T0230727

COMPACT HEAT EXCHANGERS FOR ENERGY TRANSFER INTENSIFICATION

COMPACT HEAT EXCHANGERS FOR ENERGY TRANSFER INTENSIFICATION

Low Grade Heat and Fouling Mitigation

Jiří Jaromír Klemeš • Olga Arsenyeva
Petro Kapustenko • Leonid Tovazhnyanskyy

CRC Press
Taylor & Francis Group
Boca Raton London New York

CRC Press is an imprint of the
Taylor & Francis Group, an **informa** business

MATLAB® is a trademark of The MathWorks, Inc. and is used with permission. The MathWorks does not warrant the accuracy of the text or exercises in this book. This book's use or discussion of MATLAB® software or related products does not constitute endorsement or sponsorship by The MathWorks of a particular pedagogical approach or particular use of the MATLAB® software.

CRC Press
Taylor & Francis Group
6000 Broken Sound Parkway NW, Suite 300
Boca Raton, FL 33487-2742

First issued in paperback 2017

© 2015 by Taylor & Francis Group, LLC
CRC Press is an imprint of Taylor & Francis Group, an Informa business

No claim to original U.S. Government works

ISBN-13: 978-1-4822-3259-2 (hbk)
ISBN-13: 978-1-138-74833-0 (pbk)

This book contains information obtained from authentic and highly regarded sources. Reasonable efforts have been made to publish reliable data and information, but the author and publisher cannot assume responsibility for the validity of all materials or the consequences of their use. The authors and publishers have attempted to trace the copyright holders of all material reproduced in this publication and apologize to copyright holders if permission to publish in this form has not been obtained. If any copyright material has not been acknowledged please write and let us know so we may rectify in any future reprint.

Except as permitted under U.S. Copyright Law, no part of this book may be reprinted, reproduced, transmitted, or utilized in any form by any electronic, mechanical, or other means, now known or hereafter invented, including photocopying, microfilming, and recording, or in any information storage or retrieval system, without written permission from the publishers.

For permission to photocopy or use material electronically from this work, please access www.copyright.com (http://www.copyright.com/) or contact the Copyright Clearance Center, Inc. (CCC), 222 Rosewood Drive, Danvers, MA 01923, 978-750-8400. CCC is a not-for-profit organization that provides licenses and registration for a variety of users. For organizations that have been granted a photocopy license by the CCC, a separate system of payment has been arranged.

Trademark Notice: Product or corporate names may be trademarks or registered trademarks, and are used only for identification and explanation without intent to infringe.

Visit the Taylor & Francis Web site at
http://www.taylorandfrancis.com

and the CRC Press Web site at
http://www.crcpress.com

Contents

Foreword

Compact heat exchangers, while accounting for perhaps 10% of the worldwide market for heat exchangers, have in recent years, as it has been suggested, seen their sales increase by about 10% per annum, compared to 1% for all heat exchangers. The majority of automotive heat exchangers are 'compacts', and compact heat exchangers of all types are being used increasingly where fluid inventory is an important factor, such as in refrigeration and heat pumping equipment using flammable working fluids and in chemical processes where the 'process intensification' approach leads to increased safety. Aerospace and electronics are also major uses, where small size and low weight are important. The chemical process industries were relatively slow to adopt these units because of concerns about fouling, except in areas such as cryogenics, where the aluminium plate–fin heat exchanger is difficult to improve for flexibility. Where the demands on space and efficiency are great – such as off-shore platforms for oil and gas processing – there are hundreds of compact examples such as printed circuit heat exchangers abound, as well as in other uses where fouling can be avoided or controlled. The mitigation of fouling in all heat exchanger systems, but particularly in compact units and those with enhanced features on the surface to encourage better heat transfer, remains key to their wider application. The importance attached to this topic is evident in this book.

Often the theoretical attractiveness of cycles – refrigeration/heat pumping and power generation – is countered by the practical realisation that the cycle efficiency depends upon heat exchanger approach temperatures as well as expander or compressor efficiencies. One sees a continuing trend towards the use of 'compacts' in cycles dealing with low-grade heat – vapour compression and absorption/adsorption cycles and Rankine-based organic power generation cycles. Many commercial packages in refrigeration and heat pump duties use plate heat exchangers, for example – where liquids are the heat sources or sinks. In examples where air is a source or sink, the need for air-side extended surfaces, because of modest heat transfer coefficients, provides greater challenges in the desire for compact units. This has led to much research on modifications to fins – the plate–fin characteristics detailed by Kays and London decades ago remain topical and authors such as Ralph Webb and John Hesselgreaves have tackled enhanced surfaces in single-phase and, increasingly important, in two-phase heat transfer.

While much has been written about the many types of compact heat exchangers and their design and optimisation, much less is available outside specialist organisations such as HTRI and HTFS on how such heat exchangers can be integrated into low-grade heat recovery duties (and other applications in the process sectors), and the use of process integration methodologies to implement this is one of the great strengths of the authors of this book.

I am particularly pleased to see a chapter on the integration of intensified compact heat exchangers into heat exchanger networks. Some years ago, I collaborated with Jiří Klemeš when he was at UMIST (now the University of Manchester) on a European project in this topic. Our aim was to integrate intensified unit operations (not just compact heat exchangers) into Process Integration methodologies. It is fair to say that a lack of methodologies, or the potential users' awareness about them, has been a significant hindrance to the wider take-up of intensified plant, including 'compacts', and not just in the chemical process industries. Implicit in the chapter and section headings is the extension of the methodology outside the process industries to areas such as building heating systems, where the experience of the co-authors in areas such as district heating will be brought to the fore.

Not neglected are the fundamentals, so the heat exchanger designer should have access to up-to-date correlations and data to allow the design of several types of compact heat exchangers for operation in the single phase, as well as boiling and condensing duties. The backing up of the analyses with case studies will be of inestimable value to those who are less familiar with the analytical approaches and desire to see verification of design procedures in practice.

This text encompasses all the important features associated with the successful design and implementation of many types of compact heat exchangers in a wide variety of potentially demanding and valuable applications – the emphasis on fouling mitigation, uses in heat pumping and power cycles and Process Integration gives the book a unique flavour that will ensure its value across a wide readership.

Professor David Reay
David Reay and Associates
Lancashire, United Kingdom

Preface

The book has been based on the long-term collaboration amongst the authors, which started with the British Council Know-How–funded LINK project in 1993. To exploit a synergy between the Centre of Excellence in Process Integration at the University of Manchester Institute of Science and Technology – UMIST with previously closed (secret) leading university of the former Soviet Union just opening to the world in a newly independent Ukraine proven to be very beneficial. It had been even boosted when one of the leading academics in Kharkiv started to use the accumulated know-how for a high–tech, small to medium enterprise 'Spivdruzhnist-T' closely collaborating with UMIST and a world-leading plate heat exchanger manufacturer Alfa Laval. This brought a strong industrial experience into the joint research and collaboration.

The joint research was initiated in 1993 by The British Know-How–funded REAP project 'Training educational courses on energy saving integrated processes in Ukraine' with Ukraine. Technology transfer and collaborative preparation of degree courses for Ukrainian Universities.

Another very beneficial factor boosting the research and know-how have been several successful project supported by the European Community. They should be listed and highly praised as they provided the ground for the continuous collaboration amongst the authors of this book as well some other leading partners in the field of advanced and compact heat exchangers.

They started in 1997–1999 by INTAS-96-2017, 'Integration of alternative energy sources and energy saving processes into regional energy systems', with partners from France, Ukraine and the United Kingdom and continued in 1999–2001 by the EU Programme INCO-COPERNICUS 'Sustainable Development by Retrofit and Debottlenecking for Energy Based System (REDBAS)'. Research & Technology Development Project 'PL 5046' with partners Romania, the Czech Republic, Ukraine, Belgium and the United Kingdom.

The following EU programme SYNERGY project was 'Analysis of possibilities of energy saving and application of non-traditional sources of energy, with prognostication of corresponding reduction in green-house gasses emissions in the North-East Ukraine', 2000–2001. Partners were Ukraine, Germany and the United Kingdom. In 2001–2004 followed the EU INCO-COPERNICUS project ICFP5-1999-A2PRO1 (INCO-COPERNICUS-2), 'DEMACSYS – Development and Application of Decision-making Computerised System for Environmental Impact Assessment and Pollution Reduction Management in Chemical Industrial Areas' with partners from Italy, Greece, Russian Federation, and Ukraine, and also the EO FP6 Horizontal Research Activities Including SMEs – Collaborative Research project 'SHERPA – Sustainable Heat and Energy Research for Heat Pump Applications' with partners France, Sweden, Ireland, Bulgaria, Czech Republic, Poland, Spain, the Netherlands, Austria, Italy, Ukraine, Poland and Germany.

In 2001–2004 came the TACIS/TEMPUS CD JEP-21242-20000/UKR 'Education development in environmentally safe energetics' with partners from Spain, France and Ukraine, and in 2006–2008 the EU FP6 Co-operative Research Project 'CONNECT Advanced Controllers for Economic, Robust and Safe Manufacturing Performance' with partners from Austria, Greece, Norway, Slovenia, Ukraine and the United Kingdom.

Very important for the creation of this book was the EC FP7 project 'Intensified Heat Transfer Technologies for Enhanced Heat Recovery' – INTHEAT, Grant Agreement №. 262205, in 2010–2012. Its contributing partners were Process Integration Ltd (UK), Cal Gavin (UK), Akstionerne Tovarystvo 'Spivdruzhnist-T' (Ukraine), Makatec Apparate GmbH (Germany), Oikos, svetovanjezarazvoj, d.o.o. (Slovenia), The University of Manchester (UK), the University of Bath (UK), Paderborn University (Germany), the University of Pannonia (Hungary) and EMBaffle (The Netherlands). The academic partners perform targeted research for the benefit of the involved SMEs for improving their competitiveness and active contribution to the EU's sustainability objectives.

In 2012–2015 another EC FP7 FP7-PEOPLE-2011-IRSES, MARIE CURIE ACTIONS, International Research Staff Exchange Scheme, brought opportunities for long time secondments, which were crucial for writing this book. The project 'Distributed Knowledge-Based Energy Saving Networks' – DISKNET, Grant Agreement No: PIRSES-GA-2011-294933, enabled the coordination among the following partners: the University of Pannonia (Hungary), the Centre for Research and Technology Hellas – CERTH (Greece), the University of Zagreb – Faculty of Mechanical Engineering and Naval Architecture – UNIZAG-FSB (Croatia), Aristotle University of Thessaloniki – AUTH (Greece), National Technical University 'Kharkiv Polytechnic Institute' – NTU KhPI (Ukraine), The Hashemite University (Jordan) and the University Mohammed I (Morocco).

In 2012–2015, another major contribution brought an EC project FP7-ENERGY-2011-2, 296003 'EFENIS Efficient Energy Integrated Solutions for Manufacturing Industries' with partners The University of Manchester (UK), Bayer Technology Services Gmbh (Germany), MOL plc (Hungary), VestasAircoil AS (Denmark), IPLOM, SpA (Italy), ENN Group (China), Teknologian Tutkimuskeskus VTT (Finland), AO Spivdruzhnist-T Limited Liability Company (Ukraine), ESTIA Consulting & Engineering S.A. (Greece), PannonEgyetem – University of Pannonia (Hungary), Universitaet Paderborn (Germany), Aristotelio Panepistimio Thessalonikis – Aristotle University (Greece), Università degli Studi di Genova (Italy), Univerza v Mariboru – University of Maribor (Slovenia), Hanyang University (Korea) and Alexandra Instituttet A/S (Denmark).

Also the NATO Scientific and Environmental Division Priority Area in High Technology Programme HTECH.LG 941 Creative Energy Saving Technologies Research Based on Thermodynamic Optimum Synthesis on Graphs, with Ukraine academia and industry in 1994–1998 and in 1995–1998 NATO Scientific Affairs Division Priority Area on Computer Networking programme CN.SUPPL 951064 with Ukraine academia and industry should be praised.

This book is a strong example of how researchers from different parts of Europe can closely and effectively collaborate in research and technology development and how their joint effort can be mutually beneficial, as well as benefit the wider scientific community.

The authors appreciate all these projects and collaborations that have been a major source of information for this book. They believe that the results and experience presented in this book will be of wide interest to the community and initiate even more collaboration and applicable research results in the future.

Authors

Professor Habil Jiří Jaromír Klemeš, DSc, is a Pólya professor and a head of the Centre for Process Integration and Intensification – CPI^2 at the University of Pannonia, Veszprém in Hungary. Previously, he worked for nearly 20 years in the Department of Process Integration at UMIST and after the merge at The University of Manchester, United Kingdom, as a senior project officer and honorary reader. He holds a DSc from the Hungarian Academy of Sciences, an H DSc (doctor honoris causa) from National Polytechnic University Kharkiv, Ukraine; the University of Maribor, Slovenia; and Politehnica University of Bucharest, Romania. He has many years of research and industrial experience, including research in process and heat integration, sustainable technologies and renewable energy, which has resulted in many successful industrial case studies and applications. He has extensive experience in managing major European and UK know-how projects and consulted widely on energy saving and pollution reduction. Previously, Professor Klemeš ran research in mathematical modelling and neural network applications at the Chemical Engineering Department, University of Edinburgh, Scotland. He has been a distinguished visiting professor at University Technologi Malaysia, Universiti Technologi Petronas in Malaysia, South China University of Technology Guangzhou, Tianjin University, Jiaotong Xi'an University in China, University of Maribor, Slovenia, and Brno University of Technology, the Czech Republic. He has unique success record in managing and coordinating research projects funded by the European Community FP2 to FP7, UK Know How Fund, NATO High Technology, European Training Foundation, and others. He is an editor-in-chief of *Chemical Engineering Transactions*; subject editor of *Journal of Cleaner Production* and *Energy*; executive editor of *Applied Thermal Engineering* and *Cleaner Technologies and Environmental Policies*; editor of *Resources, Conservation and Recycling*, *Integrated Technologies and Energy Saving*, and *Theoretical Foundations of Chemical Engineering*; guest editor of *Applied Energy* and several others. In 1998, he founded and has been since the president of the International Conference *Process Integration, Mathematical Modelling and Optimisation for Energy Saving and Pollution Reduction—PRES* (www.conferencepres.com). He is chair of the CAPE Working Party of the European Federation of Chemical Engineering. Among his books are *Sustainability in the Process Industry: Integration and Optimization* (McGraw-Hill), edited books *Process Integration Handbook* and *Handbook of Water and Energy Management in Food Processing* (Woodhead Cambridge/Elsevier) and *Assessing and Measuring Sustainability in Engineering* (Elsevier).

Professor Olga P. Arsenyeva, DSc, is associate professor in the Department of Integrated Technologies, Processes and Apparatuses at the National Technical University, Kharkiv Polytechnic Institute (NTU KhPI), Kharkiv, Ukraine. She graduated from the Kharkiv State Polytechnic University (Kharkiv, Ukraine), worked for the Centre for Energy Saving Process Integration (CESPI) at NTU KhPI, which was founded following the program for British Council Training and Academic Link, where she did her PhD under the supervision of Professor Leonid Tovazhnyianskyy.

In 2005, she joined Professor Petro Kapustenko in AO 'Spivdruzhnist-T' Engineering Company (Kharkiv, Ukraine). Both CESPI and AO 'Spivdruzhnist-T' had close relations with the Department of Process Integration UMIST and Centre for Process Integration – CPI^2 of Pannonia University. In 2014, she was awarded the DSc at NTU KhPI. The objects of her research are plate heat exchangers design and application and Process Integration.

Professor Petro O. Kapustenko, PhD, is professor and the deputy head, since 1994, of the Centre for Energy Saving Process Integration at National Technical University Kharkiv Polytechnic Institute (NTU KhPI), Kharkiv, Ukraine, founded with the support of DPI UMIST under British Council Know How Link project. He graduated in 1972 from the same university. In 1980, he earned his PhD at DI Mendeleev Russian Chemical Technology University in Moscow. In 2000, he became professor at NTU KhPI. In 1991, he cofounded Sodrugestvo-T engineering company and become its chief executive officer. The company has been involved in a number of projects funded by the EC and the World Bank aimed at the renovation of district heating networks and industry using Process Integration methods. They implemented advanced energy saving equipment like CHP units, compact heat exchangers, heat pumps, modern district heating substations and others, mostly in Ukraine and the Russian Federation. Professor Petro Kapustenko is also member of the Ukrainian Building Construction Academy.

Professor Leonid L. Tovazhnyanskyy, DSc, professor and head of the Department of Integrated Technologies, Processes and Apparatuses (ITPA) at the National Technical University Kharkiv Polytechnic Institute, Kharkiv (NTU KhPI), Ukraine. He graduated in 1959 from the Kharkiv V.I. Lenin Polytechnic Institute in the former Soviet Union and in 1966 was awarded a PhD. In 1988, he received his DSc at DI Mendeleev Russian Chemical Technology University in Moscow. Since 1995, he has been the director of the Centre for Energy Saving Integrated Processes and in 2000 he became the rector of the NTU KhPI and served several terms as the rector. After that he became an Honorary Rector of NTU KhPI. His research interests include heat and mass transfer processes in channels of complicated geometry, including various types of channels with corrugated walls, for industrial plate heat exchangers, optimisation of heat exchanger networks and Process Integration. Leonid L. Tovazhnyanskyy is corresponding member of the National Academy of Sciences of Ukraine.

CHAPTER 1

Introduction

The sustainable development of the modern society is impossible without the steady increase in the need for energy. Population and income growth are the key drivers behind the growing demand for energy sources. According to BP World Energy Outlook (2013), by 2030, the world's population is projected to reach 8.3 billion and additional 1.3 billion (1,300,000,000) people shall be supplied with energy. The world's income in 2030 is expected in real terms to roughly double that of the 2011 level. The world's primary energy consumption is projected to grow from 2011 to 2030 by 36%.

As is predicted by IEA (2013) the demand will grow for all forms of energy. Fossil fuels have limited natural reserves. By their nature, fossil fuels cannot serve as a basis for sustainable development in a long term. With contemporary emphasis on utilisation of low-carbon energy sources (renewables and nuclear), they will meet around 40% of the growth in primary energy demand. Nearly half of the net increase in electricity generation will come from renewables. Today's share of fossil fuels in the global mix, at 82%, is the same as it was 25 years ago. The strong rise of renewables will only reduce this to around 75% in 2035, meaning there is an increase in fossil fuel consumption by 25% as compared to that in 2011.

The use of fossil fuels for energy generation is responsible for the major part of hazardous pollutions, greenhouse gases including carbon dioxide emissions made by human activities. The projected increase of their usage will inevitably lead to deterioration of environmental situation and dangerous climate changes. Such adverse consequences makes it necessary to analyse the ways how the generated by fossil fuels energy is used. As estimated by Rattner and Garimella (2011), two-thirds of the input energy for electricity generation in the United States is lost as heat during conversion processes. In a report for the U.S. Department of Energy by BCS, Incorporated (2008), it is observed that during manufacturing processes, as much as 20%–50% of the energy consumed is ultimately lost via waste heat contained in streams of hot exhaust gases and liquids, as well as from heated product streams. In some cases, efficiency improvements resulting from waste heat recovery can increase energy efficiency from 10% to as much as 50%. For developing countries these figures can be even higher. The use of this wasted heat can significantly improve the efficiency of energy usage and decrease the combustion of fossil fuels for energy generation.

The temperature of the waste heat is playing a crucial role for estimation of the feasibility for waste heat recovery. It is significantly different for a variety of processes disposing heat to the ambient. For example, cooling water returns at the process plants have low temperatures from around 30 up to 90 °C, and temperatures of the flue gases in glass melting furnaces can be above 1,300 °C. According to the U.S. Department of Energy report (see BCS, Incorporated, 2008), waste heat temperature can be divided into groups that are defined as follows:

1. High temperatures (high-grade heat) are above 650 °C.
2. Medium temperatures (medium-grade heat) are from 230 °C to 650 °C.
3. Low temperatures (low-grade heat) are below 230 °C.

There, it is also shown that the majority of waste heat losses, about 60%, are in the low-temperature range. Though low-grade waste heat has a lower quality as a source of heat, it is available in sufficiently large quantities and its work potential exceeds that of other sources with higher waste heat temperatures.

In other literature sources, the threshold temperatures for defining low-grade heat are different. For example, Doty and Turner (2009) specified the upper limit of temperature for low-grade heat as 205 °C. Latour et al. (1982) classify heat as low grade at temperatures as low as even below 121 °C. Ammar et al. (2012) proposed another approach in their analysis with regard to the usefulness of heat within the process. By their definition, low-grade waste heat is the heat that is not feasible to utilise for heat recovery inside a specific process. Bendig et al. (2013) are defining waste heat as the sum of the exergy that is available in a process after Pinch Analysis (Klemeš et al., 2013), heat recovery, Process Integration and energy conversion (utility) integration with the help of exergy analysis. The avoidable waste heat, which can be used in a process, is better not used for a secondary application, to preserve opportunity for investments in energy efficiency.

The subject of our book is concerned with the selection of compact heat exchangers (CHEs) for low-grade energy usage with all severity of conditions intrinsic to a variety of such applications. From this point, it is assumed that low-grade heat is defined to be in the temperature range of ambient up to around 260 °C, as it is made by Law et al. (2013). Besides, according to Ammar et al. (2012), the temperature range up to 200 °C takes place in 75%–85% of all processes such as heating and cooling and inter-process heat transfer applications. For such cases, most of the results and recommendations discussed here are also applicable.

The topic covered in this book received very considerable attention in the research over the last years. Searching the papers in SCOPUS in December 2014 brought the following results:

Heat transfer intensification – 156 papers, 1,218 by title, abstract and keywords
CHEs – 551 papers, 2,732 by title, abstract and keywords
Low-grade heat – 280 papers, 5,330 by title, abstract and keywords
Fouling mitigations – 140 papers by title, 488 by title, abstract and keywords

Low-grade heat sources also include most applications of renewables, such as solar and geothermal energies, as well as sewage waters, seawater and waste to energy. The spectrum of different low-grade heat sources and the methods of their energy utilisation are briefly discussed in Chapter 2.

New challenges in efficient heat recuperation arise when integrating renewables, waste heat, polygeneration and combined heat and power units with traditional sources of heat in industry and the communal sector, as it is shown by Klemeš et al. (2010). As most of the mentioned sources can be characterised by relatively low temperatures, to utilise this low-grade heat, there is a requirement to consider minimal temperature differences in heat exchangers (HEs) of reasonable size (2010). Such conditions make it necessary to consider Compact Heat Exchangers (CHEs), which for the same duty have smaller size than traditional shell and tube HEs and smaller heat transfer area, which can be made from rather small quantities of sophisticated corrosion-resistant materials. Another advantage is small holdup volume, which enables to use less charge of expensive or dangerous fluid used as a heat agent. According to the Oxford Dictionary (1999), one of the meanings of the word 'compact' is that it is 'small and economically designed'. It is exactly what is needed when utilising low-grade heat.

Starting mainly in the 50th year of the last century, the importance and advantages of CHEs have been recognised in aerospace, automobile, gas turbine power plant and other industries. The energy crisis of 1970 gave another boost to the development of CHEs. The additional driving factors for new HE designs have been the need for reducing energy consumption with increased heat recuperation at power and process plants and minimising the capital investment in grass root and retrofitted plant projects.

In process industries, where traditional, not compact, shell and tube HEs were hitherto overwhelming, the use of plate (PHEs) and other CHEs has been increasing owing to their inherent advantages. At that time, two new types of PHEs were introduced, as well as new principle of designing plate-and-frame PHE by using in one HE plates with different corrugation pattern.

The first of these new construction types is the brazed plate heat exchanger, originally developed in 1977 at ALFA LAVAL in Sweden by engineers who later in 1983 organised the SWEP company, initially specialising on this kind of CHEs, and now widely adopted by other manufacturers. In recent times, brazed plate exchangers dominate the low- to medium (100 kW)-capacity range of refrigeration and central air conditioning equipment, practically almost completely replacing traditional shell and tube HEs.

Another derivative of PHE design made in 1970 is welded plate HE. In 1974, the cross flow welded plate HE, developed by researchers and engineers of NTU KhPI and Ukrniichemmash at Kharkiv in Ukraine, was manufactured at Pavlogradchemmash factory and commissioned at Azot chemical factory in Chirchik, Uzbekistan. This HE is designed to operate at temperature 520 °C and pressure 32 MPa in high-pressure shell of ammonia synthesis column. With the development of welding methods and especially of laser beam welding, different kinds of welded PHEs are now produced by all major PHE manufacturers. They are suitable for a wide variety of process temperatures and pressures, considerably widening the scope of PHE application in process industries.

There is renewed interest in gas microturbines, as the source for distributed power generation. As it is concluded by Shah (2005), from the cost, performance and durability points of view, the prime surface plate–type recuperators have the most potential in the microturbine applications. The modern designs of such CHE for microturbine systems use prime surfaces on both fluid sides in a form of corrugated plates welded at the side edges to form air flow passage, to prevent the leaks and mixing of the fluids. The construction principle is the same as in welded PHE, but with much smaller hydraulic diameter of the channel, in order of 1–2 mm.

The growing widespread use of heat pumps for utilisation of low-potential heat and renewable sources of energy has stipulated the development of CHEs. The trends in the development of their constructions in this area were analysed by Reay (2002).

The big potential for CHE applications is also in the utilisation of solar energy, as shown in a survey by Li et al. (2011) for a new generation of high-temperature solar receivers.

Since 1990, the advances in manufacturer's ability to produce microscale devices and systems have stipulated the development of minichannel and microchannel HEs. Microchannel HEs have been used in various engineering and scientific applications. It includes medical applications and microelectromechanical systems such as micro-HEs, inkjet printers, pumps, sensors and actuators. The needs for efficient cooling methods with high-heat-flux components are satisfied by micro-HEs. Their types also include non-metal polymer microchannel HEs and ceramic microstructure HEs. The fields of application for microchannel HEs also include heat pumps and refrigeration technologies.

Following the trend for reduction in size, weight and holdup volume of recuperators, there is also significant progress in shell and tube HEs. The manufacturers are offering tubes of diameters down to 1 mm and walls as thin as 0.04 mm from stainless steel and sophisticated alloys. There is significant progress in research and manufacturing of different tubes with enhanced heat transfer as well as tube inserts for the same purpose.

The powerful tool to improve HE performance is heat transfer intensification. In the literature, the alternative terminology is frequently used, including the phrases 'heat transfer augmentation' and the 'heat transfer enhancement'. These definitions are not fully corresponding to a more global term 'process intensification', which according to Reay et al. (2013) for heat transfer processes can be defined as follows. Heat transfer intensification is a 'process development involving dramatically smaller equipment, which leads to: …

1. Higher energy efficiency.
2. Reduced capital cost.
3. Reduced inventory/improved intrinsic safety/fast response times.'

The CHEs, 'small and economically designed', are exactly suited to this definition, and the task of heat transfer intensification for them has mainly the same meaning as the increase of compactness.

As it will be discussed in Chapter 3, there are two main methods to increase compactness of heat transfer surface, which are usually combined to obtain the best performance: first is to decrease hydraulic diameters of the channels and second is to increase overall heat transfer coefficient. Most of the researches concerned with heat transfer enhancement are concentrating on the increase of film heat transfer coefficients as the main constituents of overall heat transfer. There are different possible ways to achieve higher values of film heat transfer coefficients that can be divided into active and passive methods of heat transfer enhancement.

The active methods involve input of some additional power from the outside of HE for augmentation of heat transfer. The examples of active methods include the use of magnetic field to disturb some light particles in a flowing stream, electrostatic field, ultrasonic vibrations, fluid or surface vibrations and induced pulsations by cams or reciprocating plungers. These methods require some additional devices and external energy input and are used only in a limited number of specific applications. Active methods of heat transfer enhancement are not considered in this book.

The passive methods do not involve external power sources but use modifications of the flow channel geometry or wall surfaces to change the flow characteristics. It is resulting in redistribution across the channel of the energy delivered with the flow itself aiming to obtain heat transfer enhancement and increase in the values of film heat transfer coefficients. It can be made by inserting extra component in channels, swirl flow devices, wire inserts (see, e.g., Figure 3.19) and surface roughness or special elements inducing turbulence (see, e.g., Figure 3.18) or using surface tension forces in two phase flows and modification of the geometrical form like in coiled or in twisted tubes (see, e.g., Figure 3.20) or in channels of plate HEs (see, e.g., Figure 3.30).

As it is discussed in Section 3.5, the decrease of hydraulic diameter below some values can also lead to the increase of film heat transfer coefficient, while the flow regime can be transitional and even laminar. However, it requires much smaller length of the channel for the same allowable pressure drop. Such effect can be used where it is possible for the construction of HE and specified duty, like in microchannel HE gas–liquid (see, e.g., Figure 3.58) or liquid–liquid (see, e.g., Figure 3.63).

When HE is working in some closed circuit of a certain fluid, the heat transfer enhancement can be achieved by selection of fluid with special properties, especially high heat conductivity. In a number of researches, nanofluids have been studied as liquids with special additives for enhancing their heat transfer properties. Nanofluids are special heat transfer fluids engineered by dispersing nanoparticles with a size of < 100 nm in some base fluids. More detailed survey of researches on nanofluids can be found in a paper by Hussein et al. (2014).

Nowadays, the passive methods of heat transfer intensification are the most widely used for the development of efficient HEs for different applications, as they do not require external devices or additional energy input. Passive methods also can be used as a base for compound methods where active methods like flow or surface vibrations and magnetic fields are imposed on surfaces with passive heat transfer enhancement. To implement these methods for the development of intensified heat transfer surfaces for CHEs and methodology to use them in the systems of low-grade heat recovery, the analysis of the existing methods of heat transfer intensification, thermohydraulic design, and optimisation of CHEs with heat transfer surfaces of different geometries is required.

The requirements for a recuperative HE are generally formulated according to its position in the specific network of heat recuperation, in which the structure is determined by the nature of the technological processes in certain production system with heat utilisation. In designing heat recuperation network, the target temperatures for technology equipment are usually strictly specified by the technological processes in that equipment. Based on these data, the following requirements for heat transfer process in a specific recuperative HE is formulated:

1. The nature and thermophysical properties of heat exchanging streams
2. The flow rates of the streams
3. The inlet and outlet temperatures of both heat exchanging media
4. Allowable pressure drops at both sides of an HE

Generally, outlet temperatures and pressure drops can be changed during optimisation of the total production system and its heat recuperation part, but that constitutes separate problem considered on the stage of HE network (HEN) design.

In many industrial applications of heat transfer equipment, the surfaces washed by the streams involved in heat transfer process are susceptible to fouling, which refers to any change at the heat transfer surface, whether by deposits of fouling substances or by other means, which results in a deterioration of heat transfer intensity across that surface. As it is mentioned in Chapter 5 with the discussion of Equation 5.1, this phenomenon can significantly deteriorate the intensity of heat transfer process and HE performance by creating additional thermal resistance of fouling layer. Moreover, the decrease of the channels' cross-sectional area partly blocked by the fouling deposits can lead to significant increase of pressure drop in HE and finally to clogging of the channels.

According to an analysis of different publications presented by Crittenden and Yang (2011), conservative estimation of HE fouling leads to a conclusion that additional cost for fouling in industrialised countries are in the order of 0.25% of the gross domestic product. Fouling also causes a total equivalent anthropogenic emission of carbon dioxide of around 2.5%. In most processing industries, fouling creates a severe operational problem that compromises energy recovery and creates additional negative impact in the environment. It results in loss or reduction in production, increased consumption of energy and pressure losses, flow maldistributions, cost of antifouling chemicals, cleaning costs, etc. Inadequate detailed knowledge of the fouling mechanisms is frequently limiting the problem solution, even though the basic principles are understood for some time.

The economic and environmental importance of fouling can be classified as four main categories:

1. Capital expenditure
2. Energy cost and environmental impact
3. Production loss during shutdowns due to fouling
4. Maintenance cost

Extra capital cost involves surplus heat transfer surface area to compensate the fouling effect, heavier foundation cost, extra space for larger HEs, higher cost for transport and installation of the equipment, cost of antifouling equipment and online cleaning devices and treatment plants. In most cases, the additional surface area to compensate for fouling is calculated by incorporating one or more fouling resistances into the basic design (Equation 5.1 in Chapter 5). Such approach is having low accuracy, as fouling is developing in time and even if it is quickly reaching asymptotic value, this value can be influenced by a number of factors not accounted by numerical values for fouling resistances quoted in the literature.

As discussed by Crittenden and Yang (2011), the development of a fouling resistance reduces the thermal performance of an HE and economic fouling resistance could be selected. One problem with the extra heat transfer surface in an HE as a way of compensating for fouling is that there is too much surface area when the exchanger is clean. Bypassing fluid around the exchanger to counter the increased heat transfer rate is leading to the increase of fouling due to lower velocities through the exchanger. Energy costs and environmental impact in recuperative systems due to the increase of fouling directly involve the additional fuel required for the heat energy to produce the required temperature required by other process equipment. For example, the drop of the temperature on 1 °C at the end of the crude oil preheat train at a 100,000 bbl/d refinery leads to about $40,000 additional fuel cost and the release of about 750 t of CO_2/y. There are also significant production losses during shutdowns for cleaning of HEs, which are difficult to estimate. After shutdowns, also an additional cost after the production is restarted due to some period of out of specification product quality. Maintenance cost involves staff and other expenditures for cleaning of equipment from fouling, as well as an economic and ecological cost for disposal of cleaning chemicals.

The efficient use of heat from the variety of low-grade heat sources described in Chapter 2 requires their efficient integration into the heat supply, distribution and usage system including all other existing energy sources and heat consumers. Such system can include a number of streams that require to be heated and streams that have heat energy that can be extracted as they are cooled. These streams can exchange heat energy with the use of the HEs that as a system of devices transforming heat energy of hotter streams into heat energy of cooler streams represents HEN. With the same system of streams, the structure of HEN and heat transfer areas of constituting it HEs can vary to the great extent with significant differences in total heat energy exchanged (recuperated) and the numbers, heat transfer area and costs of HEs. The design of HEN represents the problem of finding its structure and parameters of constituting it HEs corresponding to the optimum of certain objective function. The widely adopted approaches for HEN design can be divided into two groups, namely, mathematical programming and Pinch Analysis as a basic methodology of process integration (Klemeš and Kravanja, 2013).

In practice, any decision about low-grade heat utilisation and equipment selection for this purpose is made considering economical aspect of the problem. The crucial role in taking such decision is playing the estimation of the trade-off between the payment received for energy beneficially utilised after the project implementation and the cost of investment required for the design of utilisation system, purchasing, installation and commissioning of the equipment.

ACKNOWLEDGMENTS

The authoring team used their many years' experience in the academia, research and industry to provide carefully selected and verified information in the field. The authors benefited from the information kindly provided by the companies Alfa Laval, Sweden; CAL Gavin Ltd., United Kingdom; and Spivdruzhnist-T, Ukraine. The information has been also tested and discussed by academic courses and especially further professional development training courses at the University of Manchester Institute of Science and Technology, United Kingdom; the University of Manchester, United Kingdom; the University of Pannonia, Hungary; Brno University of Technology, Czech Republic; Universiti Teknologi Malaysia and Petronas University of Technology, Malaysia; Xi'an Jiaotong University, China and finally, the National Technical University 'Kharkiv Polytechnic Institute', Ukraine, where several authors have been having academic appointments.

The information on process intensification network – www.pinetwork.org and Heat Exchanger Action Group – from www.hexag.org has been used and is gratefully acknowledged.

A very important role was played by several European Community projects and the collaborating partners: EC FP7 project ENER/FP7/296003/EFENIS 'Efficient Energy Integrated Solutions for Manufacturing Industries – EFENIS', EC FP7 project 'Distributed Knowledge-Based Energy Saving Networks' – DISKNET – Grant Agreement No: PIRSES-GA-2011-294933 and EC FP7 project 'Intensified Heat Transfer Technologies for Enhanced Heat Recovery – INTHEAT', Grant Agreement No. 262205. The authors would like to express their gratitude to the EC for the financial support as well as for creating the researcher structure enabling multinational collaboration: the University of Manchester, United Kingdom; Bayer, DE; MOL Hungarian Oil and Gas Company, HU; Vestas Aircoil, DK; IPLOM (Industria Piemontese Lavorazione Oli Minerali – Piedmontese Mineral Oil Processing Factory), IT; ENN Science and Technology Development Co. Ltd., CN; Technical Research Centre of Finland, FI; Sodrugestvo-T, UA; ESTIA Consulting and Engineering S.A., GR; University of Paderborn, DE; Aristotle University of Thessaloniki, GR; Genoa University, IT; University of Maribor, SI; Hanyang University, KR; Alexandra Instituttet A/S, DK; CAL GAVIN Ltd., United Kingdom; MAKATEC GmbH, DE; University of Bath, United Kingdom; and EMBaffle, NL.

The authors also gratefully acknowledge the financial support from the Hungarian State and the European Union under the project 'TÁMOP - 4.2.2.A-11/1/KONV-2012-0072 – Design and optimisation of modernisation and efficient operation of energy supply and utilisation systems using renewable energy sources and ICTs', which has been significantly contributing to the completion of this analysis.

REFERENCES

Ammar Y., Joyce S., Norman R., Wang Y., Roskilly A.P. 2012. Low grade thermal energy sources and uses from the process industry in the UK. *Applied Energy* 89(1): 3–20.

BCS Incorporated. 2008. Waste heat recovery: Technology and opportunities in U.S. industry. U.S. Department of Energy, Washington, DC. www1.eere.energy.gov/manufacturing/intensiveprocesses/pdfs/waste_heat_recovery.pdf (accessed 10 April 2015).

Bendig M., Maréchal F., Favrat D. 2013. Defining "waste heat" for industrial processes. *Applied Thermal Engineering* 61: 134–142.

BP World Energy Outlook. 2013. www.bp.com/content/dam/bp/pdf/statistical-review/BP_World_Energy_Outlook_booklet_2013.pdf (accessed 11 April 2015).

Crittenden B., Yang M. 2011. Technical review of fouling and its impact on heat transfer. Report on project FP7-SME-2010-1 262205/INTHEAT. intheat.dcs.uni-pannon.hu/wp-content/uploads/2011/11/D1.2.pdf (accessed 12 October 2014).

Doty S., Turner, W.C. 2009. *Energy Management Handbook*, 7th edn. The Fairmont Press, Lilburn, GA, USA.

Hussein A.M., Sharma K.V., Bakar R.A., Kadirgama K. 2014. A review of forced convection heat transfer enhancement and hydrodynamic characteristics of a nanofluid. *Renewable and Sustainable Energy Reviews* 29: 734–743.

IEA. 2013. World Energy Outlook 2013. www.worldenergyoutlook.org/pressmedia/recentpresentations/LondonNovember12.pdf (accessed 20 March 2014).

Klemeš J., Friedler F., Bulatov I., Varbanov P. 2010. *Sustainability in the Process Industry. Integration and Optimization*. McGraw-Hill, New York, USA.

Klemeš J.J., Kravanja Z. 2013. Forty years of heat integration: Pinch analysis (PA) and mathematical programming (MP). *Current Opinion in Chemical Engineering* 2(4): 461–474.

Klemeš J.J., Varbanov P.S., Kravanja Z. 2013. Recent developments in process integration. *Chemical Engineering Research and Design* 91(10): 2037–2053.

Latour S.R., Menningmann J.G., Blaney B.L. 1982. Waste heat recovery potential in selected industries. Project summary EPA-600/S7-82-030. nepis.epa.gov/Exe/ZyNET.exe/2000TU36.txt? (accessed 20 March 2014).

Law R., Harvey A., Reay D. 2013. Opportunities for low-grade heat recovery in the UK food processing industry. *Applied Thermal Engineering* 53: 188–196.

Li Q., Flamant G., Yuan X., Neveu P., Luo L. 2011. Compact heat exchangers: A review and future applications for a new generation of high temperature solar receivers. *Renewable and Sustainable Energy Reviews* 15: 4855–4875.

Rattner A.S., Garimella S. 2011. Energy harvesting, reuse and upgrade to reduce primary energy usage in the USA. *Energy* 36: 6172–6183.

Reay D., Ramshaw C., Harvey A. 2013. *Process Intensification: Engineering for Efficiency, Sustainability and Flexibility*. Butterworth-Heinemann, Oxford, UK.

Reay D.A. 2002. Compact heat exchangers, enhancement and heat pumps. *International Journal of Refrigeration* 25: 460–470.

Shah R.K. 2005. Compact heat exchangers for microturbines. Micro gas turbines. Educational notes RTO-EN-AVT-131, Paper 2. RTO, Neuilly-sur-Seine, France, 2-1–2-18.

CHAPTER 2

Low-Grade Heat

Issues to be Dealt With

2.1 WASTE HEAT FROM INDUSTRY

As it was concluded by International Energy Agency (IEA, 2008), in 2005, manufacturing industry was globally consuming the most energy, with 33% share, followed by households (29%) and transport (26%). Its carbon dioxide emissions share was also biggest, constituting 38%. From 10% to 50% of this energy is lost with waste heat to the environment. About 60% of this energy is going as low-grade heat. The typical sources of low-temperature heat are presented in Table 2.1.

The temperature and other conditions of the industrial waste heat are highly dependent of particular industry. It can be in the form of contaminated steam in petrochemical and refining industry at the level about 150 °C or cooling water about 30 °C–55 °C. In food and beverage industries, the level can be about 80 °C. Globally, the largest savings potentials can be found in iron and steel, cement, and chemical and petrochemical sectors. On average, Japan and the Republic of Korea have the highest levels of industrial energy efficiency, followed by Europe and North America. Energy efficiency levels in developing and transition countries show a mixed picture. Generally, the efficiency levels are lower than in Organisation for Economic Cooperation and Development (OECD) countries, but, where there has been a recent, rapid expansion using the latest plant design, efficiencies can be high (according to IEA, 2013). Final energy use in industry was 116 EJ in 2005. The useful energy and the lost energy in major industrial sectors of the United States are shown in Figure 2.1.

The divisions of waste heat recovery potentials between different processes most promising for heat recovery chemical and oil refining industries, according to the analysis of Pellegrino et al. (2004), are presented in Tables 2.2 and 2.3.

The chemical, petrochemical and other heavy industries are distributed unevenly between different countries in the world; food and beverages sector are present rather evenly more closely following the distribution of the world's population. There is significant potential to recover waste heat from food and beverage industries, as shown by Law et al. (2013). Only in the United Kingdom this sector has about 10.08×10^{15} J of wasted low-grade heat available for heat recovery. The use of this heat is similar

Table 2.1 Typical Sources of Low-Grade Heat in Industry and their Temperatures

Source	Temperature (°C)
Process steam condensate	55–88
Cooling water from	
Central cooling water circuit	30–55
Furnace doors	30–55
Bearings	30–90
Welding machines	30–90
Injection moulding machines	30–90
Annealing furnaces	65–230
Forming dies	25–90
Air compressors	25–50
Pumps	25–90
Internal combustion engines	65–120
Air conditioning and refrigeration condensers	30–45
Liquid still condensers	30–90
Drying, baking and curing ovens	90–230
Hot processed liquids	30–230
Hot processed solids	90–230
Commercial building exhaust	20–25
Paint booths, lab exhaust, etc.	20–25

to the use of green, low (or even neutral)-carbon energy source, as it is using what is practically a waste product that would be polluting the environment if not utilised. Therefore, when this heat is used in place of fossil fuels, the reduction in hazardous and greenhouse gas emissions would be significant, as would cost savings.

2.2 WASTE HEAT FROM BUILDINGS

In many countries, residential and commercial buildings account for about 40% of the total energy demand and 36% of the total greenhouse gas emissions and are hence considered both consumers and producers of energy (see Kapustenko and Arsenyeva, 2013). Space heating and hot tap water supply constitute the largest share of energy disposed in residential buildings. Besides heat losses through the walls, windows, doors and other construction elements, the waste heat is going out to the environment with sewage waters and ventilation air exhausting from premises, which are forming low-grade heat with temperature close to interior temperature of buildings.

2.2.1 Sewage Waters

Hot water has a number of uses inside the buildings, including showers, tubs, sinks, dishwashers and clothes washers. The different public and private pools are also discharging warm waters to sewage collectors. As it was confirmed by

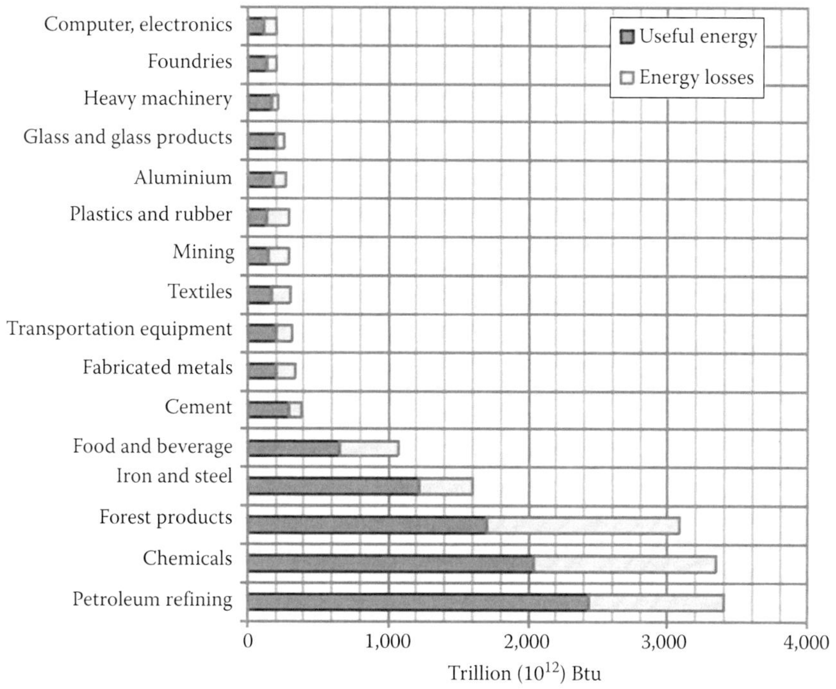

Figure 2.1 Useful energy and lost energy in major industrial sectors of the United States (1 Btu = 1.055056 kJ) (after Ferland, 2014).

Cipolla and Maglionico (2014), in most of these uses, the wastewater retains a considerable part of initially supplied hot water energy in the form of low-grade heat that could be recovered. Besides, even the cold tap water is heated when it enters inside the building at winter time, consuming some amount of energy used for space heating.

To heat water, the considerable amount of fossil fuels is spent. For example, as shown by Frijns et al. (2013) in the Netherlands, about 23% of the natural gas demand for domestic usage is consumed for heating water. Moreover, in developed countries in recent years, the considerable efforts are made for improvement of buildings' energy performance, with development and practical implementation of such concepts as 'positive energy building', 'superinsulated house' and 'zero-energy building'. A study reported by Schmid (2009) has shown that 15% of the heating energy that is supplied to buildings is going to the environment through the sewer collectors, and this value increases up to 30% in well-insulated buildings having low energy consumption. This resulted in the fact that in modern buildings sewers are one of the largest sinks for heat lost in buildings. To recover this low-grade heat, different strategies, like recovery inside the building, recovery at sewer from raw water, recovery at sewage treatment plant from cleaned water, can be used. To be implemented, all these strategies need efficient heat transfer equipment.

Table 2.2 Waste Heat Recovery Opportunities in Chemical Manufacture in the United States

Chemical Manufacture	Energy Savings (10^{15} J)
Ethylene Chain	*27.01*
Ethylene	19.66
Polyethylene – 1997	0.43
Polyvinyl chloride	0.74
Ethylene oxide	0.46
Ethylene glycol	0.45
Polystyrene	5.15
Propylene Chain	*5.56*
Propylene	3.37
Polypropylene – 1997	0.14
Propylene oxide – 1997	0.43
Acrylonitrile	0.28
Acrylic acid	0.89
Acrylic fibre	0.45
BTX (Benzene, Toluene, Xylene) Chain	*16.14*
BTX	3.59
Benzene	0.33
Ethylbenzene	1.00
Styrene	8.69
Polystyrene	1.09
Cumene	0.21
Terephthalic acid	0.62
Nylon 6.6	0.59
Nylon 6	0.21
Agricultural Chemicals – Fertilizers	*34.61*
Urea	0.23
Phosphoric acid (furnace process)	0.51
Ammonia	33.66
Soda	*16.15*
Caustic soda (chlorine/sodium hydroxide)	8.00
Soda ash (sodium carbonate)	8.16
Total Five Chains	*99.47*
Estimated Additional Savings	*211.02*
Total Industry	*310.49*

Source: After Pellegrino et al. (2004).

2.2.2 Ventilation Air Exhaust

The modern energy-efficient buildings are well insulated to save energy, but without proper ventilation, they would not be suitable for healthy living. Air within the living rooms can be much more polluted than ambient air outside. Different biological pollutants like mould, pet dander and plant pollen, as well as

Table 2.3 Waste Heat Recovery Opportunities in Refineries in the United States

Petroleum-Fired Systems	Energy Savings (10^{15} J)
Atmospheric distillation	101.54
Vacuum distillation	31.49
Solvent deasphalting	2.74
Delayed coking	7.39
Fluid coking	1.50
Flexicoking	1.41
Visbreaking	0.27
Fluid catalytic cracking	21.10
Catalytic reforming	49.63
Alkylation	31.44
Ether manufacture	3.52
Isomerisation	4.22
Catalytic hydrotreating	74.12
Catalytic hydrocracking	26.06
Lube oil Mfg	20.68
Total	*376.67*
Petroleum Steam Systems	
Atmospheric distillation	63.31
Vacuum distillation	11.61
Fluid catalytic cracking	0.00
Catalytic hydrocracking	4.22
Catalytic hydrotreating	26.38
Catalytic reforming	14.77
Alkylation	17.94
Isomers	5.28
Total	*143.49*
Total Petroleum Systems	*520.16*

Source: After Pellegrino et al. (2004).

chemicals like radon and volatile organic compounds, can create a toxic environment inside living premises. To prevent such dangerous situation, the buildings need whole-house ventilation throughout the day to maintain a healthy indoor environment.

In different countries, there is an acting legislation specifying minimal ventilation rate. For example, United States Department of Energy (USDOE) (2002) recommends that according to the American Society of Heating, Refrigerating and Air-Conditioning Engineers, Inc. (ASHRAE), the living area of a home must be ventilated at a rate of 0.35 air changes per hour or 15 ft^3/min (about 25 m^3/h) per person, whichever is greater. In winter at room temperature 20 °C and temperature outside −10 °C, it means that the incoming fresh air should receive, in an hour, about 940 kJ of energy (0.26 kW of power) for the person to be heated. The same considerable amount of low-grade thermal energy would be lost with exhaust air to the environment, if not recuperated or utilised in other ways.

Energy-recovery ventilation systems provide a controlled way of ventilating a home while minimizing energy loss. They reduce the cost of heating ventilated air in winter by transferring heat and/or humidity from the warm inside air being exhausted to the fresh (but cold) supply air. In summer, the inside air cools the warmer, incoming supply air to reduce ventilation cooling cost. There are two types of energy-recovery systems. Heat recovery ventilator is drawing fresh outside air into a home and simultaneously exhausting air from the home. During such exchange, the two airstreams pass through a heat exchanger within the unit where thermal energy from one airstream is transferred to the other. Energy-recovery ventilator is working similarly, except that it allows a portion of the moisture in the more humid airstream (usually the indoor air in winter and the outdoor air in summer) to be transferred to the dryer airstream. For heat recovery purposes, heat pumps (HPs) also can be used, which can take heat from outgoing air and transfer it to fresh air or heating elements inside the building, usually heated floor.

2.3 WASTE TO ENERGY

With the improvement of living and nutrition standards in modern society, as well as the rise in world population, the consumption of different goods is sustainably growing, inevitably leading to the increase in the amounts of wastes from households and commercial activities. It can be an important source of energy, which can be also regarded as green, because if it is disposed on the fields, it would emit the same amount of hazardous pollutants and carbon dioxide to the environment during its degradation. Waste disposal represents the primary purpose of thermal processing of waste. But during this process, heat recovery becomes more and more important and inherent part of up-to-date technologies in this area. Nowadays waste-to-energy (WtE) technologies are considered as much as possible and feasible (see, Kilkovský et al., 2014). The utilisation of waste can contribute to renewable energy production and represents an important item of energy-focused policies. Together with biomass-to-energy technologies, such approach can substitute a quite significant amount of fossil fuel. The majority of operating plants are exploiting Rankine cycle, which expands the turbine steam that drives the generator. The effectiveness of transformation from energy chemically bound in waste to its final useful forms (heat or electricity) is affected by a number of aspects including types and properties of incinerated waste, technologies being used, local conditions and current energy prices.

A typical technology for municipal solid waste treatment is presented in Figure 2.2. Waste is burned on a moving grate, and flue gas (off-gas in the case of incineration) flows to the combustion chamber where, at a sufficient temperature and time, the decomposition of even the most stable harmful compounds takes place. The operational temperature maintained in this zone is 900 °C–950 °C. According to European Union (EU) legislation, a minimum operational temperature of 850 °C is required. Burning supplementary fuel in the chamber may be used if necessary.

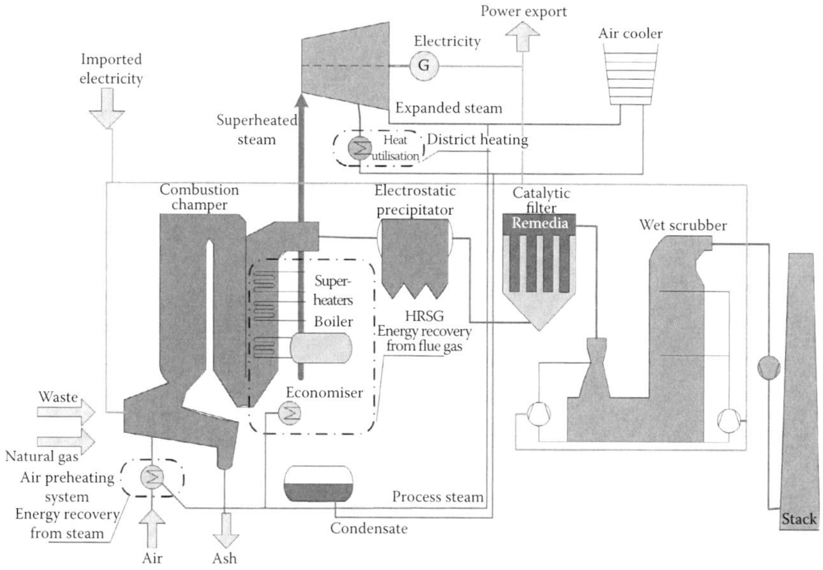

Figure 2.2 Simplified flowsheet of up-to-date waste-to-energy technology with emphasis on heat recovery (adapted from Kilkovský et al., 2014).

However, this is only done in the case of nonstandard conditions such as starting and shutting down the operation. Auxiliary energy consumption is negligible compared to the overall energy input through the waste. This part of incineration plant is called thermal system. Off-gas from the chamber then enters to the heat recovery system where in the waste heat boiler (heat recovery steam generator) its sensitive heat is utilised for producing superheated steam. The steam produced is then utilised within flexible cogeneration system comprising a condensing turbine with bleeding (extraction). The advantage of this arrangement is the ability to quickly react to changes on heat demand represented here by district heating system (heat utilisation in Figure 2.2). The district heating water is heated to a required temperature by condensing the steam extracted from certain turbine section. For this purpose, heat exchanger can be efficiently used. During reduced demand of heat (summer operation), most of the steam is utilised for electricity generation without the necessity to reduce the throughput of waste. Thermal output (heat) is reduced and electricity output is increased. Excessive low-grade heat in a form of low-pressure steam is rejected in vacuum condenser via air or water cooling system. Off-gas (with lower temperature at approx. 250 °C after giving up its energy in the heat recovery system) is then cleaned in the off-gas cleaning system. Dust is first removed mechanically in an electrostatic precipitator, which is followed by a bag-house catalytic filter of the remaining fine solid particles and dioxins and furans removal.

As can be concluded from the aforementioned description, WtE technology is producing a significant amount of low-grade waste heat. It is an intrinsic feature of all electricity generation facilities using Rankine cycle with condensing steam turbine for electricity generation with efficiency not more than 40%. Considerable amount of energy supplied to the process is discharged as latent heat of vacuum steam at the end, which is discharged to the environment with the use of vacuum condenser as low-grade heat of temperature 35 °C–40 °C. When there is a need in low grade heat of higher temperature, it can be taken as steam from intermediate section of turbine. This heat can be used for district heating or other purposes and the unit that is working in cogeneration heat and power mode. That can increase the overall efficiency up to 76% or even more.

2.4 RENEWABLE SOURCES OF HEAT ENERGY

The necessity to minimise fossil fuel consumption and environmental hazards from carbon dioxide and greenhouse gas emissions is recognised worldwide. The Directive 2009/28/EC reaffirms commitment of the EU to the unionwide development of energy from renewable sources by endorsing a mandatory target of a 20% share of energy from renewable sources by 2020. It establishes a common framework for the promotion of energy from renewable nonfossil sources, namely, wind, solar, aerothermal, geothermal, hydrothermal and ocean energies; hydropower; biomass; landfill gas; sewage treatment plant gas; and biogases. The following are considered the possibilities of solar and geothermal sources of energy to produce low-grade heat.

2.4.1 Solar Heating

The main source of external energy on earth is the solar radiation. According to Doty and Turner (2009), the intensity of it at the outer edge of atmosphere is about 1,353 W/m^2. This value is known as the *solar constant.* It somewhat varies throughout the year, being largest in January when the earth is closest to the sun. This radiation is partly reflected back to space, partly absorbed into the atmosphere and reemitted and partly scattered. Only about two-thirds of this energy comes to the surface of the earth at peak hours. This peak solar radiation on the earth surface is about 1 kW/m^2. Solar radiation data for a particular location are often given in kWh/m^2/d. Usually, it is given as average solar radiation data for fixed-plate surfaces tilted to horizontal direction at latitude degrees. For example, at different locations in the United States, this value varies from 3.5 to 6.8 kWh/m^2/d. The usage of this solar energy can be divided into four categories: (1) passive that is using design principles and building construction materials to better use the solar energy; (2) solar thermal, to produce low-grade heat; (3) solar-thermal electric to produce electricity usually using high-grade heat produced with the solar energy; and (4) solar photovoltaic to produce electricity with photovoltaic panels.

Thermal use of solar energy is currently one of the most widespread applications of renewable energy sources. According to a report by Mauthner and Weiss (2013), at the end of 2011, an installed capacity of solar thermal systems accounted for 234.6 GW, which corresponds to a total of 335.1 Mm^2 of collector area. The vast majority of the total capacity in operation was installed in China (152.2 GW) and Europe (39.3 GW), which together accounted for 81.6% of the total installed. Harnessing solar energy for heat typically is fulfilled with the use of solar thermal collectors, as the most popular method, and solar ponds. Most of these devices are producing low-grade heat. The required temperature of low-grade heat produced by solar collectors depends on the application. For space heating, a temperature in the range of 45 °C to 90 °C is required, depending on space heating technology employed. For domestic tap hot water preparation, it is from 50 °C to 60 °C. For industrial processes, it is from 60 °C to 260 °C and even higher, up to 400 °C.

2.4.1.1 Solar Ponds

Solar pond (see Figure 2.3) is the simplest form of solar thermal energy collector for a big surface area, which is simply a shallow pool several meters deep filled with salt water. The salinity of the water is increasing with depth, and the salinity gradient prevents convection currents. Solar radiation penetrates through the water of the pond to its lower layer with concentrated salt solution. The temperature there rises since the absorbed heat cannot be moved upward by convection. Therefore, solar heat is stored in the lower layer of the pond. In summer at an ambient temperature about 20 °C, solar pond can give low grade thermal energy with temperature about 80 °C.

2.4.1.2 Solar Collectors

The use of the solar thermal energy significantly varies in different regions around the world. Its main applications are hot tap water preparation, space heating and heat supply for industrial processes and cooling. The main elements, which

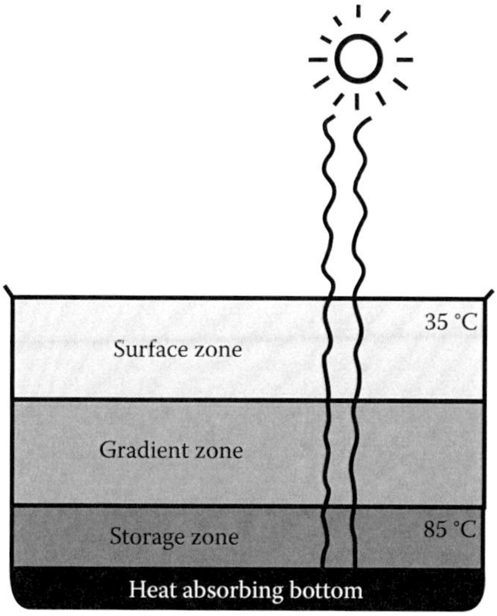

Figure 2.3 Schematic drawing of solar pond.

determine the capacity, efficiency and cost of the solar thermal system, are solar collectors. The major types of solar collectors producing low-grade heat are

- Unglazed collectors
- Flat-plate collectors (FPC)
- Evacuated tube collectors (ETC)

There are a big number of different modifications for each of these types, produced by different manufacturers. Most of them are designed to produce heat at temperatures below 95 °C for applications like district heating and hot water and swimming pool heating (see EU project Solar District Heating SDH, 2014). According to European technology platform Renewable Heating and Cooling (RHC -Platform, 2012), highly efficient vacuum-insulated FPC and ETC can produce heat with temperatures up to 250 °C for applications like industry process heating, desalination and cooling.

The heat with temperatures up to 250 °C and even much higher can be produced by advanced concentrating collectors, like linear Fresnel and parabolic trough collectors, solar dishes and solar towers as described in, for example, RHC-Platform (2012). This medium- and high-grade heat, which is out of scope of this book, are used for high-temperature processes, mainly for electric power generation with steam thermal cycles.

Concerning solar low-grade heat installations worldwide, according to the market analysis made by Mauthner and Weiss (2013), ETC dominated with a share of 63% of the total installed capacity in operation. This type of collectors counts even 82% of the newly installed capacity, but mainly because of the role of evacuated tube collectors in China, since this is so far the largest market with high growth rates. On the contrary, in Europe, most of the large solar collector installations are based on FPC, as listed in SDH (2014). The largest such facility in Europe at Marstal, Denmark, has a total area of 33,300 m^2 with a heat capacity of 23,300 kW. The largest installation with ETC in Wels, Austria, has a total area of 3,388 m^2, 10 times smaller than with FPC, and a heat capacity of 2,400 kW. All major solar heat facilities for different industrial applications (food, chemicals, car industries), listed by Vives et al. (2012), are using FPC collectors.

2.4.2 Geothermal Heat

Among the renewable energy sources, the use of low-grade heat of geothermal energy has the longest experience with district heating systems, counting for continuous operation of Chaudes-Aigues thermal station in France since the fourteenth century (Rezaie and Rosen, 2012). A comprehensive review of geothermal energy direct utilisation was made by Lund et al. (2011). The total amount of directly utilised geothermal energy in 2010 was 117,778 GWh. The leaders in geothermal heat annual energy use for district heating are Iceland, China, Japan, Turkey, Germany and France. Besides direct utilisation worldwide, 67,246 GWh in 2010 was used for electricity generation (Bertani, 2012).

There are two ways of obtaining geothermal heat. In some places, it can be obtained directly from hot springs, hot geysers, or steam going from under the earth.

In Iceland, for example, the Nesjavellir Geothermal Power Station utilises the steam that is coming spontaneously from underground. There are also some other countries besides Iceland that have reliable geothermal sources of this kind. Among them are Hungary, Kenya, the United States and Russia.

Spontaneous geothermal heat sources are rather rear exceptions. In most places, geothermal heat has to be extracted. The temperature of the earth is rising with the depth from the ground. The temperature gradient is from 20 °C/km to 30 °C/km in most areas of the world, not including areas near tectonic plate boundaries. A typical method of using geothermal heat is injecting cold water deep into the ground and pumping it back when it is heated by the underground heat. Geothermal heat is rather cheap, sustainable, reliable and environmental friendly.

2.5 HEAT PUMPS TO INCREASE HEAT POTENTIAL

One of the most important problems with utilisation of low-grade heat energy is to find the local consumer whose needs will be satisfied with this energy potential. A lot of low-grade heat sources, like sewage waters, water of industry cooling cycles, cooling water from power generation plants and some geothermal water and groundwater, have temperatures of 20 °C–40 °C or even lower. Such energy cannot be directly used in majority of applications.

Increasing the temperature and potential of low-grade heat is possible with the use of HP, which can transfer energy from one fluid with low-temperature level to another with high-temperature level using additional mechanical or heat energy. This additional energy can be regarded as a price for inverting the natural heat flow which, according to the second law of thermodynamics in nature, tends to be from a higher to a lower temperature. The efficiency of an HP is usually measured by its coefficient of performance (COP), which shows how much thermal energy is transferred to the medium to be heated for the unit of additional energy required for driving HP.

$$\mathrm{COP} = \frac{\mathrm{Thermal\_Energy\_Supplied}}{\mathrm{Driving\_Energy\_Required}} \tag{2.1}$$

The driving energy, which is required to drive HP, can be applied in different ways (using electric motor, gas engine, electricity, or any form of thermal energy). To estimate the efficiency of HP system, the primary energy ratio (PER) is used, which shows how much thermal energy is supplied to the medium to be heated for the unit of primary energy spent to drive HP system:

$$\mathrm{PER} = \frac{\mathrm{Total\_Thermal\_Energy\_Supplied}}{\mathrm{Primary\_Energy\_Spent}} \tag{2.2}$$

In nature, there are a number of physical phenomena that can be used to cool low-temperature media on one side by transferring heat to media with higher temperature on another side. According to these phenomena, different types of HPs can be constructed. The main such types are mechanical compression HP, absorption HP, ejector HP, chemical HP and thermoelectric HP. All these types of HP have different principles of construction, different forms of energy to be spent for heat transfer and ways to deliver it and different performance and economical characteristics. With limiting resources and raising cost of fossil fuel, global warming and other environmental issues becoming a focus of world attention, the interest in HP as a means of energy recovery much increased. HPs offer efficient solutions to these adverse effects by recirculating environmental and waste heat back into heat-consuming processes in applications ranging from households and commercial buildings to process industries. The positive impact on the environment depends on the HP type, energy source and efficiency. The principles of functioning and main characteristics of the most suitable HPs for low-grade heat utilisation are discussed as follows.

2.5.1 Vapour-Compression Heat Pumps

The principle of vapour-compression HP directly follows from Carnot thesis published in 1824, as is stated in a book by Reay and MacMichael (1979). The principle is the same as for vapour-compression refrigeration systems (see, e.g., Green and Perry, 2008). The ideal Carnot cycle is reversible and has temperature–entropy diagram as shown in Figure 2.4. It consists of adiabatic (isentropic due to reversible character) compression (1–2), isothermal rejection of heat during the condensation process (2–3), adiabatic expansion (3–4) and isothermal addition of heat due to evaporation (4–1). While the Carnot cycle is an unattainable ideal, it can serve as a standard of comparison and provide an approximate guide to the temperatures that characterise the effectiveness of the whole process of heat transfer from low- to high-temperature level.

The heat transferred to the hot media at the high temperature (Q_{high}) is the objective for HP operation. The amount of heat transferred in the process Q is equal to the heat taken from the low-temperature level. The input of additional energy W is added to this amount, and this sum is delivered to high-temperature level. Also according

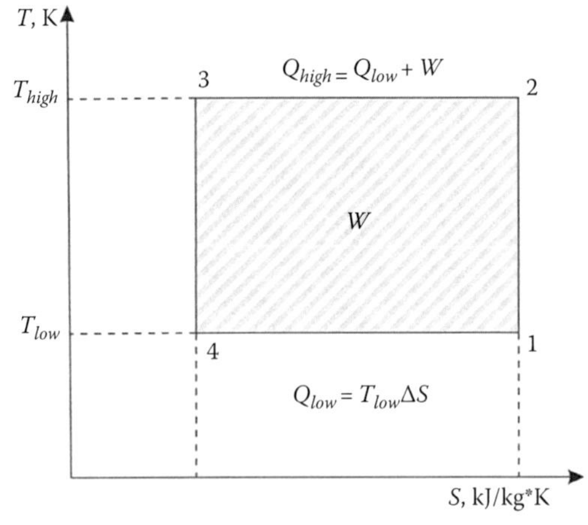

Figure 2.4 Temperature–entropy diagram of the Carnot cycle.

to the third postulate of thermodynamics, for reversible processes, like Carnot cycle, the change of entropy is $\Delta S = \Delta Q/T$. Therefore, the COP of HP operating with Carnot cycle can be calculated as follows:

$$\mathrm{COP} = \frac{Q_{high}}{W} = \frac{Q+W}{W} = \frac{T_{high}}{T_{high} - T_{low}} \tag{2.3}$$

In refrigeration applications, the objective is to take heat at lower temperature. In this case, the coefficient of performance COP_R is determined as the ratio of heat removed from the low-temperature level Q to the energy input W as follows:

$$\mathrm{COP}_R = \frac{Q_{low}}{W} = \frac{Q}{W} = \mathrm{COP} - 1 \tag{2.4}$$

For the time being, refrigeration is a more widespread and matured technology than heat pumping. As the principles of both technologies are the same, the refrigeration terminology is frequently used also in HPs, as, for example, the working medium of vapour-compression cycle is called refrigerant in both cases.

Basic methods of transferring heat to higher-temperature level by vapour-compression HPs use similar processes: evaporation of refrigerant in the evaporator, condensation in the condenser where heat is supplied to hot media and expansion of refrigerant in the expansion device. The difference between these methods is in the way how compression is done, as shown in Figure 2.5. It can be using mechanical compressor driven by electric motor, gas engine–driven mechanical compressor, pressure difference by ejector and thermal energy for absorption and desorption. As far as heat exchangers are concerned, in all these methods, at least two recuperative

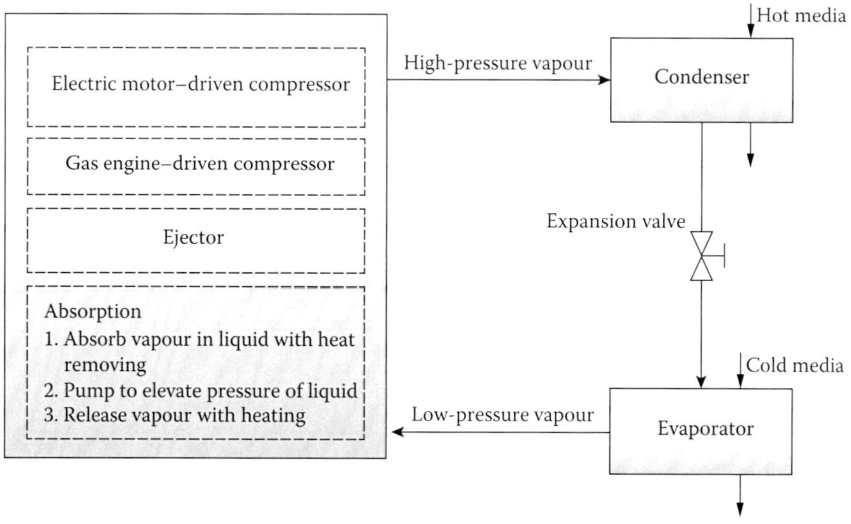

Figure 2.5 Basic methods to transform low-pressure vapour into high-pressure vapour.

heat exchangers are required to work as condenser and evaporator. The number of heat exchange position can be significantly increased with complicated schemes and also when indirect heat exchange is required. The main features of these basic methods are discussed as follows.

2.5.1.1 Mechanical Compressor Heat Pump Systems

Vapour-compression cycles are the most widely used principle in HPs. While *T–S* diagram of ideal Carnot cycle is very simple and convenient for thermodynamic analysis, in practice, the process is much more complicated. The more close to reality vapour-compression cycle (Rankine) is presented in Figure 2.6b in *T–S* coordinates. Isothermal processes are requiring isobaric evaporation and condensation at constant pressures, but in reality, flow of refrigerant in the condenser and evaporator is accompanied by pressure drop. Some subcooling in the condenser and vapour superheating at the compressor suction line are always observed, both because of the continuing process in the heat exchangers and the influence of the environment. These subcooling and superheating are usually desirable factors to ensure that only liquid enters the expansion device. Superheating is recommended as a precaution against droplets of liquid being carried over into the compressor. The operation of the HP shown schematically in Figure 2.6a is realised in the following way. The heat from low-temperature media is supplied to the refrigerant in the evaporator. The refrigerant is boiling there at pressure P_0 and corresponding saturation temperature T_0. The vapour after separation from liquid droplets in the separator enters to the compressor, where its pressure is increasing to P_C. In *T–S* diagram, it corresponds to 1–2 line. The dotted line 1–2* shows isentropic compression that cannot be practically fulfilled because of friction and irreversible heat losses. The temperature of superheated vapour after compression is becoming T_2 that is higher than saturation temperature at pressure P_C. This vapour is cooling down and, after, condensing at saturation temperature T_C in the condenser (line 2–3 in Figure 2.6b). The liquid

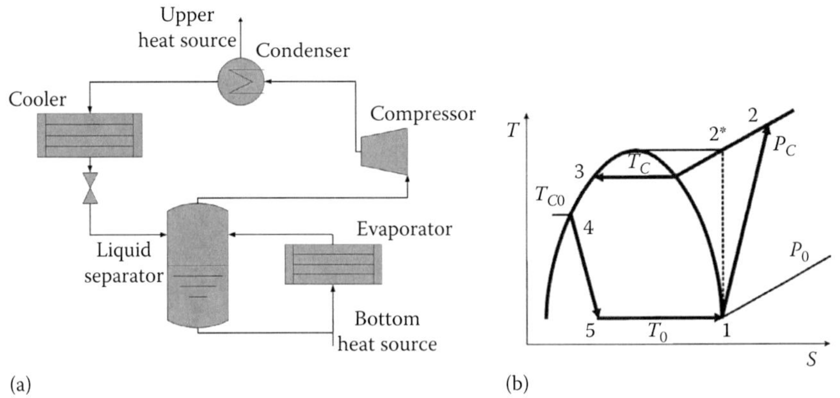

Figure 2.6 One-stage heat pump flow chart (a) and *T–S* diagram of its operation (b).

refrigerant is flowing to the subcooler (line 3–4) where it cools down to temperature T_{C0}. After that, the liquid refrigerant is expanding in the expansion valve (line 4–5), and its pressure drops to P_0 and temperature to T_0. After the liquid refrigerant enters the separator, it flows to the evaporator, and the cycle is renewed.

There are many ways to increase efficiency of the cycle performance (COP), which depend on the refrigerant and working temperatures of HP. One typical example is the use of a cooler in which the refrigerant flows to the compressor suction. It allows to cool liquid refrigerant before expansion, but also to superheat vapour before being processed in the compressor to guarantee there the absence of liquid phase. For large condensing and evaporation temperature (and pressure) difference, compression can be separated in two stages, where two compressors are used with the same refrigerant. Two levels of temperatures can be obtained using condensers on both stages. Another option is cascade system, where two systems with different refrigerants and compressors are exchanging heat through mutual heat exchanger that serve as a condenser for a system with low temperatures and as evaporator for a system with higher temperatures.

There is also a big variety of other thermodynamic cycles that can be used for construction of vapour-compression HPs like double Rankine cycle, cycles of Stirling and Brayton, and their combinations Otto–Rankine, Stirling–Rankine, Rankine–Brayton and Stirling–Brayton. These cycles have different efficiencies with different refrigerants and in practice are used only in some specific applications. But in any of such, HPs require heat exchangers to serve as evaporators and condensers.

There are different types of compressors that are used in mechanical vapour-compression HPs: reciprocating; rotary as screw, vane, scroll, etc.; centrifugal; or turbocompressors. The choice of compressor depends on the capacity and type of HP application. For the choice of heat exchangers to work with specific compressor, one of the main characteristics is the level of vibrations created in the system.

The compressor of an HP should be driven by engine. For this purpose, different kinds of driving engines can be applied. The most widely used are electric motors, which have high efficiency, can have up to 95% nominal loads and are reliable, convenient in service and friendly to the local environment. But when considering the electricity production efficiency, which for most of existing power generation facilities does not exceed 35%–40%, the primary energy usage of HP with electric motor drive can be considered with PER slightly higher than unity. When, for power generation, fossil fuel is used, the reduction in its consumption and global emissions of carbon dioxide and other pollutants is limited. For example, taking the HP with COP equal to 3.1 (as average value according to Reay and MacMichael, 1979), it follows that in the most favourable conditions PER could be equal to 1.178 (COP multiplied on efficiencies of motor 0.95 and electricity generation 0.4).

To drive compressor, internal combustion engines, like gasoline, diesel, gas or gas turbine, can be used. Modern diesel and gas internal combustion engines can have efficiency about 0.5 and even higher. To drive compressor with such engine, PER = 1.55 can be achieved for conditions of the discussed example. The internal combustion engines are producing low-grade heat which is going out with exhaust gases and waste heat released by the engine cylinder jacket. This heat can be used to increase the output of HP and to achieve higher primary energy utilisation, up to PER = 1.8–2. Such measure will require additional heat exchangers in the system. As it is confirmed in a survey published by Hepbasli et al. (2009) for experimental gas engine–driven heat pump (GEHP) systems, PER could reach 2.08 and even 2.43 for existing time gas engines with efficiency below 0.45. It means that for the same amount of produced heat energy, such HP will require two or even more times less gas, than the best condensing gas-fired boiler. Having such good primary energy efficiency, with developments in control systems, reliability and operability of gas engines, GEHP systems can be very competitive in a number of applications.

The crucial element determining construction and efficiency of an HP is the working medium which is taking heat at low temperature and supplying it to higher-temperature level, or refrigerant. The most significant refrigerants are listed in *2013 ASHRAE Handbook: Fundamentals*. More data can be found in the handbook by Green and Perry (2008). Nowadays, there are rapid changes in refrigerants, so it is important to consult the most recent data and publications.

Before, HP refrigerants were natural, like ammonia, CO_2 and SO_2. With the fast development of refrigeration, in 1930–1970, the new refrigerants emerged, such as chlorofluorocarbons (CFCs) and hydrochlorofluorocarbons (HCFCs). These refrigerants are halocarbons which contain one or more of three halogens chlorine, fluorine and bromine. They possess many advantageous qualities such as odourless, non-toxic, nonflammable, nonexplosive and compatible with the most engineering materials and have reasonably high COP.

At the end of the twentieth century, the refrigerant was recognised as extremely important for environment protection, such as the chlorine and bromine atoms from halocarbons emitted to the atmosphere. The chlorine atoms might be expected to cause the breakdown of large amounts of ozone (O_3) in the stratosphere, depleting ozone layer. Montreal Protocol entered into force in 1989, and later, the agreements ban the use of certain CFCs and halogen compounds. Year 2013 was set as the time to freeze and starting 2015 reduce the consumption and production of HCFCs. Chemical companies are working on the development of safe and efficient refrigerants for the refrigeration and HPs. Many projects are concerned with design and study of refrigerant mixtures, both azeotropic and zeotropic having desirable qualities. The attention has been returned also to natural refrigerants like ammonia, carbon dioxide, propane and butane.

The natural refrigerants are not causing harmful effect on ozone layer and have no big global warming potential, but most of them possess other undesirable qualities. Ammonia is toxic and aggressive to copper. Propane and butane are flammable. CO_2 has low COP and requires pressure in condenser 100 bar and higher. It puts forward special requirements for heat exchangers, in which the compactness and small inside volume are the most important, allowing to decrease the charge of refrigerant in HP system.

2.5.1.2 Ejector Compression Heat Pump Systems

These systems substitute an ejector for a mechanical compressor in a vapour-compression system. To drive ejector HP, there should be available source of heat with temperature higher than the target temperature in HP condenser. There is heat exchanger (generator) in which the media of high-temperature heat source (steam or liquid) evaporate liquid refrigerant supplied by the pump. The evaporated refrigerant is flowing to the ejector that utilises the momentum of a high-velocity jet (up to 1,000 m/s) to entrain and accelerate a slower-moving vapour that is evaporated in the main evaporator of the HP at lower pressure. The resulted mixture of vapours then enters the diffuser section where the velocity is gradually reduced due to the increasing cross-sectional area. The pressure in the condenser is lower than in the generator, but maintained higher than in the evaporator because of the expansion valve. This pressure kept high enough for saturation temperature in the condenser can maintain required for HP temperature level.

The principle of ejector HP is the same as of stem-jet refrigeration system described, for example, in the handbook by Green and Perry (2008). The majority of steam-jet systems being currently installed are multistage. They are being simple, rugged, reliable, requiring low maintenance and low cost and vibration free, but are not widely accepted due to the characteristics of the cycle. The same characteristic can be attributed to ejector HP, which has limited number of applications. One such application was proposed by Sun et al. (2014), designed to improve performance of district heating network. Installed at heating substation in combination with heat exchanger for radiator water heating, ejector HPs allow to increase the temperature of the building radiator circuit using the heat of the returned water at primary district heating network. Such heating substation requires at least four heat exchangers: water to water, two evaporators and condenser. Nevertheless, as is stated by the authors, it is proved efficient in some district heating systems in China.

2.5.2 Chemical Heat Pumps

Another way to transfer heat from low to higher level of temperature is to use chemical processes of some substances, endothermic decomposition and exothermic synthesis. These processes can be implemented in gas–liquid (absorption) and gas–solid (adsorption) phases. Important difference consists also in nature of the sorption processes. Physical sorption is a result of the van der Waals force where the interaction energy is very weak and adsorption energy is small, it is higher in chemical sorption and processes involving reversible endothermic and exothermic chemical reactions. A chemical HP is more environmentally friendly than more conventional vapour-compression HPs, as the compressor consumes much more electrical energy and also most natural substances are used as working media.

2.5.2.1 Absorption Heat Pump

Absorption HP uses non-mechanical principle for creating pressure difference in the condenser and evaporator, which are employed in a similar way as in mechanical compression HP. In this case, the driving force is the thermal energy which is supplied in generator to the solution consisting of two components. The refrigerant, a volatile component, is evaporating at high level of temperature to condensate after it is processed in the condenser. Another much less volatile component is called absorbent. Two main combinations of components in the solution are used: lithium bromide as absorbent with water refrigerant and water as absorbent with ammonia refrigerant. After condensation, just like in any vapour-compression HP, liquid refrigerant loses pressure in the expansion valve and evaporates in the evaporator at low-pressure level, taking heat from low-temperature media. After it flows from the generator, the solution with low content of refrigerant component enters the absorber to be enriched with the refrigerant that is flowing as the vapour from the evaporator. The liquid then is pumped to the generator.

As the liquid solution is practically incompressible, the energy for pumping it is practically negligible, and the source of driving the energy for the process is heat supplied by the generator, at which the maximum temperature of the cycle is maintained. An absorber is a component where strong absorbent solution is used to absorb the refrigerant vapour flashed in the evaporator. The absorbent is sprayed over the absorber heat transfer surface through which the higher-temperature media of an HP are heated, absorbing heat of dilution and sensible heat of solution cooling. After being heated in the absorber, high-temperature media are further heated in the main condenser (the parallel connection is also possible).

Nowadays, there are different modifications of absorption HP developed to improve its performance, which are described in more detail by Chan et al. (2013). All of them have relatively low COP compared to mechanical compression HP but can be used where electricity is scarce or expensive taking driving energy from a heat source such as natural gas, propane, solar-heated water, geothermal-heated water or low-grade heat from industry. The HPs using gas are called gas absorption HPs and are commercially produced for industry and domestic applications.

2.5.2.2 *Adsorption Heat Pump*

The basic principle of adsorption HP functioning is the same as that of absorption HP, with the difference that solid matter cannot be so easily moved like liquid. For the implementation of adsorption process, as well as desorption, the solid–gas reactor is necessary. A simple adsorption HP cycle can be fulfilled with four components. It is a condenser, an evaporator, an expansion valve (as in all vapour-compression cycles) and an adsorber, which is a certain type of reactor having heat transfer surface and filled with an adsorbent (such as zeolite, active carbon, silica gel). Adsorber can work in two modes: adsorption mode when heat generated in the process is transferred to heated media through heat transfer surface and desorption mode when heat is transferred to adsorbent through heat transfer surface and adsorbate gas is generated. As both modes cannot be realised simultaneously, the work with one adsorber is intermittent and can be considered as two cycles in a series. The first one is an HP cycle in which the adsorbate is expanded in the expansion valve and vapourised in the evaporator by taking heat from the low-level temperature source. After that, during adsorption process in the adsorber, the heat is released to heated media with intermediate temperature. This cycle can work till the saturation condition is reached in the adsorber or all adsorbate is evaporated. Then another cycle (desorption) can be started in which the heat is supplied to the adsorber from high-temperature source, and adsorbate vapour is released. During condensing in the condenser, the adsorbate transfers heat to the heated media at second intermediate temperature. When desorption process is finished, the HP cycle can be started. Such intermittent character can be avoided using two or multiple adsorber reactors or multiple bed reactors, as described in the literature (see, e.g., Demir et al., 2008).

The intermittent character of adsorption HP cycle despite its disadvantages offers possibility to store heat for a long time or move it to substantial distances without considerable losses that is discussed in the next section. The adsorption HP cycle is characterised by four temperatures: low temperature in the evaporator, where heat is taken from low-temperature media; intermediate temperature in the adsorber during adsorption process, where heat is supplied to higher-temperature media; intermediate temperature in the condenser during condensation; and high temperature at the adsorber during desorption process, where thermal heat for driving HP is supplied. The levels of these temperatures and corresponding pressures depend on the properties of adsorbent and adsorbate. Water and ammonia are the most used adsorbates. The solid–gas adsorption process has a wide range of applications in the utilisation of low-grade heat, because besides zeolite, active carbon, silica gel, etc., there are more than two hundred reactive salts on which base can be constructed adsorbents for temperature levels of 60 °C–300 °C, as shown in the paper by Ma et al. (2009). Ammoniates, hydrates and metal hydrides can also be used as reactants that were tested with HPs prototypes by the number of authors. The survey of these works is presented by Cot-Gores et al. (2012) with classification of tested substances as shown in Figure 2.7.

2.6 STORAGE AND TRANSPORT OF THERMAL ENERGY

Many low-grade heat sources, especially industrial waste heat and solar energy, have various time characteristics in energy production. Such character is also featuring the demand for thermal energy. In sunny days, it can be a surplus of solar energy for heating the building. But at night and on rainy days, when heating is needed mostly, there is no supply from the sun. To overcome this obstacle, thermal energy should be stored during higher-production periods to be used when it is mostly needed. Another problem is thermal energy transportation to the places of its consumption, as the consumer of energy can be situated at considerable distance of low-grade energy sources.

The main purpose of thermal energy storage systems development is to synchronise the time and rate of energy available for supply with its demand. The requirements to energy storage systems depend on the type of application and especially on the amount of energy to be stored and the time length of storage cycles. There are different ways of how the low-grade thermal energy can be stored.

The simplest and most frequently used in different applications is sensible heat storage. The certain volume of material (usually liquid) is kept in thermally insulated space being heated to a required temperature to store thermal energy.

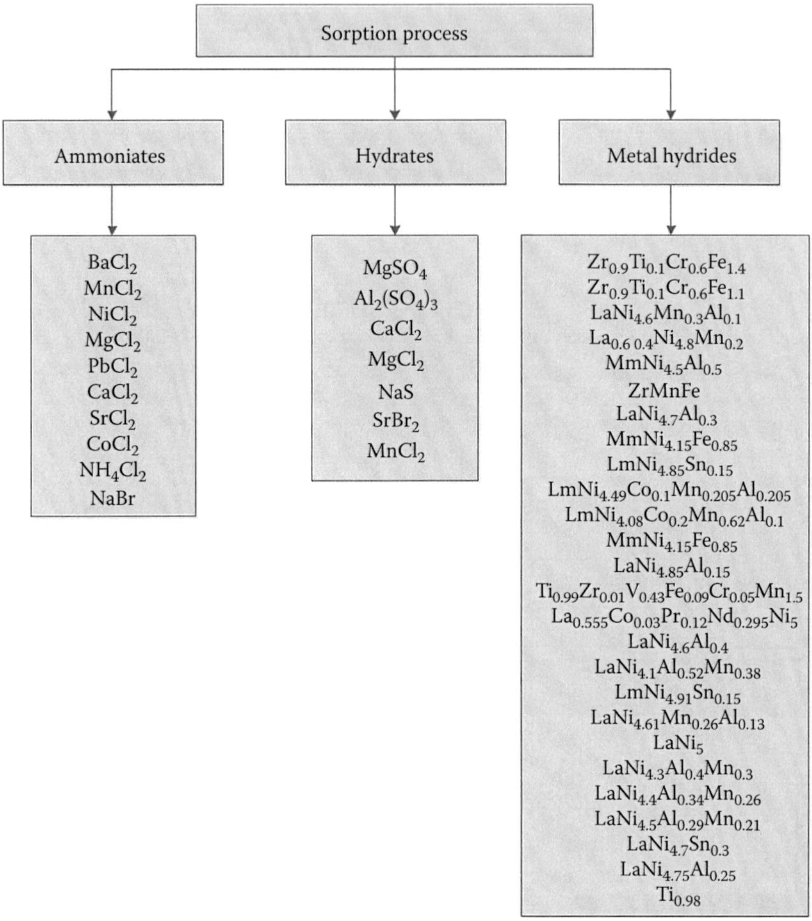

Figure 2.7 Overview of working pairs tested in prototypes (after Cot-Gores et al., 2012).

The thermal energy density of storage is equal to the product of the specific heat of this material by the difference of maximal and ambient temperatures. Hot water storage (hydro-accumulation), underground thermal energy storage (aquifers, boreholes, caverns, pits) and rock-filled storage are the main examples of sensible heat storage implementations in practice. This type of heat storage has the lowest energy density compared to other storage types, as can be seen from comparison in Figure 2.8. It requires big volumes of storage material and amounts of insulation to minimise energy losses.

Latent thermal energy storage systems use the effect of latent heat during phase change between solid and liquid states, as, for example, cold storage water/ice and heat storage by melting paraffin waxes or other phase change material (PCM). Latent thermal energy storage units have bigger energy density than sensible storage units

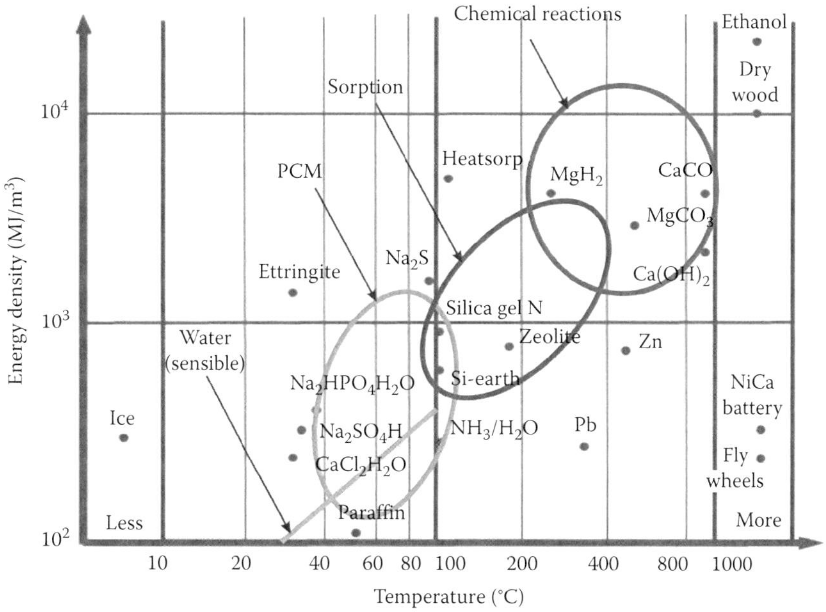

Figure 2.8 Energy density of energy storage methods (adapted from N'Tsoukpoe et al., 2009).

and are generally smaller. Their exterior surface is also smaller, so it needs less insulation or to be insulated to prevent energy losses better than in the case of sensible heat storage. The review of modern methods for thermal energy storage with PCM is published by Nkwetta and Haghighat (2014).

The chemical thermal energy storage includes sorption and thermochemical reactions. In thermochemical energy storage, the endothermic dissociation reaction is used in which the heat is charged in storing media to be stored for future use when it is recovered in a chemically reverse reaction. Both cycles are the same as in chemical HPs described in Section 2.5. The reactants for sorption processes are also the same as shown in Figure 2.7. The chemical storage based on sorption method has much higher energy density than sensible and latent heat storage, as illustrated in Figure 2.8. It requires much less volume for the storage of the same amount of thermal energy. The storing energy media can be kept at ambient temperature to eliminate heat losses, and thermal energy can be stored for much longer periods, such as storing thermal energy up to half a year from winter to be used in summer. The detailed discussion of methods for long-term energy storage is published by Xu et al. (2014).

The use of thermochemical reactions as a means to store thermal energy exhibits the highest energy density, as can be judged on comparison in Figure 2.8. The examples are reforming of methane with steam or carbon dioxide, ammonia dissociation and synthesis and others mostly requiring high- or medium-range temperatures above 500 °C. Very few reactants have reaction temperatures below 260 °C corresponding to the range of low-grade heat utilisation. Methanol can be considered as one of the most promising candidates as the chemical reaction storage medium for temperature about 150 °C–200 °C. Methanol is clean and relatively cheap. The chemical reaction is as follows:

$$CH_3OH \leftrightarrow CO + 2H_2 \quad \Delta H_{298K} = 95 \text{ kJ/mol } (CH_3OH) \tag{2.5}$$

The temperatures of methanol reactions are from 150 °C to 200 °C with the appropriate catalysts, for example, copper–aluminium alloy with weight content 50% of each.

At the stage of storage charging, the reaction of methanol dissociation according to (2.5) should be used directly. When extracting the heat, a two-step liquid-phase methanol synthesis is proposed. It enables to increase heat recovery efficiency. The first step is the carbonation of methanol to methyl formate:

$$CH_3OH + CO \rightarrow CH_3OOCH \quad \Delta H_{298K} = 38.1 \text{ kJ/mol } (CH_3OH) \tag{2.6}$$

The second in sequence is hydrogenation of the formate:

$$CH_3OOCH + 2H_2 \rightarrow 2CH_3OH \quad \Delta H_{298K} = 62.8 \text{ kJ/mol } (CH_3OH) \tag{2.7}$$

To implement these reactions in practice, it is required rather complex chemical reaction unit described by Ma et al. (2009). It consists of the two catalytic chemical reactors involving two distillation columns. Both distillation and catalytic reactions are involved, that is, the ion-exchange resin and the copper–aluminum alloy. The significant drawbacks of such system are rather higher temperatures of thermal energy to be stored and the necessity to store large amounts of hydrogen. At the same time, it can be used for transport of industrial low-grade heat for a long distance with transportation of the reagents by pipelines from energy source to the place of its usage. The transportation efficiency for a distance of 30 km is more than two times higher than the efficiency of transportation using steam or water.

Nowadays, water is the most widespread medium to transport low-grade heat from source of its generation to final users. In liquid state, it is used in the temperature range from ambient to 150 °C by most of district heating systems. As a steam, it is used in some district heating also, but more widely in industry at refineries, distilleries and chemical and other production sites. This type of thermal energy transport can have considerable energy losses to ambient, especially at higher, temperatures. It requires a good thermal insulation of transport pipelines and is limiting economically viable distances.

To minimise losses of thermal energy during its transport for long distances, in most technologies, thermal energy storage can be used. In choosing the best method of transportation, a detailed economy analysis is required, which accounts for the costs of energy and facilities for thermal energy transport and gains of losses prevention.

The latent heat thermal energy storage method can be modified for transportation of the accumulated low-grade thermal energy to distant user. For this purpose, the container with the PCM is charged with the low-grade waste heat using closed circuit and heat exchanger. Then, it is transported by trucks, trains or ships to release heat at the site of its usage. The energy density of PCM storage is higher than that of sensible storage. The outer surface of PCM container is also small compared to pipelines with the same volume and can have good thermal insulation to prevent heat losses. The container should be specially designed to improve the efficiency of the closed circuits with the heat exchangers for charging and discharging of transported thermal energy. According to the survey by Ma et al. (2009) in Japan, two kinds of PCM TransHeat containers for thermal energy transportation are developed: one using sodium acetate trihydrate ($NaC_2H_3O_2{\cdot}3H_2O$ melting point 58 °C, latent heat of fusion 264 kJ/kg) and the other erythritol ($CH_2OH(CHOH)_2CH_2OH$ melting point 126 °C, latent heat of fusion 340 kJ/kg) as PCMs with supply temperatures 50 °C and 110 °C. Such container system can be developed as a networking system for thermal energy distribution.

The solid–gas adsorption process offers much high energy density in storing thermal energy and more opportunities for energy transport. A big number of reactive salts can be used to achieve desired level of temperatures. The heat losses to the environment are practically negligible. The container previously described for PCM energy transport can be very efficient with gas adsorption process.

The hydrogen-absorbing alloys also have been studied for transportation of low-grade heat, as they can have reversible reactions with hydrogen. Such alloy consists of metals, some of which are rare, for example, $Ti_{0.4}Zr_{0.6}Cr_{0.8}Fe_{0.7}Mn_{0.2}Ni_{0.2}Cu_{0.03}$. The heat transportation system with hydrogen-absorbing alloys includes mainly two locations of hydrogen-absorbing alloys, at the source site and the user site, and a pipeline between the two sites. Usually, the alloys at the source site and the user site are different to match the temperature condition. Nasako et al. (1998) have reported the results of their research on low-grade heat transport. Two different metal alloys $LmNi_{4.55}Mn_{0.25}Co_{0.2}Al_{0.1}$ and $LmNi_{4.4}Mn_{0.2}Co_{0.1}Sn_{0.1}$ were used at the heat generation side and utilisation side, respectively. The hydrogen was delivered between the sides by two pipelines with different pressures. The temperature of transported heat was 90 °C. The process is effective, but not cost competitive mostly because of the cost of the alloys.

2.7 LOW-GRADE HEAT TO POWER

One of the main problems in efficient utilisation of low-grade heat is to identify the customer needing available or upgraded thermal energy potential for the price that includes the costs of delivery and upgrading. The kind of energy that can be conveniently transported through the existing distribution network and has a high demand is electricity. To convert heat into mechanical work and ultimately generate electricity, it is possible with different thermodynamic cycles, among which the most widespread is Rankine cycle. It is used to generate about 80% of electricity worldwide. The simplified diagram of Rankine cycle is presented in Figure 2.9, where vapour generated in the evaporator using external heat produces mechanical energy in a turbine. The most frequently used working medium is water, but for temperatures below 370 °C, such process is not economically viable. For that reason, for low-grade heat

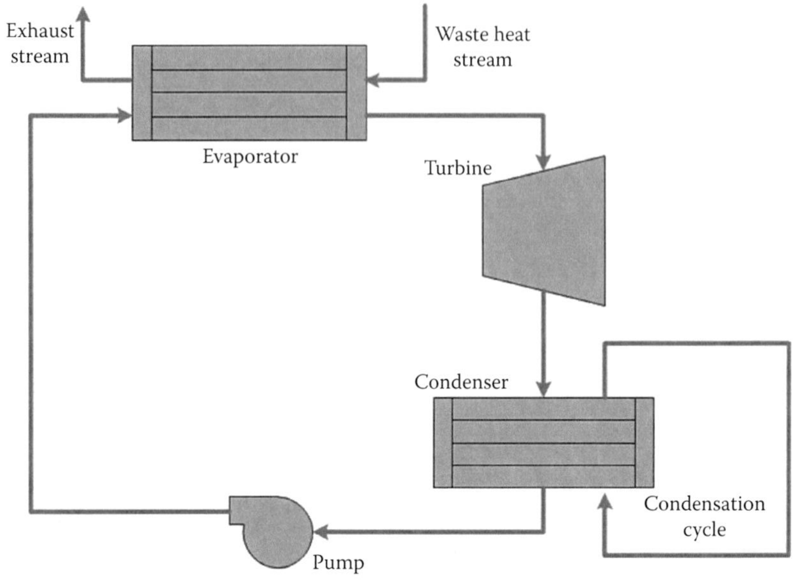

Figure 2.9 Simplified flow diagram of Rankine cycle.

utilisation, different organic chemicals are used as working media, and such cycle is called organic Rankine cycle (ORC). With increased interest in low-heat utilisation in the last decades, a number of different thermodynamic cycles have been investigated for the generation of electricity using low-grade heat. The major ones are ORC, supercritical Rankine cycle (SRC) and Kalina, Goswami and trilateral flash cycle. Some of these cycles are described as follows.

2.7.1 Organic Rankine Cycle

Steam Rankine cycles usually work at temperatures above 673 K, and the ORC, sometimes called as the subcritical ORC, is a modification of the traditional steam Rankine cycle, illustrated in Figure 2.9, with the water substituted by some of organic substance as working fluid. The working fluid should have low boiling point, to recover heat at lower temperatures between 373 and 673 K.

The principles of ORC operation are illustrated in Figure 2.9. The working fluid is pumped from low pressure in the condenser to high pressure in the evaporator as liquid with only little energy input. As an option, a recuperative heat exchanger can be added to improve the ORC efficiency, where the higher-temperature fluid after being processed in the turbine is used to preheat the lower-temperature fluid before it enters in the evaporator. In the evaporator, the liquid is heated at constant pressure by an external low-grade heat source and evaporates becoming a dry saturated vapour. This vapour expands in the turbine which through the generator produces electric power. The temperature and pressure of the vapour there decreases, and it then condensed at a constant temperature in the condenser to become saturated liquid before it is pumped to the evaporator.

The economic characteristics and operating efficiency of an ORC are highly determined by working fluid, as outlined by Tchanche et al. (2011), who also analysed the number of potential working fluids and presented their characteristics. There is a big variety of potential substances identified for ORCs: hydrocarbons, hydrofluorocarbons, HCFC, CFC, perfluorocarbons, siloxanes, alcohols, aldehydes, ethers, hydrofluoroethers, amines, fluid mixtures (zeotropic and azeotropic) and inorganic fluids. Examples of fluids in operating ORC plants are R-114, HFE-30, isobutene and n-pentane (R601). They also described potential applications of ORC for solar thermal power systems, desalination and ocean thermal energy conversion systems, waste heat recovery systems and biomass power plants. The conclusion is that a growing number of ORC manufacturers and installers propose machines that are easily adaptable to existing heat sources for on-site power generation at different power sizes.

2.7.2 Supercritical Rankine Cycle

One of the requirements for fluids of ORC is that they must have critical temperatures higher than the temperatures of the working cycle. There are a lot of substances with relatively low critical temperatures and pressures. These fluids can be compressed directly to their supercritical pressures and heated to their supercritical state before expansion and in such a way better satisfy the process conditions. The configuration of an SRC is practically the same as ORC, illustrated in Figure 2.9, but it does not need a superheater for vapour before the turbine or recuperative heat exchanger after it that is required in ORC to improve performance. The working fluids can be CO_2, H_2O or organic (hydrocarbons) that are heated directly from a liquid state into a supercritical state bypassing the two-phase region. It allows a better thermal match with the heat source resulting in less exergy destruction and negating the need to use phase separation equipment, as discussed in the analysis of SRC published by Chen et al. (2010). Both the ORC and SRC have advantages of their own. Although the SRC allows obtaining a better thermal match than the ORC, the SRC normally needs high pressure, which may lead to difficulties in operation and a safety concern.

2.7.3 Kalina Cycle

The driving force for power generation with turbine is the pressure difference of moving its vapour and the pressure on the turbine outlet, which significantly influences turbine efficiency. In ORC, the low pressure at the turbine outlet is created by condensing the vapour in the condenser. The intensity of condensation process and finally its pressure is determined by the temperature of cooling media and efficiency of heat transfer process in the condenser. Another method of creating low pressure at the turbine outlet was proposed by Kalina in the late 1970s and early 1980s. In Kalina cycle, the low pressure is created by absorption process of volatile component in binary mixture of two fluids, typically water and ammonia. Ammonia with some steam content is used as a vapour to drive the turbine, while water is absorbing it in

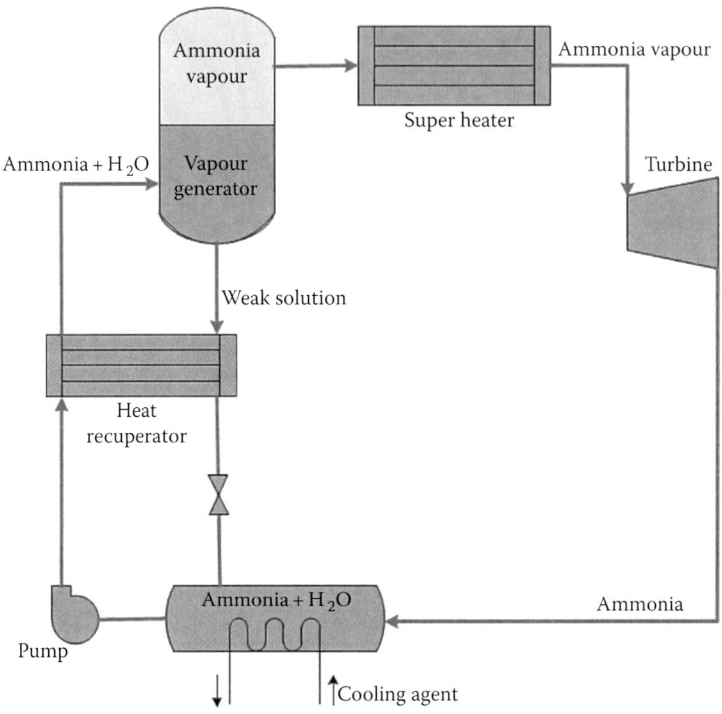

Figure 2.10 Flow diagram of the basic Kalina cycle.

the absorber, and the resulting solution is used to generate vapour mixture in the generator with application of the low-grade heat. It has the advantage that the boiling of the working fluid mixture occurs over a range of temperatures, as discussed by Law et al. (2013). This allows a greater degree of heat recovery from the heat source as, in a countercurrent evaporator, and the heat source can be cooled to a lower temperature. The ratio between water and ammonia varies in different parts of the system to decrease thermodynamic irreversibility and therefore increase the overall thermodynamic efficiency. Several modifications of Kalina cycle have been proposed in the last decades based on different applications. A basic configuration of the Kalina cycle is shown in Figure 2.10. It is claimed that the novel features of the Kalina cycle lead to an increase in power output of up to 20% compared to conventional ORC, but in practice, this increase may be as small as 3%.

2.8 REQUIREMENTS FOR HEAT TRANSFER EQUIPMENT WHEN UTILIZING LOW-GRADE HEAT

In this chapter, the presented analysis of different low-grade heat sources shows that in vast majority of applications, the heating media of the heat source cannot be used directly or mixed with the media delivering heat to the final user;

the transferring heat between these media must be made with recuperative heat exchangers. Besides, all the methods of low-grade heat upgrading or transforming it to power require recuperative heat exchangers for the implementation of heat transfer processes: gas–gas, gas–liquid, liquid–liquid, evaporation and condensation. The conditions of all these processes differ on temperatures, pressures, available pressure drops, stream's thermal and physical properties and tendencies for fouling and corrosion of heat transfer surfaces. All variety of these process conditions cannot be in the feasible way satisfied by one kind of heat exchangers, for example, the most widespread conventional shell-and-tube heat exchangers, even from technical viewpoint. Another consideration is economical, as heat transfer equipment can constitute considerable share in capital cost of many installations for utilisation of low-grade heat.

Nowadays, the development of heat transfer equipment with the aim to satisfy the requirements of different applications in industry, mining, transport, aerospace and communal sector has resulted in creation of a various types of heat exchangers. A lot of these types can be used in different applications of low-grade heat utilisation that put forward some specific requirements. The main such requirements are formulated as follows.

2.8.1 Small Temperature Differences

As can be judged from the analysis made in first sections of this chapter, a majority of low-grade heat sources in industry, communal sectors and renewable geothermal and solar sources have rather limited temperature levels. For the direct utilisation of this heat, the temperature difference between the source and consumer is usually limited. When upgrading of the heat with HPs is used, any additional difference in temperature in the evaporator or condenser is leading to decrease the COP of the total system and deterioration of the system performance.

The powerful tool for increasing energy efficiency in different applications, including utilisation of low-grade thermal energy, is process integration. The current state of this methodology is described in the book edited by Klemeš (2013). The advantages of process integration tools for the development of low-grade heat utilisation systems have been demonstrated in a number of publications, for example, integration of an HP in a food industry waste heat usage by Kapustenko et al. (2008) and buildings' heating systems by Boldyryev et al. (2013). One of the main parameters in process integration is the minimal temperature difference (MTD) in the heat recuperation system. The lower MTD results in better heat recuperation and increased efficiency of energy usage, but with increased cost of heat transfer area for recuperative heat exchangers. When using conventional shell-and-tube heat exchangers, the lower limit of the MTD is about 10 K with the lowest value of 5 K, counting on their construction features. Using the lower values of MTD without considerable increase of capital costs would be beneficial for energy efficiency of recuperative heat exchangers system. It is possible with other types of heat exchangers based on heat transfer enhancement technologies.

2.8.2 Close Temperature Approach

When two fluids are exchanging heat in a heat exchanger, the heated fluid is receiving all thermal energy transferred from the fluid which is being cooled, except the heat losses to the environment, which are usually small and can be practically eliminated with insulation at certain types of heat exchangers. Meanwhile, the useful part of thermal energy, called exergy, is lost as it is infeasible to heat cold fluid to the temperature with which the heating hot fluid is supplied. These losses of exergy can be minimised by heating cold fluid as closer as possible to the supply temperature of hot fluid and to make temperature approach minimal.

2.8.3 Fouling Mitigation

In a number of applications, low-grade heat is produced with substances that have high content of materials that form deposits on heat transfer surfaces causing adverse effect on heat transfer. It concerns many industrial waste heat sources, geothermal and sewage waters, etc. Fouling formation is causing the rise in pressure drop and increased energy consumption for pumping the media. The decrease of overall heat transfer coefficient lowers the amounts of energy recuperated in the system and requires that the surplus surface area should be installed to avoid this. One of the measures to mitigate fouling is the use of enhanced heat transfer surfaces. For example, the fouling rate in plate heat exchangers as a rule should be about ten times less than in conventional shell and tubes for the same conditions. It requires the reliable and accurate methods to account for fouling in heat exchanger design, which are limited for the time being. Another method is periodical cleaning of heat transfer surface. The construction of heat exchanger must allow the cleaning procedures for the stream side with fouling formation.

2.8.4 Compactness and Limited Cost when Using Expensive Materials for Heat Transfer Surface

There are large numbers of low energy sources in which thermal energy is transferred by streams of highly corrosive nature, especially waste heat in chemical industry, coke oven plants, sewage waters and some geothermal applications. A lot of heat is wasted with steam outgoing to the environment in flue gases, which as condensed form acid solutions absorbing oxides from a flue gas. In all of these cases, heat transfer surface must be produced from materials consisting of metal alloys and other materials of higher corrosion resistance. It can make such heat exchanger very expensive and even prohibit heat utilisation on economic grounds. The problem can be solved with compact enhanced heat transfer surfaces, which can require much smaller amounts of expensive material for their production leading to sufficient reduction of heat exchanger cost.

Depending on specific applications that are numerous in low-grade heat utilisation, other requirements for heat exchangers can arise. Therefore, to select the proper type and needed design, all the variety of available heat transfer equipment should be analysed to find the best solution on the market or to improve it with further development for the case of interest.

REFERENCES

ASHRAE Standard Committee. 2013. *ASHRAE Handbook Fundamentals*. www.ashrae.org (accessed 10 April 2014).

Bertani R. 2012. Geothermal power generation in the world. 2005–2010 update report. *Geothermics* 41: 1–29.

Boldyryev S.A., Garev A.O., Klemeš J.J., Tovazhnyansky L.L., Kapustenko P.O., Perevertaylenko O.Yu., Arsenyeva O.P. 2013. Heat integration of ammonia refrigeration cycle into buildings heating systems in buildings. *Theoretical Foundations of Chemical Engineering* 47(1): 39–46.

Chan C. W., Ling-Chin J., Roskilly A. P. 2013. A review of chemical heat pumps, thermodynamic cycles and thermal energy storage technologies for low grade heat utilisation. *Applied Thermal Engineering* 50(1): 1257–1273.

Chen H., Goswami D.Y., Stefanakos E.K. 2010. A review of thermodynamic cycles and working fluids for the conversion of low-grade heat. *Renewable and Sustainable Energy Reviews* 14: 3059–3067.

Cipolla S.S., Maglionico M. 2014. Heat recovery from urban wastewater: Analysis of the variability of flow rate and temperature in the sewer of Bologna, Italy. *Energy Procedia* 45: 288–297.

Cot-Gores J., Castell A., Luisa F., Cabeza L.F. 2012. Thermochemical energy storage and conversion: A-state-of-the-art review of the experimental research under practical conditions. *Renewable and Sustainable Energy Reviews* 16: 5207–5224.

Demir H., Mobedi M., Ulku S. 2008. A review on adsorption heat pump: Problems and solutions. *Renewable and Sustainable Energy Reviews* 12: 2381–2403.

Directive 2009/28/EC of the European Parliament and of the Council on the promotion of the use of energy from renewable sources and amending and subsequently repealing Directives 2001/77/EC and 2003/30/EC. *Official Journal of the European Union* L 140: 16–47, June 5, 2009.

Doty S., Turner W. C. 2009. *Energy management handbook*. The Fairmont Press, Inc., Lilburn, GA, USA.

Ferland K., Papar R. 2007. Industrial waste heat recovery. texasiof.ceer.utexas.edu/PDF/Documents_Presentations/Roundtables/RoundTable_0307.pdf (accessed 10 April 2014).

Frijns J., Hofman J., Nederlof M. 2013. The potential of (waste) water as energy carrier. *Energy Conversion and Management* 65: 357–363.

Green D.W., Perry R.H. 2008. *Perry's Chemical Engineers Handbook*, 8th edn. McGraw-Hill, New York, USA.

Hepbasli A., Erbay Z., Icierc F., Colakd N., Hancioglu E. 2009. A review of gas engine driven heat pumps (GEHPs) for residential and industrial applications. *Renewable and Sustainable Energy Reviews* 13: 85–99.

IEA. 2008. Worldwide trends in energy use and efficiency. www.iea.org/publications/freepublications/publication/Indicators_2008.pdf (accessed 20 March 2014).

IEA. 2013. World energy outlook 2013. www.worldenergyoutlook.org/pressmedia/recentpresentations/LondonNovember12.pdf (accessed 20 March 2014).

Kapustenko P.O., Arsenyeva O.P. 2013. Process integration for energy saving in buildings and building complexes. In Klemeš J.J. (Ed.), *Handbook of Process Integration (PI)*. Woodhead Publishing Limited, Cambridge, UK, 938–965, Chapter 31.

Kapustenko P.O., Ulyev L.M., Boldyryev S.A., Garev A.O. 2008. Integration of a heat pump into the heat supply system of a cheese production plant. *Energy* 33(6): 882–889.

Kilkovsky B., Stehlik P., Jegla Z., Tovazhnyansky L. L., Arsenyeva O., Kapustenko P. O. 2014. Heat exchangers for energy recovery in waste and biomass to energy technologies–I. Energy recovery from flue gas. *Applied Thermal Engineering* 64(1): 213–223.

Klemeš J. (Ed.). 2013. *Handbook of Process Integration (PI): Minimisation of Energy and Water Use, Waste and Emissions*. Woodhead Publishing/Elsevier, Cambridge, UK.

Law R., Harvey A., Reay D. 2013. Opportunities for low-grade heat recovery in the UK food processing industry. *Applied Thermal Engineering* 53: 188–196.

Lund J.W., Freeston D.H., Boyd T.L. 2011. Direct utilization of geothermal energy 2010 worldwide review. *Geothermics* 40: 159–180.

Ma Q., Luo L., Wang R.Z., Sauce G. 2009. A review on transportation of heat energy over long distance: Exploratory development. *Renewable and Sustainable Energy Reviews* 13: 1532–1540.

Mauthner F., Weiss W. 2013. Solar heat worldwide. Markets and contribution to the energy supply 2011. Steinhuber Infodesign, Graz, Austria. IEA SHC. www.aee-intec.at/0uploads/dateien932.pdf (accessed 10 April 2014).

Nkwetta D.N., Haghighat F. 2014. Thermal energy storage with phase change material—A state-of-the art review. *Sustainable Cities and Society* 10: 87–100.

N'Tsoukpoe K.E., Liu H., Le Pierres N., Luo L. 2009. A review on long-term sorption solar energy storage. *Renewable and Sustainable Energy Reviews* 13: 2385–2396.

Pellegrino J., Margolis N., Miller M., Justiniano M., Thedki A. 2004. Energy Use, Loss and Opportunities Analysis: US Manufacturing and Mining. US Department of Energy.

Reay D., MacMichael D. 1979. *Heat Pumps: Design and Applications*. Pergamon Press, Oxford, UK.

Rezaie B., Rosen M.A. 2012. District heating and cooling: Review of technology and potential enhancements. *Applied Energy* 93: 2–10.

RHC-Platform. 2012. Strategic research priorities for solar thermal technology. European technology platform on renewable heating and cooling. www.rhc-platform.org (accessed 10 April 2014).

Schmid F. 2009. Sewage water: Interesting heat source for heat pumps and chillers. Energy-Engineer FH, Swiss Energy Agency for Infrastructure Plants, Zürich, Switzerland, 10 pp. www.bfe.admin.ch/php/modules/publikationen/stream.php? Extlang=en&name=en_ (accessed 10 April 2014).

SDH. 2014. Solar district heating. www.solar-district-heating.eu/ServicesTools/Plantdatabase.aspx (accessed 10 April 2014).

Sun F., Fu L., Sun J., Zhang S. 2014. A new ejector heat exchanger based on an ejector heat pump and a water-to-water heat exchanger. *Applied Energy* 121: 245–251.

Tchanche B.F., Lambrinos Gr., Frangoudakis A., Papadakis G. 2011. Low-grade heat conversion into power using organic Rankine cycles – A review of various applications. *Renewable and Sustainable Energy Reviews* 15: 3963–3979.

USDOE. 2002. Whole house ventilation systems. USDE. web.ornl.gov/sci/roofs+walls/insulation/fact%20sheets/whole%20house%20ventilation%20systems.pdf (accessed 10 April 2014).

Vives M., Schaefer D., Salom J. 2012. Suitable solar thermal collectors for large scale industrial applications. IREC-TR-00003. Catalonia Institute for Energy Research IREC, Sant Adrià de Besòs, Catalonia, Spain.

Xu J., Wang R.Z., Li Y. 2014. A review of available technologies for seasonal thermal energy storage. *Solar Energy* 103: 610–638.

CHAPTER 3

Compact Heat Exchangers

3.1 MAIN DEVELOPMENTS IN COMPACT HEAT EXCHANGERS

The advantageous features of compact heat exchangers (CHEs) have been recognized in aerospace, car and gas turbine power plants since the start of the development of efficient transport vehicles. A further boost in the development of CHEs gave the energy crisis of the 1970s. The additional driving factors for new heat exchanger (HE) designs have been the need for reducing energy consumption with increased heat recuperation at power and process plants with limited increase of space and material consumption and minimizing the capital investment in grass root and retrofitted plant projects. In process industries the use of plate heat exchangers (PHEs) and other CHEs has been increasing owing to their inherent advantages of small size and high heat transfer efficiency. Two new types of PHEs, brazed and welded, were introduced, with the new principle of designing plate-and-frame PHE using one HE plate with different geometries of corrugation pattern.

Brazed Plate Heat Exchanger (BPHE), originally developed in 1977 at Alfa Laval in Sweden by engineers who later in 1983 organized the Sweden SWEP company, initially specializing on this kind of CHEs, is now widely adopted by other manufacturers. In recent times brazed plate exchangers dominate the low to medium (100 kW) capacity range of refrigeration and central air-conditioning equipment, almost completely replacing traditional shell and tube HEs in these applications.

Another derivative of PHE design made in the 1970s is welded PHE. In 1974 the cross-flow welded PHE, developed by researchers and engineers of NTU KhPI and Ukrniichemmash at Kharkiv in Ukraine, was manufactured at Pavlogradchemmash factory and commissioned at Azot chemical factory in Cherchick, Uzbekistan. This HE is designed to operate at a temperature of 520 °C and pressure of 32 MPa in high-pressure shell of ammonia synthesis column. With the development of welding methods, and especially of laser beam welding, different kinds of welded PHEs are now produced by all major PHE manufacturers. They are suitable for a wide variety of process temperatures and pressures, considerably widening the scope of PHE application in process industries.

There is renewed interest in gas microturbines, as the source for distributed power generation. As it is concluded by Shah (2005), from the cost, performance and durability points of view, the prime surface plate-type recuperators have the most potential in microturbine applications. The modern designs of such CHE for microturbine systems use prime surfaces on both fluid sides in a form of corrugated plates welded at the edges to form an airflow passage, to prevent the leaking and mixing of the fluids. The construction principle is the same as in welded PHE, but with much smaller hydraulic diameter of the channel, in the order of 1–2 mm.

The growing widespread use of heat pumps for utilization of low potential heat and renewable sources of energy was analyzed by Reay (2002). CHE applications have a great potential in solar energy utilization, as shown in a survey by Li et al. (2011) for a new generation of high-temperature solar receivers.

The advances in manufacturer ability to produce microscale devices and systems have stipulated the development of minichannel and microchannel HEs. Microchannel HEs have been used in various engineering and scientific applications. It includes medical applications and microelectromechanical systems (MEMS) such as micro HEs, ink-jet printers, pumps, sensors and actuators. Their types also include nonmetal polymer microchannel HEs and ceramic microstructure HEs. The fields of application for microchannel HEs also include heat pumps and refrigeration technologies.

There is also significant progress in shell and tube HEs toward increase of their compactness. Heat transfer tubes of diameters down to 1 mm and wall as thin as 0.04 mm from stainless steel and sophisticated alloys are now available on the market. There is significant increase in research and manufacturing of tubes with enhanced heat transfer as also tube inserts for the same purpose.

In the chemical engineering field, the requirements of process intensification lead to the development of multifunctional devices such as HE/reactors. They enable to perform in one piece of equipment chemical reactions simultaneously with heat supply for endothermic reactions or heat withdrawal for exothermic reactions. Such equipment is made on the same principles of construction as HEs described in this chapter but this field is out of scope here. The comprehensive survey of publications on concepts and technology of HE/reactors can be found in the paper by Anxionnaz et al. (2008).

To analyze the trends in the development of CHEs and ways to increase their effectiveness in low-grade heat utilization, let us first consider basic principles and terminology of compactness and its impact on heat transfer and hydraulic performance for flow inside channels.

3.2 BASIC PRINCIPLES AND TERMINOLOGY OF COMPACTNESS

The main parameter often quoted as a measure of compactness is surface area density. In a first edition of their book on CHEs, Kays and London (1984) introduced two parameters characterizing surface area density. The first one is the ratio of heat transfer surface area for one side, occupied by one fluid in HE (hot or cold), to the volume on that side:

$$\sigma_i = \frac{F_i}{V_i} \quad \mathrm{m}^2/\mathrm{m}^3 \tag{3.1}$$

where $i = 1$ for hot fluid and $i = 2$ for cold fluid.

Another parameter is the ratio of heat transfer surface area on one side to the total volume occupied by HE:

$$\psi_i = \frac{F_i}{V} \quad \mathrm{m^2/m^3}, \tag{3.2}$$

Defined in this way, both parameters characterize compactness of heat transfer surface area on one side of HE, with the difference in referencing volume at that side in case of β or total volume of HE occupied by heat transfer walls in case of ψ. The relationship between these two parameters can be expressed through the share of the volume of the specific side to the total volume of fluids on both sides of HE:

$$v_i = \frac{V_i}{V_T} \tag{3.3}$$

The volume occupied by heat transfer wall materials V_W is accounted by introducing HE porosity:

$$p = 1 - \frac{V_W}{V} = \frac{V_T}{V} \tag{3.4}$$

The relationship is as follows:

$$\psi_i = \sigma_i \cdot v_i \cdot p \tag{3.5}$$

The definition of compactness on one side of HE β given by Equation 3.10 is directly linked to the hydraulic diameter D_h of the channels for fluid on that side:

$$D_h = \frac{4 \cdot f_{ch}}{\Pi_{ch}}, \tag{3.6}$$

where

f_{ch} is the cross-section flow area of the channel, $\mathrm{m^2}$
Π_{ch} is the perimeter washed by the fluid, m

For the channel of length L, even if flow cross-section area varies along the channel, the hydraulic diameter is

$$D_h = \frac{4 \cdot V_{ch}}{F_{ch}}, \tag{3.7}$$

where

V_{ch} is the volume of fluid inside the channel, $\mathrm{m^3}$
F_{ch} is the surface area in the channel washed by the fluid, $\mathrm{m^2}$

Comparing with Equation 3.1 one can get

$$\sigma_i = \frac{4}{D_{hi}} \quad \mathrm{m^2/m^3}, \tag{3.8}$$

The most cited book on CHEs written by Kays and London in its first edition at 1962 was based on research of intensified heat transfer surfaces of transport HEs (including HEs for automobiles, aerospace and other vehicles). It is concerned mostly of extended surface HEs of plate–fin and tube–fin types. HEs of these types were considered as compact in contrast to HEs of shell and tube type. The development of all types of HEs, including shell and tubes, to be more compact and efficient raised the need of a quantitative characteristic.

The HE is regarded as compact for surface area density greater than a certain level. Shah and Sekulic (2003) have defined this level for gas-to-fluid HE as 700 m^2/m^3 for heat transfer surface area density σ_G on a gas side, which corresponds to the hydraulic diameter D_h less than 6 mm. For operating in a liquid or phase change stream, HE is referred as compact when the heat transfer area density σ_L is greater than 400 m^2/m^3, which corresponds to the hydraulic diameter D_h less than 10 mm.

Estimating the compactness at the specific side of the HE using parameter σ is important for an engineer designing HE or selecting an HE type for a specific duty. But for the engineer selecting equipment for a specific process plant, it is important to estimate the size of the HE as a whole piece. It is directly possible with a parameter ψ characterizing the total volume occupied by the heat transfer area of the HE. To estimate total compactness ψ through compactness on one side of HE σ, Equation 3.5 can be utilized.

For primary heat transfer surface HEs, like PHEs, shell and tube, and spiral, surface areas on both sides can be taken as equal. Assuming that for liquid–liquid duties the volumes on both sides are approximately equal ($v_i \approx 0.5$) and for thin heat transfer wall porosity is close to unity, we can estimate that HE can be regarded as compact for total compactness ψ bigger than 200 m^2/m^3. This figure was recommended by Hesselgreaves (2001) for industrial HEs. The size of the HE for the given duty depends not only on its heat transfer surface area but also on the heat transfer and hydraulic characteristics of the channels on both sides. HE with enhanced heat transfer characteristics can require much less heat transfer area than traditional ones.

3.3 HEAT TRANSFER ASPECTS OF COMPACTNESS

Based on the averaged characteristics, the heat transfer surface area of HE for a specific duty can be determined by the following equation:

$$F = \frac{Q}{U \cdot \Delta T_m}, \tag{3.9}$$

where

Q is the specified heat transfer load, W
U is the average overall heat transfer coefficient, $W/(m^2 \cdot K)$
ΔT_m is the mean temperature difference between heat exchanging streams, K

It can be expressed through logarithmic mean temperature difference (LMTD or $\Delta T_{\ln}$) with the introduction of a correction factor C_{FT}:

$$\Delta T_m = C_{FT} \cdot \Delta T_{\ln}, \tag{3.10}$$

$$\Delta T_{\ln} = \frac{\Delta T_b - \Delta T_s}{\ln\left(\Delta T_b / \Delta T_s\right)}, \tag{3.11}$$

where ΔT_b and ΔT_s are the bigger and smaller temperature differences of hot and cold streams at the ends of HE, K. In the limiting case when $\Delta T_b = \Delta T_s$, the uncertainty limit is $\Delta T_{\ln} = \Delta T_b = \Delta T_s$.

Let us consider the primary heat transfer surface HE. The overall heat transfer coefficient can be determined by the following equation:

$$U = \frac{1}{\frac{1}{h_1} + \frac{1}{h_2} + \frac{\delta_w}{\lambda_w} + R_{f1} + R_{f2}}, \tag{3.12}$$

where

- h_1 and h_2 are the film heat transfer coefficients for hot stream (index 1) and cold stream (index 2) in HE, $W/(m^2 \cdot K)$
- δ_w is the wall thickness, m
- λ_w is the wall thermal conductivity, W/(mK)
- R_{f1} and R_{f2} are the thermal resistances of fouling on both sides, $m^2 \cdot K/W$

In Equation 3.9, the heat transfer load Q, LMTD $\Delta T_{\ln}$ and thermophysical properties of streams are determined by specific process conditions that must be satisfied by HE. For the same process requirements, the value of the heat transfer area is determined by construction features of HE, which are determined by film heat transfer coefficients h_1 and h_2 and LMTD correction factor C_{FT} and also influenced by the thermal resistance of fouling. Hence to minimize heat transfer area and make HE smaller, one needs to keep the highest value of C_{FT} (the maximum $C_{FT} = 1$ is possible for countercurrent flow arrangement) and to achieve the highest overall heat transfer coefficient U.

In denominator of Equation 3.12, there is a sum of five terms. Each of these terms corresponds to a certain thermal resistance. To maximize the overall heat transfer coefficient, these terms should have minimal possible values.

The material of the heat transfer wall and its thickness δ_w are usually determined by other requirements than just heat transfer. First, the material must withstand the corrosive or aggressive action of both stream fluids. The strength characteristics of such material are determined by the wall thickness, ensuring its ability to withstand pressure difference between streams. The minimization of wall thickness is concerned mostly with the need to save material and consequently cost and weight of HE, but also is beneficial for heat transfer. For CHEs made from metals, the share of wall thermal resistance in overall resistance to heat transfer is rather small. For example, most of the contemporary plate-and-frame HEs are made from stainless steel with rather low metal thermal conductivity (about 16 W/mK) and have wall thickness of 0.5 mm, with thermal resistance approximately 0.000031 $m^2 \cdot K/W$. Brazed PHEs (BPHEs) have wall thickness even less, down to 0.3 mm, with thermal resistance of the wall about 0.000019 $m^2 \cdot K/W$. For nonmetals, like plastics or ceramics, the wall thermal resistance is considerably higher and can significantly influence overall heat transfer performance of HE.

The thermal resistance to heat transfer between walls and streams in HE $1/h_1$ and $1/h_2$ can be reduced by increasing velocities in channels, which in most cases leads also to reduction of fouling deposits and of their thermal resistances R_{f1} and R_{f2}. But increase of flow velocity leads to increase of pressure drop in channels, which can be calculated by the following equation:

$$\Delta P_i = 2 \cdot f_i \cdot \frac{L_i}{D_{hi}} \cdot \rho_i \cdot W_i^2 \quad i = 1, 2 \tag{3.13}$$

where

f_i is the Fanning friction factor for flow in respective channels
L_i is the channel length, m
D_{hi} is the hydraulic diameter of the channel, m
ρ_i is the density of the fluid, kg/m^3
W_i is the average velocity in the channel, m/s

At the same flow rate, the increase in pressure drop results in higher power to pump the stream through HE channels. In a majority of practical applications, the allowable pressure drops ΔP_i^o, which must not be exceeded, are specified. When HE can be regarded as a part of the system, which includes also pumps, the pressure drop can be found as a result of solving the optimization problem with some of the cost objective functions.

The problems arising on designing and selecting HE can be solved with reliable thermohydraulic models accounting for different types of construction and their features. The base for thermal design is equations that enable to calculate film heat transfer coefficients in channels of specific type. The film heat transfer coefficient in dimensionless form can be represented by the Nusselt number:

$$\text{Nu} = \frac{h \cdot D_h}{\lambda} \tag{3.14}$$

where λ is thermal conductivity of the fluid, $W/(m \cdot K)$.

In compact surface terminology, the Colburn j factor is also frequently used:

$$j = \frac{\text{Nu}}{\text{Re} \cdot \text{Pr}^{1/3}} = \text{St} \cdot \text{Pr}^{2/3} \tag{3.15}$$

where

$$\text{Re} = \frac{W \cdot D_h}{\nu} \text{ is the Reynolds number} \tag{3.16}$$

$$\text{Pr} = \frac{c_p \cdot \rho \cdot \nu}{\lambda} \text{ is the Prandtl number} \tag{3.17}$$

$$\text{St} = \frac{h}{W \cdot \rho \cdot c_p} \text{ is the Stanton number} \tag{3.18}$$

ν is the cinematic viscosity, m^2/s
c_p is the specific heat of the flowing fluid, $J/(kg \cdot K)$

The analysis of the influence of different factors on compactness in general form is possible only with a lot of assumptions and simplifications. It gives general directions in the development of CHEs and is based on equations of heat balance, which for cold stream can be written in the following form:

$$Q = U \cdot \Delta T_m \cdot F = G_2 \cdot c_{p2} \cdot \left(T_{22} - T_{21}\right), \tag{3.19}$$

where
G_2 is the mass flow rate of the cold stream, kg/s
T_{21} and T_{22} are the temperatures of the cold stream at the inlet and outlet of HE, K

The analysis is made for countercurrent one-pass HE ($\Delta T_m = \Delta T_{\ln}$). The main assumptions are that in Equation 3.12, the thermal resistances of the wall, fouling and heat transfer from the hot stream are negligible compared with the thermal resistance to heat transfer in the cold stream:

$$\frac{\delta_w}{\lambda_w} \ll \frac{1}{h_2}, \quad R_{f1} \ll \frac{1}{h_2}, \quad R_{f2} \ll \frac{1}{h_2}, \quad h_2 \ll h_1 \tag{3.20}$$

These assumptions can be justified for the gas side of the gas–liquid HE, for which they were initially introduced by Kays and London and much less realistic for HE liquid–liquid. But such approach enables to analyze the thermal behavior at one side of the HE by prepositioning that $U = h_2$. Following Hesselgreaves (2001) in this case from Equation 3.19, one can write

$$\frac{h_2 \cdot F}{G_2 \cdot c_{p2}} = \frac{T_{22} - T_{21}}{\Delta T_{\ln}} = \text{NTU}, \tag{3.21}$$

where NTU is the number of heat transfer units for the cold stream side of the HE.

From Equations 3.14 through 3.16 can easily received

$$\frac{h_2 \cdot D_{h2}}{\lambda_2} = \frac{W_2 \cdot D_{h2}}{\nu_2} \cdot j_2 \cdot \mathrm{Pr}_2^{1/3} \tag{3.22}$$

The velocity of the cold stream is

$$W_2 = \frac{G_2}{\rho_2 \cdot F_{cs2}}, \tag{3.23}$$

where F_{cs2} is the cross-sectional area of side 2 of the HE, m^2.

Accounting for Equations 3.17, 3.22 and 3.23, Equation 3.21 can be expressed as

$$j_2 = \frac{F_{cs2}}{F} \cdot \mathrm{Pr}^{2/3} \cdot \mathrm{NTU}. \tag{3.24}$$

From the definition of hydraulic diameter, it follows that

$$D_{h2} = \frac{4 \cdot F_{cs2}}{\Pi_2} = \frac{4 \cdot F_{cs2} \cdot L}{\Pi_2 \cdot L} = \frac{4 \cdot F_{cs2} \cdot L}{F}, \tag{3.25}$$

where Π_2 is the perimeter of the channels washed by fluid on the cold side of the HE, m.

Equation 3.24 determines the value of the Colburn j factor, which satisfies heat transfer process conditions, specified by the number of heat transfer units NTU. Accounting for Equation 3.25 it can be rewritten as

$$j_2 = \frac{D_{h2}}{4 \cdot L} \cdot \mathrm{Pr}^{2/3} \cdot \mathrm{NTU}. \tag{3.26}$$

The pressure drop in HE is determined by Equation 3.13. Substituting there W_2 from Equation 3.23 and rearranging,

$$\frac{2 \cdot \rho_2 \cdot \Delta P_2}{G_2^2} = f_2 \cdot \frac{4 \cdot L}{D_{h2} \cdot F_{cs2}^2} \tag{3.27}$$

Excluding D_{h2}/L from Equations 3.26 to 3.27 and rearranging,

$$\frac{2 \cdot \rho_2 \cdot \Delta P_2}{G_2^2} = f_2 \cdot \frac{\mathrm{Pr}^{2/3} \cdot \mathrm{NTU}}{j_2 \cdot F_{cs2}^2} \tag{3.28}$$

Introducing the mass velocity

$$g_2 = \frac{G_2}{F_{cs2}} \quad \mathrm{kg/(m^2 s)}, \tag{3.29}$$

the 'core velocity equation', named so by London (1983), takes the following form:

$$\frac{g_2^2}{2\cdot\rho_2\cdot\Delta P_2}=\frac{j_2/f_2}{\mathrm{Pr}^{2/3}\cdot\mathrm{NTU}} \tag{3.30}$$

For specified heat transfer process conditions (Pr_2, NTU, ρ_2 and ΔP_2 are known), g_2 is a function only of j_2/f_2. London (1964) described parameter j/f as the 'flow area goodness factor'. Based on the assumption that j/f is only a weak function of Reynolds number (j/f in the range from 0.2 to 0.3), Hesselgreaves (2001) has made some important conclusions.

First is based on Equation 3.30, in which g_2 should not change much, as j/f is not considerably changing. As in Equation 3.29, G_2 is fixed by process conditions, than for F_{cs2} as follows:

- Flow area is largely independent of hydraulic diameter.

In view of this, from Equation 3.27 follows:

- Flow length decreases as hydraulic diameter decreases.

Both conclusions are rather vaguely describing trends. We can check these influences in a more detailed examination of the correlations of j and f for some heat transfer surfaces with different forms of channels.

3.4 THERMAL AND HYDRAULIC PERFORMANCE OF DIFFERENT HEAT TRANSFER SURFACES

In recently published literature there is a lot of correlations available for different forms of channels with different types of heat transfer enhancement. The most complete data are for flows inside straight smooth pipes, which were investigated much longer and thoroughly than any other heat transfer surfaces. For the friction factor in pipes is used single equation proposed by Churchill (1977), which can be applied for laminar, transitional and turbulent flow in smooth tubes (roughness is zero):

$$f=2\cdot\left[\left(\frac{8}{\mathrm{Re}}\right)^{12}+\frac{1}{(A+B)^{\frac{3}{2}}}\right]^{\frac{1}{12}} \tag{3.31}$$

$$A=\left[2.457\cdot\ln\left(\frac{1}{\left(\frac{7}{\mathrm{Re}}\right)^{0.9}+0.27\cdot\frac{0}{D}}\right)\right]^{16}\qquad B=\left(\frac{37{,}530}{\mathrm{Re}}\right)^{16}$$

For calculation of film heat transfer coefficient in the turbulent flow regime, the equation proposed by Gnielinski (1975) is

$$\mathrm{Nu} = \mathrm{Nu}_T = \frac{f \cdot \mathrm{Pr} \cdot (\mathrm{Re} - 1{,}000)}{2 \cdot \left[1 + 12.7 \cdot \sqrt{f/2} \cdot \left(\mathrm{Pr}^{2/3} - 1\right)\right]} \quad \text{for } \mathrm{Re} > 4{,}000 \tag{3.32}$$

In laminar and transitional flow regimes are used the equations proposed by Gnielinski (2013).

For $\mathrm{Re} \le 2{,}300$,

$$\mathrm{Nu} = \mathrm{Nu}_L = \left[3.66^3 + 0.7^3 + \left(1.615\sqrt[3]{\mathrm{Re} \cdot \mathrm{Pr} \cdot \frac{D}{L}} - 0.7\right)^3 + \left(\frac{2}{1 + 22 \cdot \mathrm{Pr}}\right)^{\frac{1}{6}} \cdot \left(\mathrm{Re} \cdot \mathrm{Pr} \cdot \frac{D}{L}\right)^{0.5}\right] \tag{3.33}$$

For $4{,}000 \ge \mathrm{Re} > 2{,}300$

$$\mathrm{Nu} = \mathrm{Nu}_{L.2,300} + \frac{\left(\mathrm{Nu}_{T.4,000} - \mathrm{Nu}_{L.2,300}\right) \cdot (\mathrm{Re} - 2{,}300)}{4{,}000 - 2{,}300} \tag{3.34}$$

The influence of temperature difference between bulk of the stream and wall is neglected here.

To analyze the effects of heat transfer enhancement, let us consider two heat transfer surfaces with different types of heat transfer intensification. One type is the flow inside tubes having a relatively small scale, compared to tube diameter, elements causing turbulence in the near wall region. Another type is the flow inside the PHE channel, where big scale corrugations on forming adjacent channel plates have multiple contact points. The size of corrugations is comparable to the equivalent diameter of the channel.

Kalinin et al. (1990) reported about their experiments with round tubes having small inside elements formed as diaphragms of special shape with some distance between them. They have found that the relative increase of the Nusselt number compared to the smooth tube was about the same as the relative increase in the friction factor and was equal to about two times. Similar figures for the relative increase of the Nusselt number and friction factor was observed also by Kukulka and Smith (2013) in their experiments with Vipertex 1EHT enhanced heat transfer tubes for the heating of propylene glycol in the range of Reynolds numbers from 250 to 18,000. For the examples in Chapter 3, illustrating the effects of such type of intensification, use the same Equations 3.31 through 3.34 as for smooth tubes with the results for the Nusselt number and friction factor multiplied by 2.

The comparison with tubes and channels having small-scale turbulence promoters (see, e.g., Kays and London, 1984) has shown that the increase in the Nusselt number for most of them is smaller than the increase in the friction factor.

Results of analysis for such channels are somewhere between the results chosen here for enhanced pipes and for channels with another type of enhancement, such as channels of PHEs.

The generalized correlation for the friction factor for the main corrugated field of criss-cross-flow channels of PHEs was proposed by Arsenyeva (2010). The validity of the correlation for different forms of corrugations was confirmed later by Arsenyeva et al. (2011) for corrugations with an inclination angle β from 14° to 72° and double height to pitch ratio γ from 0.52 to 1.02 in the range of Reynolds numbers from 5 to 25,000. This correlation is as follows:

$$f = 2 \cdot \left[\left(\frac{12 + p2}{\mathrm{Re}} \right)^{12} + \frac{1}{(A+B)^{\frac{3}{2}}} \right]^{\frac{1}{12}}; \tag{3.35}$$

$$A = \left[p4 \cdot \ln \left(\frac{p5}{\left(\frac{7 \cdot p3}{\mathrm{Re}} \right)^{0.9} + 0.27 \cdot 10^{-5}} \right) \right]^{16}, \quad B = \left(\frac{37{,}530 \cdot p1}{\mathrm{Re}} \right)^{16} \tag{3.36}$$

where $p1, p2, p3, p4, p5$ are the parameters defined by the channel corrugation form:

$$p1 = \exp(-0.15705 \cdot \beta); \quad p2 = \frac{\pi \cdot \beta \cdot \gamma^2}{3}; \quad p3 = \exp\left(-\pi \cdot \frac{\beta}{180} \cdot \frac{1}{\gamma^2} \right); \tag{3.37}$$

$$p4 = \left(0.061 + \left(0.69 + tg\left(\beta \cdot \frac{\pi}{180} \right) \right)^{-2.63} \right) \cdot \left(1 + (1-\gamma) \cdot 0.9 \cdot \beta^{0.01} \right); \quad p5 = 1 + \frac{\beta}{10}.$$

where

$\gamma = 2 \cdot b/S$ is the corrugation doubled height to pitch ratio

β is the angle of corrugation to the main flow direction

To calculate the film heat transfer coefficients in channels of PHEs, the analogy between heat and momentum transfer can be used, as it is demonstrated by Arsenyeva et al. (2012). Based on modified Reynolds analogy, the following equation was proposed:

$$\mathrm{Nu} = 0.065 \cdot \mathrm{Re}^{6/7} \cdot (\psi \cdot \zeta_s)^{3/7} \cdot \mathrm{Pr}^{0.4} \cdot \left(\frac{\mu}{\mu_w} \right)^{0.14} \tag{3.38}$$

Here ψ is the share of pressure loss due to friction on the wall in total loss of pressure.

The value of ψ is estimated by the following equation:

$$A_1 = \frac{380}{\left[tg(\beta)\right]^{1.75}}; \quad \text{at Re} > A_1 \quad \psi = \left(\frac{\text{Re}}{A_1}\right)^{-0.15\cdot\sin(\beta)}; \quad \text{at Re} \le A_1 \quad \psi = 1 \qquad (3.39)$$

In further calculations at this chapter, the influence of the temperature difference between the bulk of the stream and wall is neglected ($\mu/\mu_w \approx 1$).

The validity of Equation 3.5 for criss-cross flow types of PHE channels was confirmed for Reynolds numbers from 100 to 25,000 and Prandtl numbers from 1.9 to 9.

The graphical comparison of friction factor dependence of Reynolds numbers for three different heat transfer surfaces is presented in Figure 3.1. The calculations are made for water with an average temperature of 30 °C. The geometrical parameters of the PHE channel are $\beta = 60°$ and $\gamma = 0.625$. It shows that friction factors in the PHE channel are much higher compared to the other two. The friction factor above that for smooth tube increases to 5–80 times with the increase of Re, while for enhanced tube, just 2 times in all shown range of Reynolds numbers.

It can be expected that such dramatic increase in friction factor for PHE channels can be compensated by increase in heat transfer coefficient. In Figure 3.2 the calculations for the same conditions of Nusselt numbers compared to heat transfer surfaces are presented. In the region of Re < 50, Nusselt number takes approximately the same value, as for smooth channel with parallel walls Nu = 7.54. After that it starts to increase compared to smooth straight tubes and channels up to Re = 2,300, which corresponds to the beginning of the transition regime in smooth tubes. This biggest increase of Nusselt number for examined PHE channel is about 20 times. After that

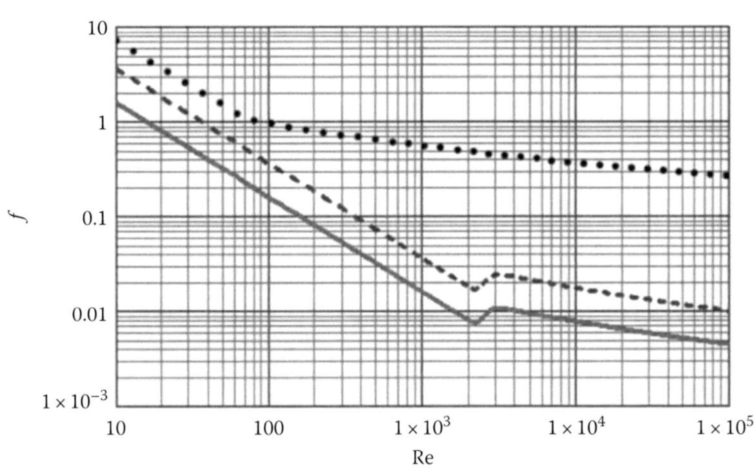

Figure 3.1 The friction factor dependence from Re number of compared heat transfer surfaces: solid curve is for smooth straight tube (Equation 3.31); dashed curve is for tube with heat transfer enhancement and dotted curve is for PHE channel.

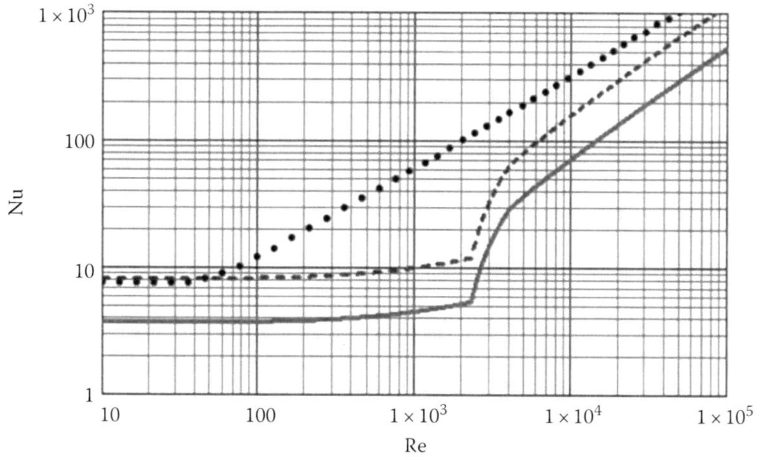

Figure 3.2 Nusselt number dependence from Re number of compared heat transfer surfaces: solid curve is for smooth straight tube (Equations 3.32 through 3.34); dashed curve is for tube with heat transfer enhancement and dotted curve is for PHE channel.

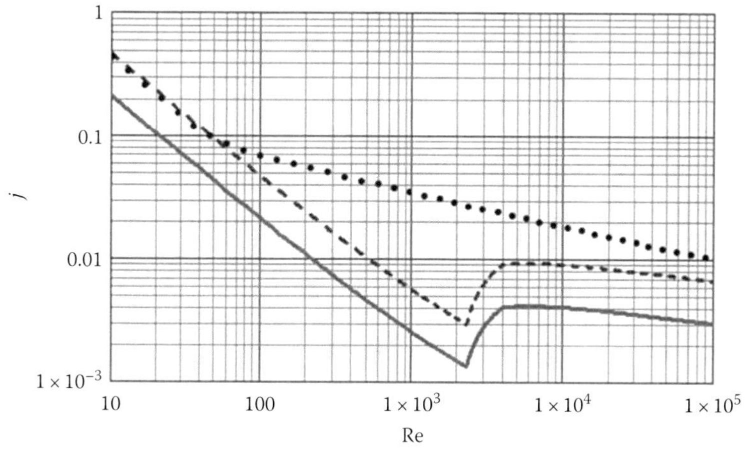

Figure 3.3 Colburn *j* factor dependence from Re number of compared heat transfer surfaces: solid curve is for smooth straight tube; dashed curve is for tube with heat transfer enhancement and dotted curve is for PHE channel.

it diminishes becoming about four times for developed turbulent flow in tubes at Re > 10,000. The same character is observed for Colburn *j* factor (see Figure 3.3).

To estimate the efficiency of compared heat transfer surfaces, first we can easily calculate the 'flow area goodness factor' *j*/*f*. This comparison is presented in Figure 3.4.

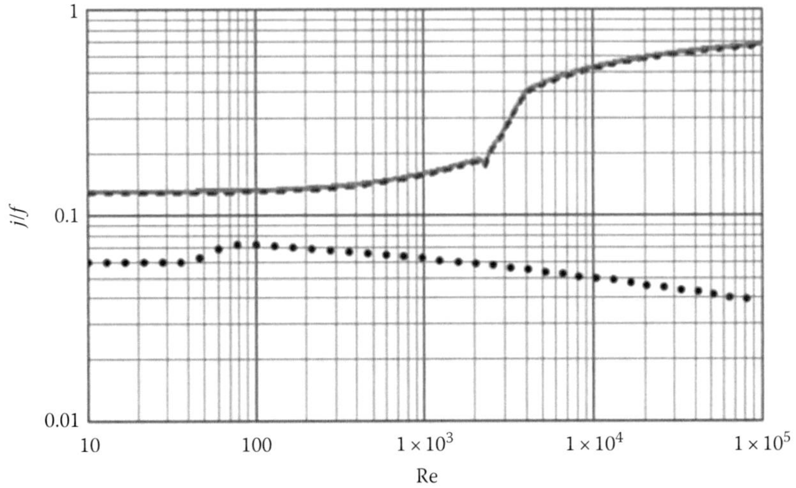

Figure 3.4 Factor *j*/*f* dependence from Re number of compared heat transfer surfaces: solid curve is for smooth straight tube; dashed curve is for tube with heat transfer enhancement and dotted curve is for PHE channel.

Considering the comparisons of the curves presented in Figure 3.4, we can make the following conclusions:

1. The factor *j*/*f* for different heat transfer surfaces significantly differs, in our case from 0.7 for tubes to 0.035 for PHE channel. Therefore to make any analysis based on the assumption about this factor, the constant value would not be justified. Even for tubes this factor is changing from 0.12 to 0.7, which is much more than those assumed by Hesselgreaves (2001). Such assumption can be made only for one type of flow regime: laminar or turbulent.
2. The flow area goodness factor *j*/*f* for PHE channels is much lower than that for smooth tubes and tubes (and channels) with small-scale promoters of turbulence. This fact shows that the channels of criss-cross flow type are not the best solution for transport radiators or other extended heat transfer surface applications. To compare performances of different heat transfer surfaces in other applications, more detailed analysis is needed, as presented in section 3.5.

3.5 INFLUENCE ON COMPACTNESS OF HEAT TRANSFER SURFACE GEOMETRICAL FORM AND ITS SCALING FACTOR

With the wide range variation of factor *j*/*f*, as shown in Figure 3.4, it is not possible to draw any general conclusions. But for given conditions and specified hydraulic diameter D_h, the ‘core velocity’ is Equation 3.30 with correlations for prediction of Colburn *j* factor and Fanning friction factor mathematically can be presented as one algebraic equation with one variable. Using Equations 3.23 and 3.28 with substitution of Equations 3.14 through 3.16 into Equation 3.30 gives

$$\frac{W_2^2 \cdot \rho_2}{2 \cdot \Delta P_2} = \frac{\dfrac{h_2 \cdot D_{h2}}{\lambda_2}}{\dfrac{W_2 \cdot D_{h2}}{\nu_2} \cdot \dfrac{c_{p2} \cdot \rho_2 \cdot \nu_2}{\lambda_2} \cdot f_2 \cdot \text{NTU}} \tag{3.40}$$

After rearranging

$$W_2 = \sqrt[3]{\frac{2 \cdot \Delta P_2 \cdot h_2\left(W_2\right)}{c_{p2} \cdot \rho_2^2 \cdot \text{NTU} \cdot f_2\left(W_2\right)}} \tag{3.41}$$

In this equation, $h_2(W_2)$ and $f_2(W_2)$ for specified D_h are functions of flow velocity W_2. These functions are expressed through respective correlations, in our case by Equations 3.31 through 3.39. The analytical solution of Equation 3.41 is not possible, but it is easily solved on PC by iterations. The obtained result of flow velocity W_2 corresponds to the heat transfer surface that simultaneously satisfies the required thermal and hydraulic conditions. All other parameters of its surface can be easily calculated using equations of the previous section. Let Reynolds number be Equation 3.16. The required heat transfer surface area from Equations 3.19 and 3.21 is

$$F = \frac{Q}{\Delta T_{\ln} h_2}, \text{ m}^2 \tag{3.42}$$

The ratio of heat transfer surface length to hydraulic diameter L/D_{h2} from Equation 3.13 is

$$\frac{L}{D_{h2}} = \frac{\Delta P_2}{2 \cdot W_2^2 \cdot \rho_2 \cdot f_2} \tag{3.43}$$

This value also satisfies the thermal conditions of Equation 3.27.

Such estimation of heat transfer surfaces can be made for any specified process conditions. Let us, for example, consider the heating of water from T_{21} = 5 °C to T_{22} = 55 °C by some vapor with constant temperature $T_{11} = T_{12}$ = 60 °C. The thermophysical properties of water are taken for its average temperature 30 °C: ρ_2 = 996 kg/m^3; $\nu_2 = 0.805 \times 10^{-6}$ m^2/s; λ_2 = 9.618 W/(mK); c_{p2} = 4,174 J/(kg · K). Prandtl number is Pr = 5.43. The required heat load is Q = 600 kW and the allowable pressure drop is ΔP_2 = 20,000 Pa. The number of heat transfer units required by this process conditions is NTU = 2.40.

The influence of hydraulic diameter D_h on the characteristics of the heat transfer surface in a best way satisfying process conditions can be estimated by solving Equation 3.41 for some fixed D_h value. Here the results of such calculations for the range of D_h from 0.00002 to 0.08 m with the step 0.00002 m are presented.

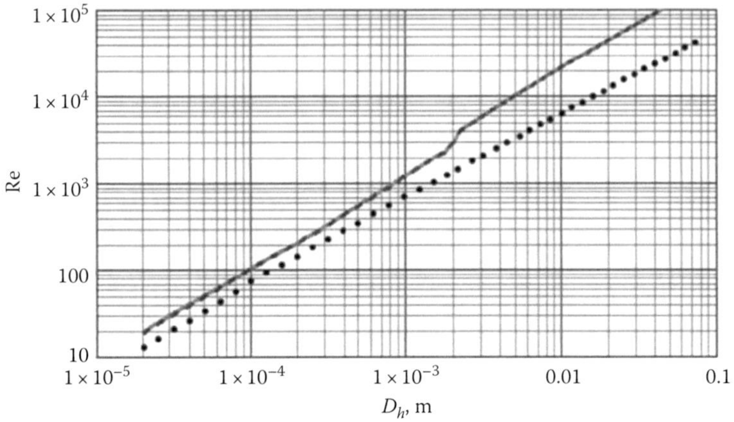

Figure 3.5 The change of required Re number with hydraulic diameter D_h of compared heat transfer surfaces (NTU = 2.40, ΔP_2 = 20 kPa): solid curve is for smooth straight tube; dashed curve is for tube with heat transfer enhancement and dotted curve is for PHE channel.

Reynolds number (dimensionless velocity) is considerably changing with D_h, as shown on the graph in Figure 3.5. For smaller D_h, which corresponds to high compactness, lower Reynolds number (and velocity in channels) is observed down to 10. As the values of Re for the same D_h are very close to each other in straight tubes and are considered here as enhanced tubes, for PHE channels Re is much lower (from 4 times at $D_h = 0.05$ m to 1.5 times at $D_h < 0.0005$ m).

Hydraulic diameter directly characterizes the compactness of the heat transfer surface area σ (see Equation 3.8), but HE with a smaller surface area will have a smaller size and therefore more compact at the same σ. The calculated heat transfer surface area, required by process conditions at different hydraulic diameters, is shown by the graph given in Figure 3.6. The picture becomes much more different for smooth tubes, enhanced tubes and PHE channels. It is influenced by change of flow regimes from turbulent to through transitional to laminar, with the change of Reynolds number illustrated in Figure 3.5.

For smooth tubes in conditions of example under consideration (solid curve in Figure 3.6) for hydraulic diameter D_h bigger than 2.5 mm, the heat transfer area F changes with D_h not more than 20% from 3.3 to 4 m². It corresponds to Re > 4,000 (see Figure 3.5) or the upper part of transitional and developed turbulent flow in smooth tubes. In this range of Re numbers, factor j/f is from 0.35 to 0.55, which is close to assumptions made by Hesselgreaves (2001), and can justify conclusions made on these grounds. A similar trend is observed for enhanced tubes considered here (dashed curve in Figure 3.6), which are increased compared to smooth tubes of Nusselt number that is close to the increase in friction factor. But the required heat transfer area becomes about two times smaller ($F = 1.6 - 2$ m²).

The one side compactness σ of both surfaces, determined by Equation 3.1, will increase with decrease of hydraulic diameter D_h, as follows from Equation 3.8. It is

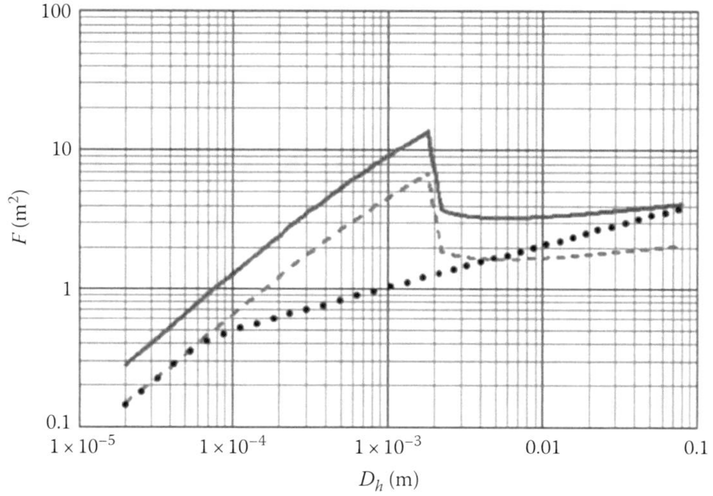

Figure 3.6 The change of required heat transfer surface area F with hydraulic diameter D_h of compared heat transfer surfaces (NTU = 2.40, ΔP_2 = 20 kPa): solid curve is for smooth straight tube; dashed curve is for tube with heat transfer enhancement and dotted curve is for PHE channel.

determined by simple geometrical characteristics of the surfaces when the scaling factor is applied and can be called 'geometrical compactness'. At the same time the heat transfer surface area with enhanced tubes is two times smaller and the HE is also smaller in size than HE without enhancement. To account for this fact, the new parameter 'heat compactness' χ is introduced here. It reflects what size of heat transfer area is required for the heat transfer process in HE with smooth straight tubes of 20 mm inner diameter that would correspond to the volume of HE. Heat compactness can be expressed as follows:

$$\chi = \frac{\sigma \cdot F_{st20}}{F}. \tag{3.44}$$

Here F_{st20} is the heat transfer area of shell and tube HE with tubes of 20 mm inner diameter, required for the same process conditions.

In example considered here for $D_h > 2.5$ mm for smooth tube HE, heat compactness is very close to the geometrical compactness $\chi \approx \sigma$ (see Figure 3.7). For enhanced tubes HE, χ is two times high than σ; it reflects the decrease of required heat transfer area for enhanced surface (see Figure 3.6). For lower D_h down to 1.9 mm, the required heat transfer area increases up to four times for both smooth and enhanced tubes (see Figure 3.6), with consequent decrease in compactness in Figure 3.7. It corresponds to the lower part of the transition flow regime 2,300 < Re < 4,000 (see Figure 3.5). After transition to the laminar flow regime, the heat transfer surface area decreases again with decrease of D_h, but reaching down its values for turbulent flow only for

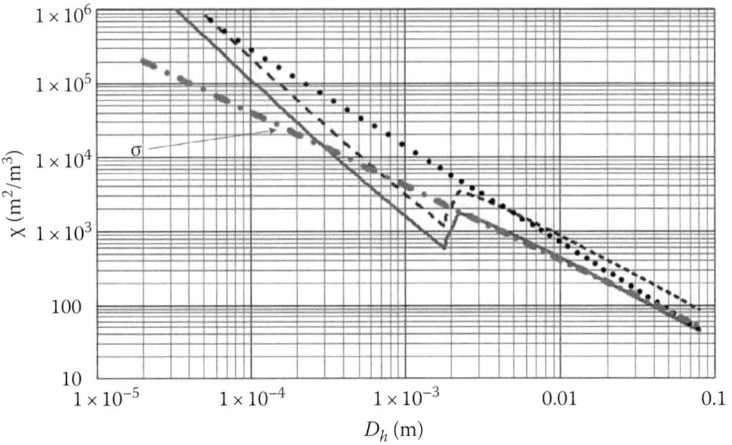

Figure 3.7 The change of heat compactness χ with hydraulic diameter D_h of compared heat transfer surfaces (NTU = 2.40, ΔP_2 = 20 kPa): solid curve is for smooth straight tube; dashed curve is for tube with heat transfer enhancement; dotted curve is for PHE channel and dot-dash line for geometrical compactness σ.

hydraulic diameter lower than 0.25 mm, which corresponds to Re about 200 – 300. In this region heat compactness is also lower than the geometrical one (see Figure 3.7).

A different picture is observed for PHE channel (dotted curves in Figures 3.6 and 3.7) because their transition from laminar to turbulent flow is not so emphasized like in tubes. For high hydraulic diameter about 80 mm, there is no advantage even over smooth tubes. But the required heat transfer area steadily decreases with decrease of hydraulic diameter, and heat compactness also becomes considerably higher than the geometrical one. At hydraulic diameter about 8 mm, the heat transfer surface area and heat compactness of the PHE surface become equal to that of enhanced tubes and, for lower D_h values, supersede these parameters. Mostly at D_h = 1.9 mm, where the required heat transfer area is 10 times smaller than for smooth tubes and half of that for enhanced tubes. Similar compactness becomes 10 or 5 times relatively higher.

The discussion earlier shows that beside increase in geometrical compactness, the reduction of the hydraulic diameter for different heat transfer surfaces can reduce also to those required for specific process heat transfer area and increase heat compactness. Not counting for cases of intensive fouling or channel clogging, from the construction view point, the main obstacle of using smaller hydraulic diameters is the difficulty to arrange the surface area that is big enough for the required heat load. With the reduction of the hydraulic diameter, the length of the heat transfer area should also be reduced, as shown in Figure 3.8. For the considered example with heat load Q = 600,000 W at D_h = 1 mm, the required heat transfer area of PHE is 1 m^2 (see Figure 3.6) and its length L = 0.05 m (Figure 3.8). The perimeter of the surface then is about 20 m. For primary surface countercurrent HE, it is not possible to organize such construction. But for D_h = 5 mm, the required heat transfer surface

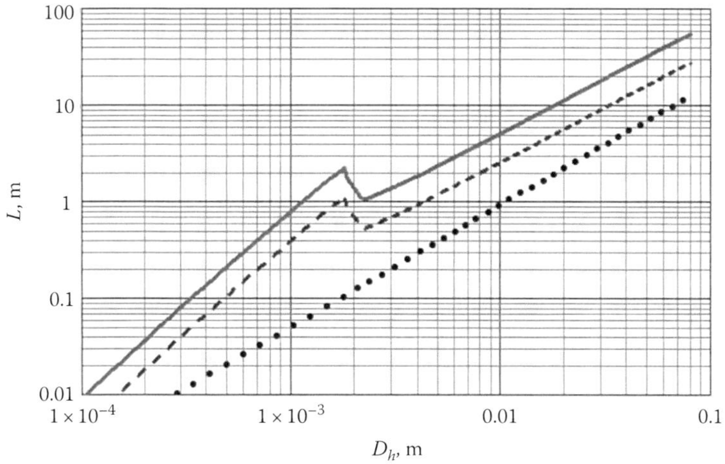

Figure 3.8 The required length of compared heat transfer surfaces (NTU = 2.40, ΔP_2 = 20 kPa): solid curve is for smooth straight tube; dashed curve is for tube with heat transfer enhancement and dotted curve is for PHE channel.

area is $F = 1.65$ m^2 and length is 0.4 m (see Figure 3.8). The perimeter of the surface is about 4.1 m. It can be easily organized in PHE with plates having width 0.2 m or even less. A more detailed discussion for this question is discussed in Chapter 5. On the other hand for heat load Q = 6,000 W, the required heat transfer area at D_h = 1 mm would be 0.01 m^2 and surface perimeter 0.2 m and it could be constructed as a minichannel HE. The same heat transfer surface area can be obtained also for smooth and enhanced tubes, but with microchannels with a hydraulic diameter of the order of 0.1 mm (see Figure 3.6).

For heat load Q = 600,000 W, the shell and tube HE with smooth tubes of 5 mm diameter requires $F = 3.23$ m^2 heat transfer surface area (see Figure 3.6). The required tube length is 2.35 m. The heat transfer surface area of one tube is 0.037 m^2 and 88 tubes are required. For enhanced tubes of 5 mm diameter, the length is 1.18 m and heat transfer area 1.63 m^2. The heat transfer surface area of one such enhanced tube is 0.0185 m^2 and also 88 tubes are required. HE is two times shorter with the same number of tubes. But to organize clear countercurrent flow in such HE, it is rather difficult because of flow distribution in the shell and possible edge effects.

For inner tube diameter of 10 mm, the required heat transfer surface area is almost the same: 3.3 m^2 for smooth and 1.65 m^2 for enhanced tubes. The tube length is 5.1 and 2.55 m and the heat transfer area of one tube is 0.16 and 0.08 m^2. Therefore 21 tubes are required in each HE. If the length of the HE is not crucial, such HE with enhanced tubes of 10 mm will be even better than with enhanced tubes of 5 mm. It is also easier to organize countercurrent flow arrangement in it. Some producers are using this effect, making enhanced countercurrent shell and tube HEs with length needed for a specific duty, as illustrated in the next section (see Figure 3.16).

For PHEs, the required length of the heat transfer area is usually much smaller, as illustrated in Figure 3.8, and shorter PHE plates can be used. To create one-pass PHE, plates with different geometry in one frame are usually used. This will be discussed in more detail in Chapter 5.

All of the discussions earlier were concerned with clean heat transfer surfaces. But in many cases of CHE practical applications, the negative impact on the overall heat transfer coefficient can cause fouling. It must be accounted when designing and selecting HE for a specific duty. In some cases the adverse effect of fouling can even eliminate advantages of hydraulic diameter reduction and heat transfer enhancement. The models for a majority of fouling mechanisms (scaling, crystallization, precipitated solids, crude oil fouling and some types of biological fouling) account for shear stress on the heat transfer surface, which increases fouling. In their experimental study on crystallization fouling Crittenden et al. (2013) have confirmed the general effect that fouling rates can be reduced by increasing the surface shear stress through surface enhancement. The change of the wall shear stress with the hydraulic diameter for the different heat transfer surfaces is shown in Figure 3.9. For enhanced tubes with the same hydraulic diameters of 10 mm, the wall shear stress is about two times high than in smooth tubes. It is about five times higher for PHE heat transfer surface with hydraulic diameter 5 mm and steadily increases with reduction of the hydraulic diameter. For both tubes there is some reduction of shear stress in transition and laminar flow regimes. It can be concluded that there is considerable potential for fouling mitigation on enhanced heat transfer surfaces by adjusting their geometrical characteristics. It is discussed in more detail in Chapter 6.

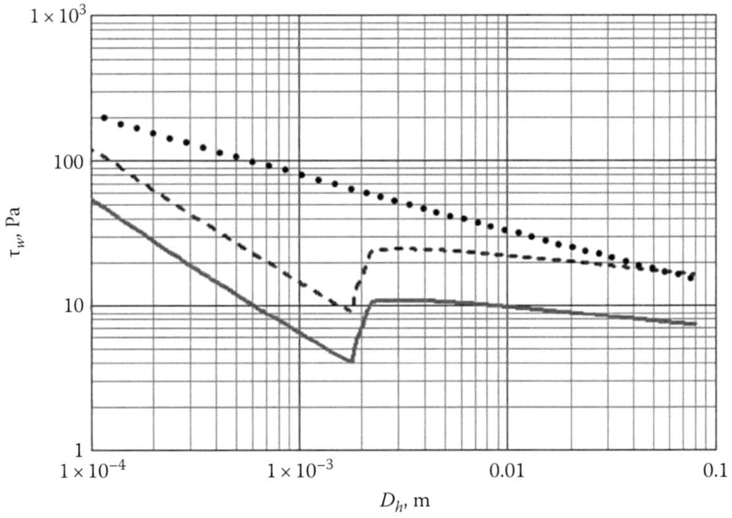

Figure 3.9 The surface shear stress of compared heat transfer surfaces (NTU = 2.40, ΔP_2 = 20 kPa): solid curve is for smooth straight tube; dashed curve is for tube with heat transfer enhancement and dotted curve is for PHE channel.

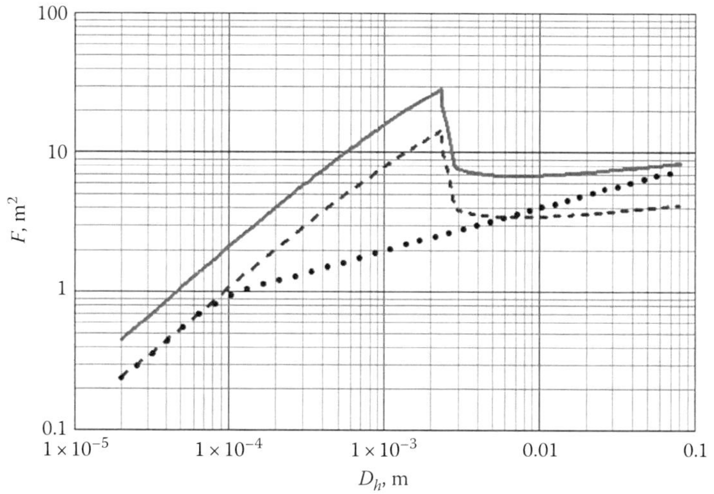

Figure 3.10 The change of required heat transfer surface area F with hydraulic diameter D_h for compared heat transfer surfaces (NTU = 3.93, ΔP_2 = 20 kPa): solid curve is for smooth straight tube; dashed curve is for tube with heat transfer enhancement and dotted curve is for PHE channel.

The presented analysis of CHE heat transfer surfaces is made for one particular example of process parameters. When process parameters are changing, the absolute values of heat transfer surface geometrical parameters are changing too, but the trends remain the same. Figure 3.10 shows the graph of the heat transfer surface area for the same conditions but the temperature of heating media is equal to 56 °C. The close temperature approach like 1 °C can be required to increase heat recuperation and to maximally use the potential of low-grade heat. It requires to increase the number of heat transfer units to NTU = 3.93. The required heat transfer area is also increased. For inner tube diameter of 10 mm, the required heat transfer surface area is about double of the previous case: 6.8 m^2 for smooth and 3.43 m^2 for enhanced tubes. For PHE channel at D_h = 5 mm, F = 3.22 m^2. At the same time the required length of heat transfer surface increases (see Figure 3.11). For smooth tubes it is becoming 8.08 m, which makes it practically impossible to make one-pass HE in one shell. For enhanced tubes L = 4.06 m and for PHE L = 0.60 m. There is a possibility to decrease the hydraulic diameter of PHE to 3 mm, when L = 0.32 m, which can reduce the surface area to 2.76 m^2. But for enhanced tubes, such possibility does not exist.

Decreasing the required heat transfer surface area is possible by increasing the allowable pressure drop in HE with appropriate changing of hydraulic diameter. For example, if for the same temperature program (NTU = 3.93) the pressure drop increases to 100 kPa, it is possible to get the same heat transfer area as for NTU = 2.40 but with a small pressure drop of 20 kPa (see Figure 3.12). For inner tube diameter of 10 mm, the required heat transfer surface area is 3.27 m^2 for smooth and 1.66 m^2 for enhanced tubes. For PHE channel at D_h = 3 mm, F = 1.60 m^2. At the same time the required length of heat transfer surface increases to 9.06 m for smooth

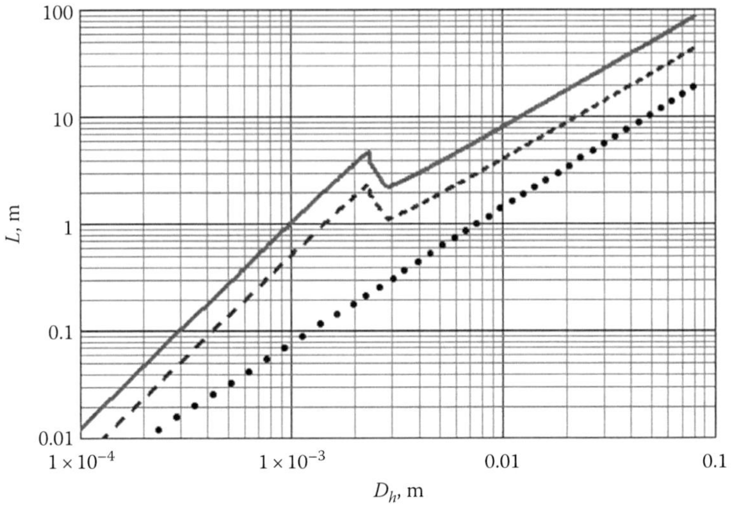

Figure 3.11 The required length of compared heat transfer surfaces (NTU = 3.93, ΔP_2 = 20 kPa): solid curve is for smooth straight tube; dashed curve is for tube with heat transfer enhancement and dotted curve is for PHE channel.

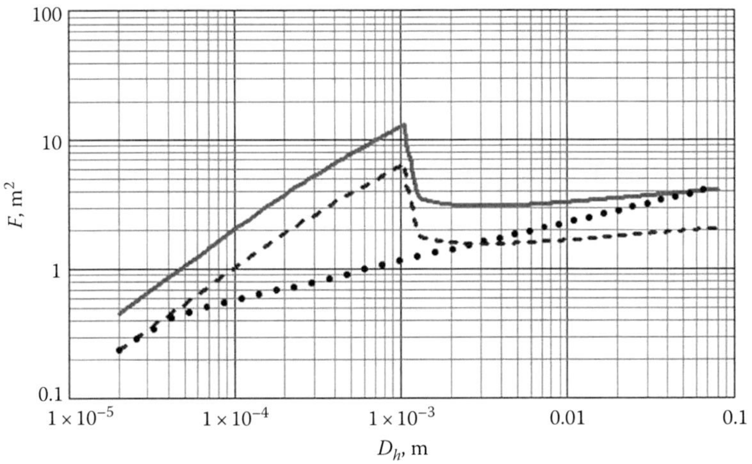

Figure 3.12 The change of required heat transfer surface area F with hydraulic diameter D_h for compared heat transfer surfaces (NTU = 3.93, ΔP_2 = 100 kPa): solid curve is for smooth straight tube; dashed curve is for tube with heat transfer enhancement and dotted curve is for PHE channel.

tubes, 4.57 m for enhanced tubes, and 0.39 m for PHE (see Figure 3.13). It opens the prospects for optimisation of heat transfer surface for HEs as part of the system including also pumps.

All the discussions earlier were for liquid–steam (also applicable for liquid–liquid) HEs with primary heat transfer surface. But the assumptions (3.20) are more

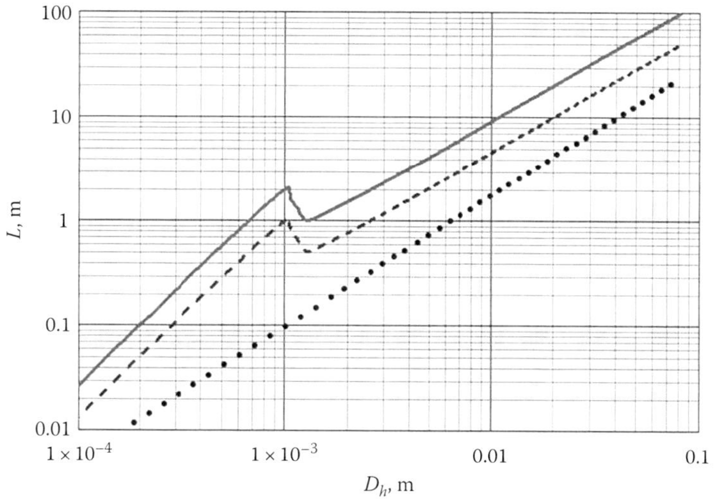

Figure 3.13 The required length of compared heat transfer surfaces (NTU = 3.93, ΔP_2 = 100 kPa): solid curve is for smooth straight tube; dashed curve is for tube with heat transfer enhancement and dotted curve is for PHE channel.

valid for gases with relatively small film heat transfer coefficient h_2. The comparison of the surfaces' heat transfer area for airflow heating from 37 °C to 43 °C in the condenser at temperature of 45 °C at pressure drop of 20 Pa is shown in Figure 3.14a. The reduction of the hydraulic diameter can significantly reduce the surface area. The Reynolds numbers at hydraulic diameter 2 mm is about 200 and at 10 mm about 2,000. The required length (see Figure 3.14b) is becoming as small as from 10 to 25 mm for enhanced channels at hydraulic diameter D_h = 2 mm. But for extended heat transfer surface, it can be achieved as is discussed in Section 3.7.7.3.

Based on the previous discussion, the following conclusions can be made:

1. To increase compactness of HE working at a specified duty, in other words to make it smaller and economically designed, it is necessary to increase not just the amount of heat transfer surface area in the unit of HE volume (geometrical compactness) but the amount of heat transferred by this surface (heat compactness) as well.
2. The correct selection of heat transfer surface for CHE for a specified duty must be made based on thermohydraulic design for which correlations for heat transfer and friction factor should be known.
3. The use in HE surfaces with enhanced heat transfer can significantly reduce the required heat transfer surface area for the same duty. At different hydraulic diameters for the same duties, the relative performance of different enhanced surfaces can significantly vary, which is significantly influenced by the flow regimes in channels (laminar, transitional, or turbulent) and types of correlations for film heat transfer coefficients and friction factors.

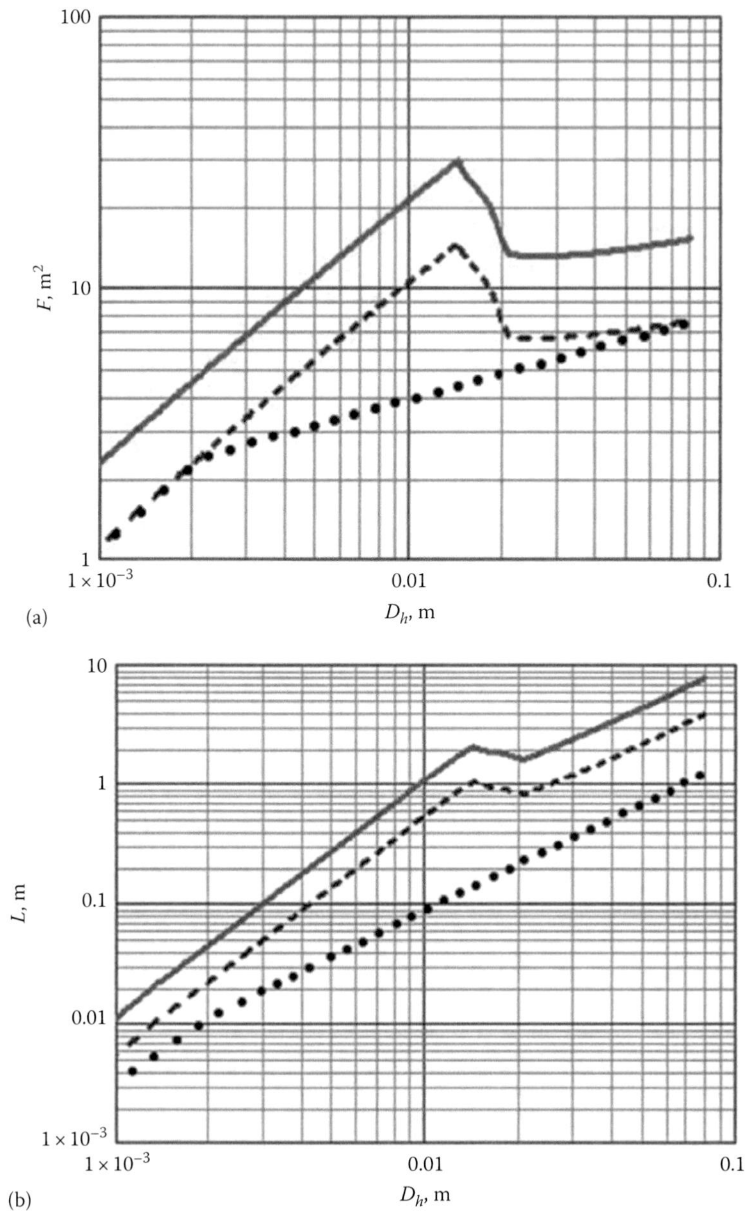

Figure 3.14 The change of parameters with hydraulic diameter D_h for heating air from 37 °C to 43 °C (solid curve is for smooth straight tube; dashed curve is for tube with heat transfer enhancement; dotted curve is for PHE channel): (a) heat transfer surface area and (b) required length of the surface.

4. As a rule the decrease of the hydraulic diameter leads not only to increase in geometrical compactness but also to smaller required heat transfer surface area (excluding transitional flow regime for channels with abrupt transition between laminar and turbulent flow regimes).
5. The decrease of the hydraulic diameter leads to the decrease of Reynolds number for heat transfer process at the same duty and to the decrease of the required length of heat transfer surface area. For enhanced heat transfer surfaces with higher friction factor correlations for the same duty, a smaller length of heat transfer area is required.
6. From the construction point of view, the choice of heat transfer surface and its hydraulic diameter for a specific duty is determined by the possibility to organize flow distribution for a required flow cross-section area and length of the surface, as well as countercurrent stream flow arrangement.
7. The increase of heat load of the HE as a rule requires an increase of the hydraulic diameter for the surface of the same geometrical form.
8. The fouling tendencies of HE streams significantly influence the performance of the heat transfer surface in HE. For most types of fouling mechanisms, the fouling tendencies on enhanced surfaces are significantly lower than on smooth surfaces because of higher levels of shear stress on the wall.

These conclusions can help in the understanding of construction principles and difference in types of CHEs, which are considered in the next two sections of this chapter.

3.6 CLASSIFICATION OF RECUPERATIVE COMPACT HEAT EXCHANGERS

3.6.1 According to the Hydraulic Diameter of Channels

The hydraulic diameter is the parameter that most clearly characterizes the geometrical compactness of the heat transfer surface. It also mainly determines the flow regime in channels at which heat transfer surface of specific geometry satisfies the required duty in the best way. For conventional HEs with hydraulic diameter more than 10 mm working with fluids like water and air, the most preferable is turbulent flow regime. The methods of turbulent heat transfer enhancement give the best results. Such flow regimes and heat transfer enhancement mechanism is mostly characteristic for the channels in the upper range of hydraulic diameters of CHEs (from 10 to 3 mm), which can be called compact flow passages, following Mehendale et al. (2000). With decrease of hydraulic diameter, required for a given duty Reynolds numbers are decreasing and the flow can be in transitional or laminar regimes. In such flows the mechanisms of artificial turbulence promotion or laminar flow disruptions can be employed. When the hydraulic diameter is measured in tens of μ, the flow became steady laminar in channels of all geometries. The main mechanism of heat transfer enhancement in this region becomes the decrease of the thickness of the flow channel.

Heat transfer and fluid flow in channels with hydraulic diameter less than 3 mm are discussed in detail in the book by Kandlikar et al. (2005). The following classification of channels by their hydraulic diameter is made:

- Conventional channels >3 mm
- Minichannels 3 mm ≥ D_h > 200 μm
- Microchannels 200 μm ≥ D_h > 10 μm
- Transitional microchannels 10 μm ≥ D_h > 1 μm
- Transitional nanochannels 1 μm ≥ D_h > 0.1 μm
- Nanochannels 0.1 μm ≥ D_h

With the decrease of the hydraulic diameter starting from a lower range of microchannels, there can be a deviation from conventional models of continuum fluid flow, like slip, transition and free molecular flow. Such channels and flows are very rare for HEs and are not considered in this book. The primary concern is the HEs with conventional compact channels (hydraulic diameters from 10 to 3 mm), which are as a rule most suitable for practically important amounts of low-grade heat utilization.

From the start of the new millennium, considerable developments were made for HEs with hydraulic diameters less than 3 mm. These developments were boosted by the needs for smaller-size HEs for refrigeration, microturbines, transport systems, electronics cooling, etc. There are still some differences in terminology for channels with hydraulic diameter from 3 to 0.2 mm. In the review paper of Mehendale et al. (2000), these size channels are called mesochannels, Balasubramanian et al. regarded these channels as microchannels with hydraulic diameter about 0.6 mm, and Tsuzuki et al. (2009) used the microchannel definition for channels with hydraulic diameters of 0.59 and 3.36 mm.

In this book we make the follow classification of CHEs by the hydraulic diameter of channels, in contrast to conventional HEs with D_h > 10 mm:

- Conventional CHEs with 10 mm ≥ D_h > 3 mm
- Minichannels and microchannels CHEs 3 mm ≥ D_h

3.6.2 According to Flow Arrangements of Heat Exchanging Streams through the Unit

Flow arrangements can be

- Parallel or concurrent flow
- Countercurrent flow arrangement
- Cross flow
- Combination of parallel flow, countercurrent flow and cross flow

3.6.3 According to the Aggregate State of Heat Carriers

- Liquid–liquid HEs
- Vapor–liquid

- Gas–liquid
- Gas–vapor
- Gas–gas HEs

3.6.4 According to the Number of Streams in One Unit

- Two-stream HEs
- Multistream HEs

3.6.5 According to Construction Principles of Heat Transfer Surface

- Compact shell and tube HE
  - With smooth tubes of small diameter
  - With enhanced heat transfer tubes having small turbulence promoters on the wall
  - With twisted tubes
  - With tubes having inserts for heat transfer enhancement
- PHE
  - Plate-and-frame PHE with gaskets
  - Welded PHE
  - Semiwelded PHE with twin plates
  - Condensing duties special design PHE
  - Evaporator special design PHE
  - BPHE
  - Fusion bonded PHE
  - Nonmetallic PHE
- Plate-and-fin heat exchanger (PFHE)
- Tube-and-fin heat exchanger (TFHE)
- Spiral heat exchanger (SHE)
- Lamella heat exchanger (LHE)
- Microchannel heat exchanger (MCHE)
  - Printed circuit MCHE
  - Matrix MCHE
  - Miniscale MCHE
- Nonmetal CHEs
  - Polymer CHE
  - Ceramic CHE

3.7 EXAMPLES OF INDUSTRIAL COMPACT HEAT EXCHANGERS

3.7.1 Compact Shell and Tube Heat Exchangers

The demand for compactness has stimulated the new developments in design of shell and tube HEs. One of the trends is to make this by just reduction of scale with smaller diameters of tubes and newer technologies of tubes fixing on the tube sheet. The example of such approach is brazed shell and tube HE for oil cooling (see Figure 3.15).

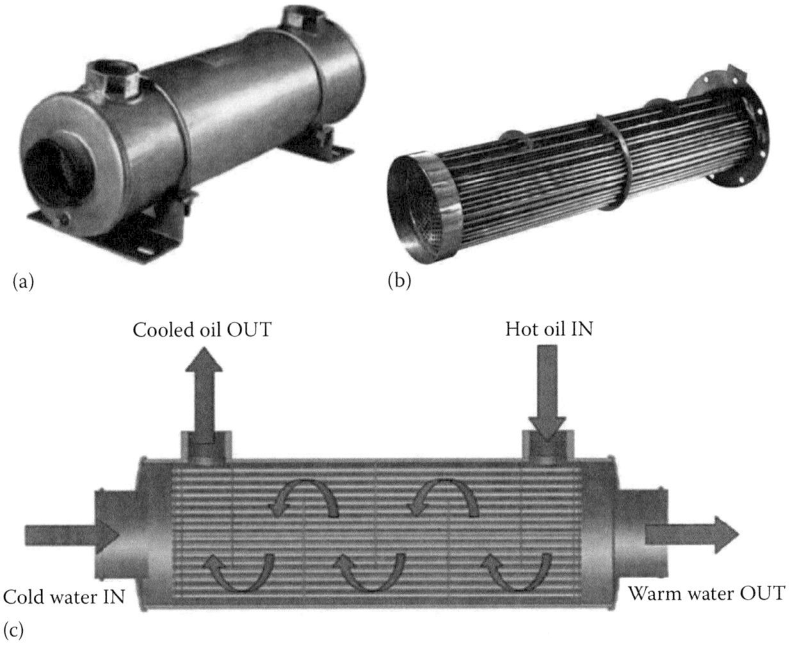

Figure 3.15 **(See colour insert.)** Compact brazed shell and tube HE: (a) assembled HE, (b) tube bundle and (c) the movement of heat exchanging streams in HE (courtesy of OAO Alfa Laval Potok, Korolev, Russian Federation).

It consists of either a fixed, brazed in place, or removable bundle. The bundle is made of copper tube sheets and either copper or 90/10 copper–nickel tubes with diameter of either 4.8, 6.35 or 8 mm. General applications include the following:

- Powertrain – diesel engine, power generation and transmissions
- Heavy-duty vehicles – on- and off-road and military vehicles
- Marine – gas and diesel inboard and outboard engines
- Specialty vehicle – fire truck, ambulance, motor coaches and transit buses
- Commercial – process fluids, irrigation, hydraulic oil, lube oil heat reclamation and process gas cooling

As can be judged from Figure 3.15, this HE represents a small replica of tubular exchanger manufacturers association (TEMA)-style shell and tube HE. One drawback of such design is a difficulty to achieve clear countercurrent flows due to the end effects and areas of cross flow on the shell side. It can make difficulties to achieve close temperature approach down to 1 K. To avoid this some manufacturers are making long, compare to the shell diameter, one-pass HEs, where clear countercurrent flows can be arranged. As it can be estimated from Figure 3.8, the use of tubes with enhanced heat transfer reduces the required length of the HE tubes and one-pass HE can be made of reasonable length. For example, HEs of TTAI® type (see Figure 3.16) produced by Teploobmen LLC have stainless steel tubes of diameter 6–10 mm with

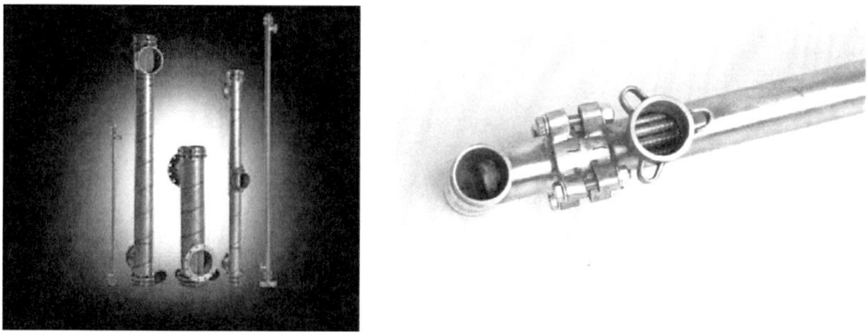

(a) (b)

Figure 3.16 Compact stainless steel HEs TTAI produced by Teploobmen LLC in Sevastopol, Ukraine: (a) example of different TTAI HEs; (b) view of HE connections (after Teploobmen, TTAI – High-efficient compact heat exchangers, 2014, www.ttai.biz/eng, accessed 20 February 2014).

small-scale cross tube corrugations, which enhance heat transfer inside the tube and also outside (shell) the channels.

A similar approach is taken by Krones company (United Kingdom) in the production of compact shell and tube HEs for safe thermal product treatment systems with maximized energy economies. The cross-corrugated stainless steel tubes are used for the manufacturing of compact countercurrent HEs, which can be assembled in modules according to production line requirements, as shown in Figure 3.17.

There are a lot of different types of tubes with small-scale turbulence promoters, including different cross-corrugated tubes, tubes with specially oriented fins and grooves, enhancing heat transfer tube inserts, which are used not only for single-phase heat transfer but also for condensers and evaporators. Quite recently a comprehensive review on passive heat transfer enhancement in tubular HEs is published by Liu and Sakr (2013). Many of these techniques, made on a smaller scale, are applicable in CHEs. For example, Vipertex 1EHT stainless steel tubes (see Figure 3.18) are produced by Rigidized® Metals Corporation (Buffalo, NY, USA) in diameters up from 9.525 mm. Counting for high heat transfer performance of these tubes, as reported by Kukulka and Smith (2013), they can be used for enhanced heat transfer CHEs at different applications.

To increase the performance of shell and tube HEs, different tube inserts can be used, which promote turbulence and can help adjust HE for required duty. For example, CalGavin company (United Kingdom) is offering such HE solutions by inserting in tubes specially calculated and designed for the specified duty HiTRAN Matrix Element tube insert (see Figure 3.19).

Another way to induce turbulence and at the same time to organize better flow distribution on the shell side is the use of twisted tubes. Produced by Koch Heat Transfer Company, the tube bundle of such HE (see Figure 3.20a) has helix-shaped tubes, which are arranged in a triangular pattern. Tubes are frequently supported by adjacent tubes, as shown in Figure 3.20b, yet fluid swirls freely along its length. Complex swirl flow on the shell side induces the maximum turbulence to improve heat transfer and also high tube side turbulence is achieved.

Figure 3.17 **(See colour insert.)** Side view of the Krones shell and tube HE module (after Krones, 2014).

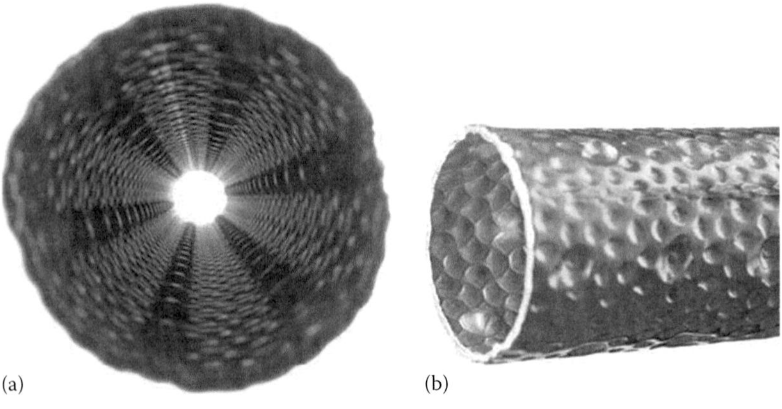

(a) (b)

Figure 3.18 (a) Cross-sectional view showing details of the inner surface of Vipertex 1EHT (type 304 L stainless steel) enhanced tube and (b) details of the outer surface of the Vipertex 1EHT. Produced by Vipertex Heat Transfer division of Rigidized® Metals Corporation, Buffalo, New York, USA (after Kukulka and Smith, 2013).

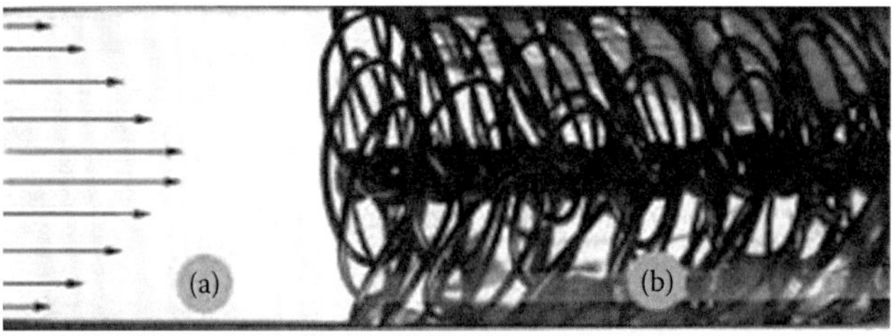
(a)
(b)

Figure 3.19 **(See colour insert.)** HiTRAN Matrix Element tube insert promoting turbulence and enhancement of heat transfer: (a) smooth tube; (b) part of tube with insert (after Calgavin, 2014).

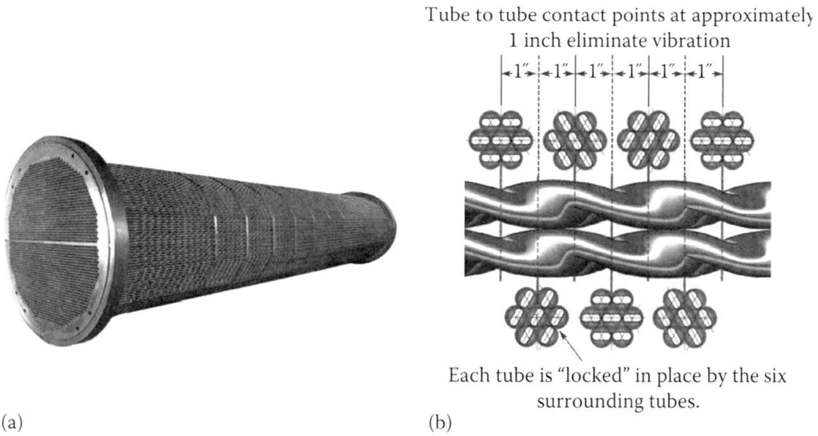
Tube to tube contact points at approximately 1 inch eliminate vibration
1″ 1″ 1″ 1″ 1″ 1″
Each tube is "locked" in place by the six surrounding tubes.
(a)
(b)

Figure 3.20 **(See colour insert.)** (a) Koch Heat Transfer Company's TWISTED TUBE® bundle and (b) scheme of tube arrangement (see Koch Heat Transfer Company, 2014).

As described earlier, shell and tube HEs are designed to work in developed turbulent flow regime or with turbulent promoters at lower Reynolds numbers. But as it was discussed in Section 3.5, similar or even better results on heat transfer area and much high compactness can be achieved for laminar flows in channels of much smaller hydraulic diameter and length. Saji et al. (2001) described the construction of microtube HE for low-temperature (less than 20 K) refrigeration cycle. HE consists of 4,800 stainless steel microtubes with an inner diameter of 0.5 mm and wall 0.1 mm. The countercurrent HE has a tube length of 310 mm and has high compactness and good satisfaction to the required duty.

The shell and tube HEs have the longest history of their development. But notwithstanding some achievements of this HE type in the area of compactness, some other types of HE, which were initially developed as compact, still have considerable advantages.

3.7.2 Plate Heat Exchangers

Because of simple geometry, round pipe at the same inside volume has minimal perimeter of the surface compared to any other form of cross section. It is best suited to transport the heated medium with minimal heat losses, at the same insulation of the walls. For HE, the task is just opposite, and the medium should be cooled (or heated) through the wall. Nevertheless from the time of industrial revolution till ours, tubular HEs are the conventional type of heat transfer equipment in many applications due to the rather simple, well-developed and reliable construction and the level of contemporary manufacturing technologies.

What is perhaps the first patent for PHE was granted by the United States Patent Office to Joseph Nason of New York, patent no 55,149 dated 29 May 1866. According to Wang et al. (2007), the other early patents for HEs consisting from plates were granted in the last three decades of the nineteenth century in Germany. But the first commercial PHEs were produced only in 1923 in England, invented by founder of APV International Richard Seligman, and around 1930 in Sweden, developed by Bergedorfer Eisenwerk of Alfa Laval. These were milk pasteurizers, in which was implemented the possibility to open heat transfer surfaces for mechanical cleaning. In later decades the growing demand for energy efficient, more compact and economical heat transfer equipment has stipulated the application of PHEs in many industries. It required the developments in their constructions, which became possible for practical applications with the development of manufacturing technologies in stamping, welding, brazing and production of new materials for plates and gaskets.

Nowadays PHEs are a wide class of heat transfer devices with primary heat transfer surfaces made as a pack of plates. The principles of PHE construction and operation (mostly for plate-and-frame PHEs) are described in practically all contemporary textbooks and handbooks on HEs, as well as in books on CHEs. In more detail, PHEs, their operation, and design are specially discussed on this type of HE and published in the books by Wang et al. (2007) and in Russian by Tovazhnyansky et al. (2004). These books can be consulted for the history and evolutions of PHE developments (as also in an earlier book in Russian by Baranovskyy et al., 1971). Here we are more concerned with contemporary state-of-the-art and advantageous features of modern PHEs. These features are as follows:

1. The cross-corrugated plates assembled in PHE are creating a big number of contact points and form a strong rigid matrix that can withstand high pressure difference between channels at relatively small plate thickness.
2. In PHE the heat transfer process takes place in channels of intricate complex geometry. The permanent changing of flow direction with separation effects induces high levels of turbulence and swirl or vortex flows even at relatively low Reynolds numbers and leads to high levels of film and overall heat transfer coefficients. Such heat transfer enhancement mechanism has no abrupt transition between turbulent and laminar flow regimes (as discussed in Section 3.4; see Figures 3.1 and 3.2).

3. High heat transfer coefficients of PHEs enable to make them much smaller in weight and volume than conventional shell and tube HEs, up to four and five times. They are compact and need a small area for maintenance.
4. Smaller and thinner heat transfer surface area of PHE allows economical use of sophisticated materials in cases of highly corrosive media.
5. PHEs can operate at very close temperature approach (down to 1 K) owing to high heat transfer coefficients and true counterflow arrangement.
6. The possibility to adjust thermal and hydraulic characteristics of PHE channels by changing the geometrical form of corrugations is much helpful for PHE optimization. By assembling plates with different corrugation patterns in one HE, it is possible to obtain one-pass arrangement of PHE closely (error not more than one plate) satisfying the required duty.
7. PHE have low hold-up volume and weight, and lower cost are required for handling, transportation and foundations. It usually needs less capital cost due to less metal consumption.
8. It is possible to treat more than two fluids in one PHE, by installing intermediate dividing sections or just rearranging pass partitioning plates. This can reduce the complexity of process plant and make it more compact.
9. Because of thin channels the volume of fluid contained in PHE is small, which is important when using dangerous or expensive fluids. Counting also for low weight of PHE, it is easier to control the required process parameters.
10. Plate-and-frame PHE with gaskets can be easily disassembled for mechanical cleaning and maintenance. It has the ability of quick change of performance parameters by just adding or extracting the number of plates in the unit in the range of the existing frame design.
11. Plates are exposed to ambient air only by its periphery edges. It much reduces heat losses even without insulation.

Historically the first PHEs were of plate-and-frame type. During the nine decades of their application and developments, their construction has undergone considerable modifications directed to increase the range of working temperatures and pressures, reliability, maintenance and operation convenience. Newer subtypes, like welded PHEs and BPHEs not requiring pressure frame and special designs, have been derived.

3.7.2.1 Plate-and-Frame PHE

The sketch drawing of plate-and-frame PHE with gaskets is presented in Figure 3.2. It consists of thin corrugated plates made by stamping from corrosion-resistant alloys, which form two channels for two heat carriers (hot and cold) providing the heat transfer process between streams in them. The plate pack is assembled between a fix frame plate and a movable pressure plate and is compressed by tightening bolts. The fixed frame plate has four connections: two for inlet and two for outlet of hot and cold streams. A movable pressure plate can have two or four connections or no connections at all. These construction features are dependent on the unit application and its design.

The connections of flange or screw type are used depending on the heat exchange type and heat carriers. Frame plates, the carrying bar and tightening bolts can be made from high-strength carbon steel, painted or zinc coated. For sanitary applications the stainless steel can be used, or carbon steel with stainless steel lining. For connections the carbon steel lining with stainless steel or sometimes titanium is used.

Each plate is fitted with a gasket from heat-resistant elastomers, which can withstand an aggressive action of streams, sealing the interplate passages and directing the fluids into alternate channels. The hot and cold mediums are always separated from each other by gaskets. Flowing along its own channel, each medium has its sealing. The gasket design minimizes the risk of internal leakage. The failure of a gasket in a PHE results in leakage to the atmosphere, which is easily detectable on the exterior of the unit. The mixing of heat carriers is impossible. Nowadays mostly glue-free gaskets are used. The mechanical cleaning of plates and change of gaskets can be performed on site, sometimes even without removing the plate from the frame, all of which contributes to reduced service costs and downtime. The exact number of plates, the form of their corrugation, and size are determined by the flow rate, physical properties of the fluids, pressure drop and temperature program. The plate corrugations promote fluid turbulence and support the plates against differential pressure.

The working principle of single-pass PHE is shown in Figure 3.21. Channels are formed between the plates and the corner ports are arranged so that the two media flow through alternate channels, formed in interplate spaces of adjacent pates. The gaskets between the plates are assembled to ensure the fluid movement into alternate channels. The heat is transferred through the plate between the channels, and complete countercurrent flow is created for the highest possible efficiency.

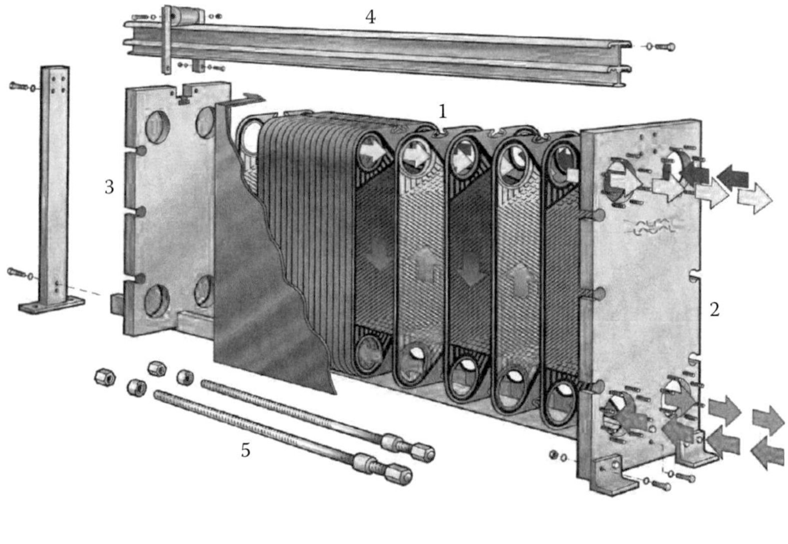

Figure 3.21 **(See colour insert.)** Schematic drawing of one-pass frame-and-plate PHE: 1, pack of plates with gaskets; 2, fixed frame plate; 3, movable frame plate; 4, carrying bar and 5, tightening bolts (courtesy of OAO Alfa Laval Potok).

The corrugation of the plates provides the passages between the plates, supports each plate against the adjacent one, and enhances the turbulence, resulting in efficient heat transfer. The plates' corrugation type, their number and channel geometry are designed for optimal heat transfer performance.

The package of channels for heat carrier movement in one direction is called the pass. PHEs can be single-pass, two-pass, three-pass and multipass units. The assembling and number of passes in one unit are defined from heat carriers' properties, temperature program and allowed pressure drops for overcoming of hydraulic losses. The total number of plates in the unit is equal to the summarized number of channels for cold and hot mediums and plus one more plate. In cases when the flow rates and (or) allowable pressure drops for hot and cold streams are significantly different, the units with nonsymmetrical design can be used, when the number of cold and cold streams passes is unequal. The working principle of two-pass HE is provided in Figure 3.22. Here the number of passes for the cold and hot side is the same. This design provides the heat carriers' movement along the channels up to some special plate, which does not have two connection holes (they are closed or absent). It forces heat carriers to channels, where they change the direction of movement. For multipass HEs the plate packs are always divided by plates with closed holes. Such plates are called pass partition plates. The heat carrier from the first plate pack moves along the collector till the pass partition plate and after it flows to the channels of the second plate pack. The flow direction in this pack is opposite to its flow direction in the first pack. Such design allows for increased flow length and bigger number of heat transfer units with the same plate.

But multipass flow arrangement has considerable drawbacks:

1. Even at equal number of passes (symmetrical pass arrangement), it will be some concurrent flow over the plates between passes that will decrease the average temperature difference and jeopardize the efficiency of PHE. At unsymmetrical pass arrangement, the correction factor for LMTD can be much more significant.

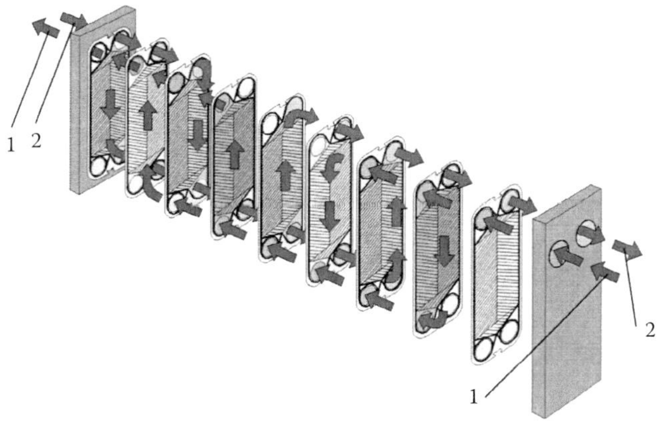

Figure 3.22 **(See colour insert.)** Schematic drawing of two-pass frame-and-plate PHE: 1, hot stream and 2, cold stream (courtesy of OAO Alfa Laval Potok).

2. One-pass PHE allows to having all four connections on a fixed frame plate (see Figure 3.21). The disassembling of plate-and-frame PHE in that case does not require to dismount any connecting pipes, which is convenient for maintenance. It is not possible with multipass PHE, which must have connections on both frame plates (see Figure 3.22).
3. At the same allowable pressure drop and hydraulic diameter, the velocity in multipass PHE should be smaller, which makes less intensive heat transfer and decrease fouling mitigation effect. For symmetrical arrangement the increase of the pass numbers analogous to the increase of the length of the heat transfer surface following the discussion in Section 3.5 (see Figures 3.8 and 3.11) requires a bigger hydraulic diameter, which jeopardizes the compactness and heat transfer area of PHE.

The method of PHE design for the specified process conditions by adjusting pass numbers was dominant until the 1980s. Having only one corrugated plate type of certain size with hydraulic diameter 6 – 9 mm or even bigger is required. But at the same hydraulic diameter and plate spacing, it is possible to obtain different heat transfer and hydraulic performance in channels between plates with different corrugation angles like those shown in Figure 3.23. The introduction of the design principle with a combination of plates with different corrugation patterns in one PHE (first described in literature by Marriot, 1977, of Alfa Laval) has allowed making one-pass PHEs for most duties (for more details see Chapter 4). It also allowed decreasing of hydraulic diameter and plate spacing, making plates geometrically stronger to be stamped from thinner metal (now down to 0.4 – 0.3 mm). At the same time it required to stamp a bigger number of plates with different corrugation patterns that became possible with advances in stamping technology. All major manufacturers of PHEs are producing a big number of plates, the combination of which can in the best way satisfy the required process duty. In Figure 3.24, are shown a fraction of plate types manufactured by Alfa Laval, Sweden.

Plates can be produced from any metal alloy with good stamping properties. The most common are stainless steels AISI 304 and AISI 316. For higher corrosive media, Alloy 254SMO, Alloy C-276, titanium, palladium-stabilized titanium, Incoloy and Hastelloy are used. The more expensive metal alloy, the more economically attractive is PHE compared to shell and tube HE with heat transfer surface of the same material.

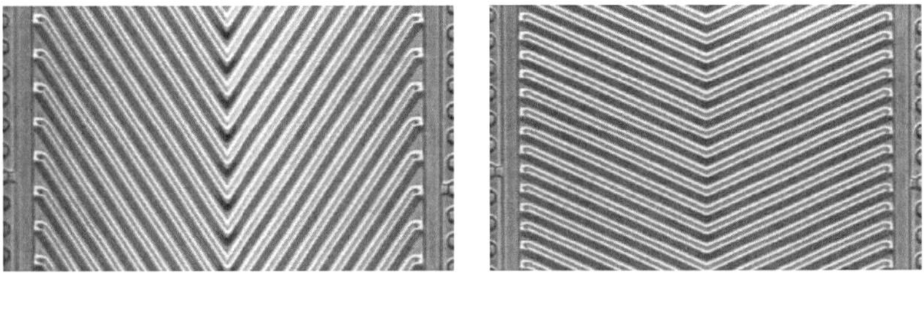

Figure 3.23 Corrugated fields with different angles of corrugations for plates that can be assembled in one single-pass PHE. Selecting exact numbers of different plates enables close satisfaction to required duty (courtesy of OAO Alfa Laval Potok).

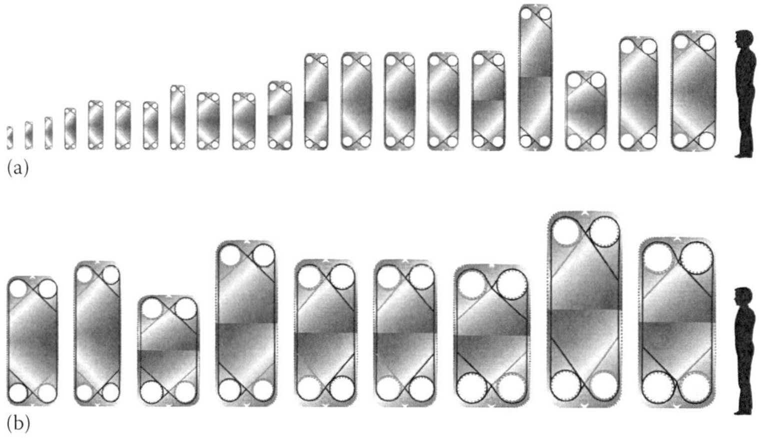

Figure 3.24 Range of plates manufactured for plate-and-frame PHEs by Alfa Laval, Sweden (courtesy of OAO Alfa Laval Potok).

The most common gasket materials are nitrile rubber (NBR), ethylene propylene diene monomer rubber (EPDM) and their different modifications.

NBR is compatible with water and fat substances and has good chemical resistance to oils. The sulfur-cured grades of this rubber have an operating temperature about 110 °C, when peroxide-cured up to 140 °C. Standard grades NBRB, NBRS, NBRP and NBRF(F) are suitable for different food and pharmaceutical processing. Some special grades of hydrogenated nitrile butadiene rubber (HNBR) has proven continuous operating temperatures up to +175 °C in crude oil applications, according to Trelleborg Sealing Solutions (2009).

EPDM is a general-purpose rubber that can be used in a wide range of chemical applications like processing of some organic solvents, strong alkalis and some acids of certain concentrations. Generally it can work at higher temperatures than NBR, showing good resistance to steam and hot water at temperatures up to 160 °C. Special-grade EPDMP withstands up to 180 °C. While generally EPDM has poor performance in fats and oils and not used in food applications, special-grade EPDMF has food approvals.

Another elastomer for gaskets is fluorocarbon material FKM. It has good performance in handling acids and aqueous solutions. The grade FKMT can be used up to 180 °C, while others for much lower temperatures, below 130 °C.

There are also a lot of other elastomers that can be used for PHE gaskets: Natural rubber (NR), Butyl rubber (IIR), Chloroprene rubber (CR), Silicon rubber, Compressed asbestos fiber, Butadiene styrene (SBR) and some others (see Gillham and Sennik, 1984).

The choice of gasket material depends mostly on fluids compositions, temperatures and pressures. The temperature limits may depend on the working pressure and also on the construction of plate and gasket grooves. Besides the exact composition of elastomer and technology, its production can significantly influence the lifetime and reliability of gaskets. The final selection of gasket for a specific application must be made by a PHE manufacturer.

The corrugated plates and gaskets are the main components of plate-and-frame PHE. The production of plates requires powerful hydraulic presses and tools for stamping, to produce gaskets; specialized equipment and technology is also required. Having the stock of plates and gaskets, even small company can calculate and assemble PHE in the best way satisfying required duty. Now most of the major manufacturers of plate-and-frame PHEs has local small affiliated or partner companies (SMEs) distributing PHEs that can manufacture and in a short time supply PHE to local customers. Such SMEs are also making support in installation of equipment, after service, and some of them produce heating modules equipped with pumps, controls and valves. For example, Figure 3.25 shows the District Heating module made with PHE produced locally from Alfa Laval supplied plates by AO Spivdruzhnist-T LLC, which is a distributor and assembly partner of Alfa Laval since 1995. Such approach improves service quality and market competitiveness and supports local SMEs and employment.

Plate-and-frame PHEs are a reliable choice in many applications, but parameters of their work are limited by pressures, generally up to 25 bar, and temperatures up

Figure 3.25 **(See colour insert.)** District heating substation module with plate-and-frame PHE produced with Alfa Laval plates by AO Spivdruzhnist-T LLC in Kharkiv, Ukraine.

to 180 °C. Moreover, the limiting temperatures and fluid of some compositions can require expensive gaskets (if any possible), the use of which can be uneconomical. It stipulated the development of PHEs without gaskets between plates, such as welded and BPHEs or semiwelded PHEs with a limited number of gaskets.

3.7.2.2 Welded PHE

Excellent experience and demonstrated advantages with plate-and-frame PHE in dairy and food applications have stipulated their use in other industries, such as chemical, petrochemical, fertilizers and power generation. It required the further development of PHE construction to make wider the range of PHE working temperatures and pressures. The first welded PHEs were developed and manufactured in the late 1960s. The design was similar to plate-and-frame PHE with modifications allowing existing welding technologies. Some of such plates developed in Ukniichemmash (Ukraine) are shown in Figure 3.26. The plates were welded in blocks and welded to special design welded manifolds for fluid flow. Resistance seam roller welding and arc welding were used. The significant drawback of production was the absence of process automation, so welding part was completely handmade.

The different design approach was used in Ukrniichemmash jointly with Kharkiv Polytechnic Institute for the development of welded PHEs produced from 1974 in Ukraine by Pavlogradchemmash manufacturer (Figure 3.27). The welding technology was modernized. With cooperation of Paton Welding Institute in Kiyv, the semiautomated line for production was developed. It ensured the quality of PHE. Being incased in a high-pressure shell of ammonia catalytic synthesis column, such PHEs are working at pressures up to 32 MPa and temperatures up to 520 °C in a number of ammonia synthesis plants at former Soviet Union countries. They are installed to recuperate the heat of after reaction gas mixture having the temperature up to 520 °C with cold (about 100 °C) gas mixture going to catalytic ammonia synthesis reaction. Compared to previously used shell and tubes, it saves much space, which is filled by catalyst to increase column capacity. The temperature program in ammonia

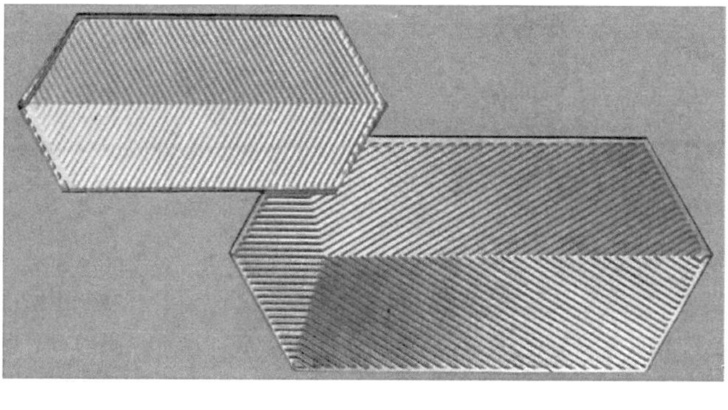

Figure 3.26 Plates for welded PHE designed in Ukrniichemmash (Ukraine) and produced for the chemical industry in the late 1960s.

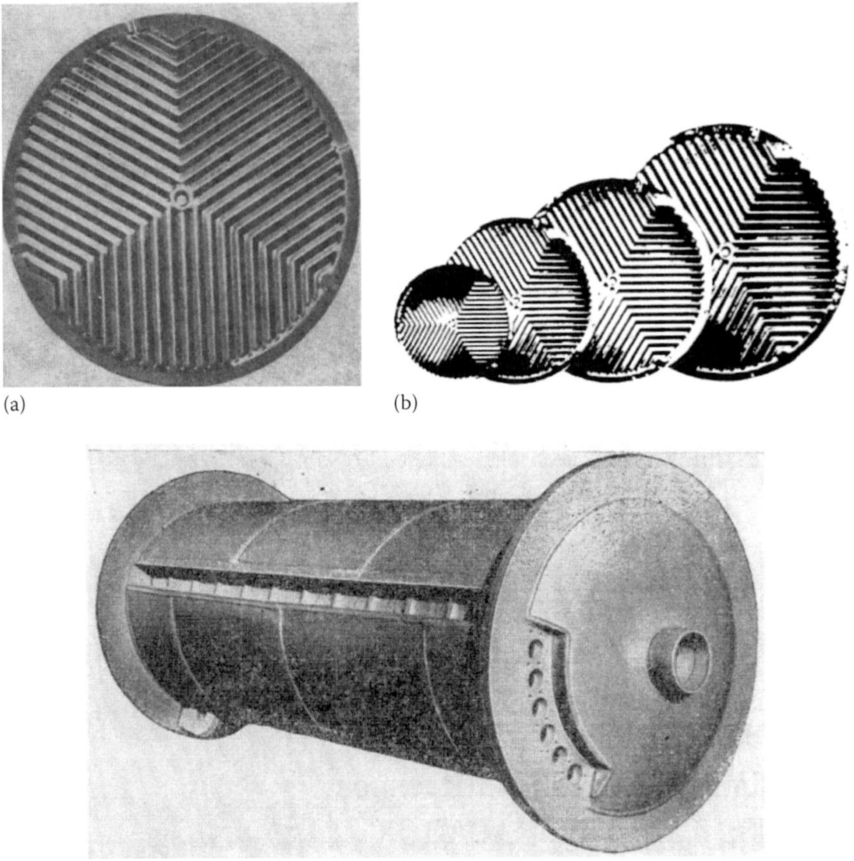

(a) (b) (c)

Figure 3.27 Welded PHE for ammonia synthesis columns produced by Pavlogradchemmash factory in Ukraine: (a) plate for column 600 mm in diameters, (b) range of plates for columns from 600 to 1,200 mm in diameter and (c) welded PHE for column 600 mm in diameter.

condensation column is much easier (0 °C–40 °C), but the pressure in both cases is up to 32 MPa. Such HEs are also working methanol column converter preheaters.

The main feature of heat transfer in this PHE is the combination of the cross flow for hot and cold fluids in one pass with total multipass counterflow system in the whole unit. It requires the accurate calculation of mean logarithmic temperature difference correction factor and special approach in the case of condensation. The cross flow made it possible to minimize pressure losses at the plate entrance and exit zones, which can be significant for countercurrent plates of plate-and-frame PHEs, where flow should be distributed from collector part at the plate corners, which have much smaller cross-section area than the main corrugated field.

By adjusting the pass number, the length of heat transfer surface could be changed to satisfy required duty.

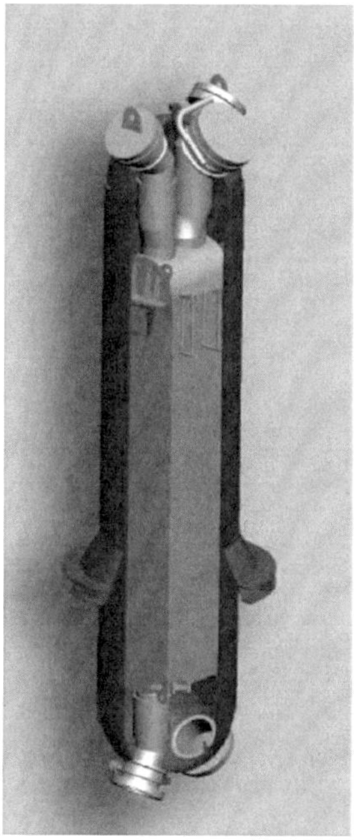

Figure 3.28 (See colour insert.) Drawing of Alfa Laval Packinox welded PHE (courtesy of OAO Alfa Laval Potok).

To exchange heat in catalytic conversion processes of chemical, petrochemical, oil and gas industries, Packinox PHE was developed (see Figure 3.28). It consists of welded block of long plates with one-pass flow arrangement that is comprised in shell. The first standard combined feed/effluent HE Packinox was supplied to a catalytic reforming unit in 1982. It operated at temperatures from 80 °C (cold end) to 530 °C (hot end) and pressure up to 45 bar depending on the process. Produced now by Alfa Laval Packinox, these PHEs are installed in hydrotreating units, in methanol production plants, in xylene isomerization units, paraxylene production plants and others. The operating ranges of Packinox depend on the plate and shell material selected, the range of temperatures between –200 °C and 700 °C, and pressures of up to 300 bar. Differential pressures are up to 100 bar. The heat transfer surface area in a single Packinox can be up to 16,000 m^2 and dry weight as high as 150 t. It is about half of that with shell and tube HEs for the same duty; the significant economy and reduction in greenhouse gases emissions are also achieved. Alfa Laval Packinox PHEs are also used in concentrated solar power plant at thermal storage system where heat from solar power is stored in molten salt, thereby enabling electricity production even on rainy days or at night.

The need of industry in efficient general-purpose CHEs for heavy duties has stipulated further development of welded PHEs. The most obvious solution is just to replace gaskets with welding following the advances in welding techniques. For laser beam welding Alfa Laval producing AlfaRex PHE (see Figure 3.29).

Figure 3.29 **(See colour insert.)** Alfa Laval AlfaRex welded PHE with cut of welded plate pack (courtesy of OAO Alfa Laval Potok).

Modifying construction of sealing elements on the plate for welding and developing automated welding appliances are required. The operating limits on pressure are up to 40 bar and temperature range from –50 °C to +350 °C.

API Heat Transfer (United States) is using for Schmidt SIGMA**WIG** all-welded PHE tungsten inert gas (TIG) welding, which required another construction of sealing, as shown in Figure 3.30. Now all major PHE manufacturers are producing PHEs made on these principles with different welding techniques and plate

Figure 3.30 The cross section of welded plate pack in Schmidt SIGMAWIG all-welded PHE (after API Heat Transfer, 2003).

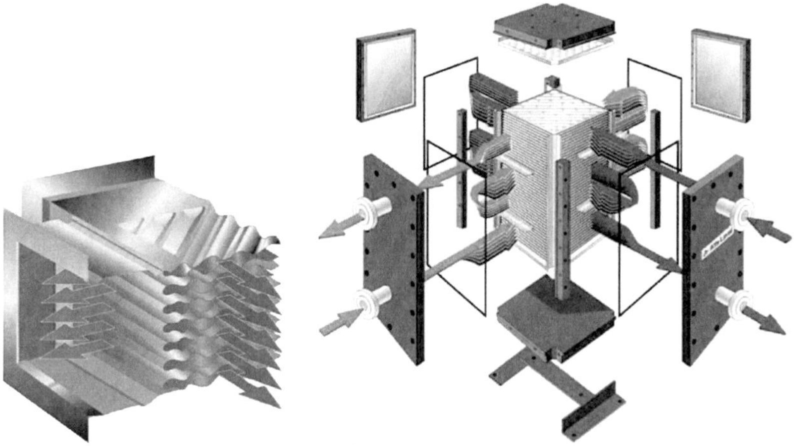

(a) (b)

Figure 3.31 **(See colour insert.)** Compabloc™ PHE of Alfa Laval production: (a) schematic drawing of a plate stack and (b) exploded view showing cross-countercurrent arrangement (courtesy of OAO Alfa Laval Potok).

constructions. But with significant widening of the range of operating temperatures and pressures, such construction prohibited any excess to the heat transfer surface for inspection and cleaning. Such welded PHEs can be used only for nonfouling streams or streams where fouling deposits can be removed by washing with certain chemicals.

Vicarb of France in 1985 has started the production of Compabloc™ welded PHE. It consists of a stack of pressed squired plates automatically welded with laser beam welding machine at alternative edges to provide cross-flow configuration as shown in Figure 3.31a. The HE can be disassembled as shown on exploded view in Figure 3.31b, and all sides of the plate pack are available for inspection and cleaning by washing with water under pressure. As can be seen on Figure 3.31b, multipass arrangement of streams is possible to arrange by baffle assembles between plate stack and panels. The pressure losses on entering and leaving the plates of the pass are minimal and flow evenly distributing across the channel width. It contributes to good thermal and hydraulic performance of such PHE. The overall counterflow makes it possible to achieve close temperature approach. Alfa Laval, after acquiring Vicarb S.A. in 1999, is producing Compabloc™ with seven different plate sizes from CP15 to CP120, and the range of temperatures is from –50 °C to 400 °C and pressures from vacuum to 42 bar (e.g., see Figure 3.32).

Now other PHE manufacturers are also producing similar by principle welded PHEs, for example, Funke is manufacturing FunkeBlock FPB, GEA producing GEABloc®, Mersen in France producing Heatex® with different plates having spacers, and SPX Corporation providing APV Hybrid-Welded HE, which history started in 1981, with different types of plate corrugation and different arrangements of plate stack, as shown in Figure 3.33.

Figure 3.32 **(See colour insert.)** Compabloc™ PHE of Alfa Laval production installed by AO Spivdruzhnist-T LLC at CHP unit in Sochi, Russian Federation.

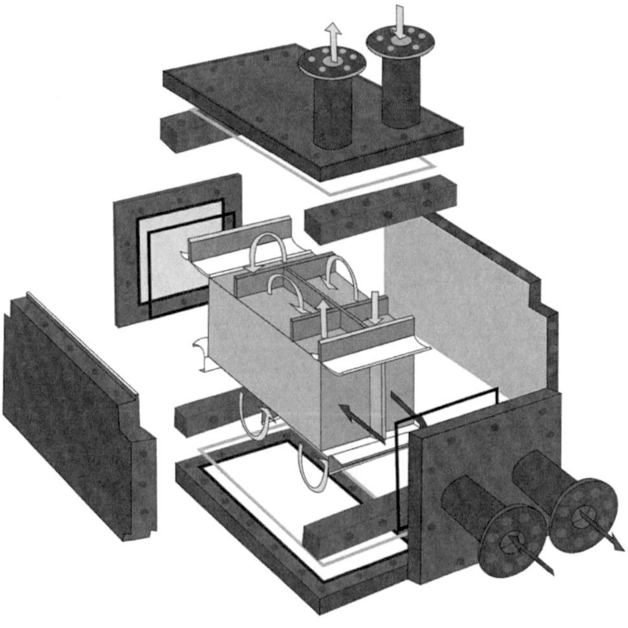

Figure 3.33 **(See colour insert.)** APV Hybrid PHE: exploded view showing cross-countercurrent arrangement (after SPX, 2012).

Another trend in welded PHE production is Plate and Shell–type HE. Founded in 1990, Vahterus Oy company in Finland developed welded PHE with registered trademark Plate & Shell® (see Figure 3.34a). This exchanger consists of a pack of welded circular corrugated plates with round holes of ports for one fluid close to the ends of the diameter, as shown in Figure 3.34b. This fluid comes through the manifold and the port into the channel between the pair of plates welded together on the periphery. There it is distributed over the plate due to inclined plate corrugations and exits through the opposite port. The round ports are stamped in such a way that plates of adjacent pairs are welded together at port perimeters, making two isolated volumes for fluids. Another fluid is coming through the port in a cylindrical shell and going through the plate pack in counterflow opposite the port in a shell. Another flow configuration can be achieved by baffles. The shell can be made capable to be disassembled to remove the plate pack and inspect or wash it on the shell side with water under pressure. The construction is rather simple for production to be economical, and the cylindrical shell is capable to withstand high pressure. Manufacturers claim that for special design by request, it is possible to have pressure up to 100 bar and temperature from −164 °C to 899 °C. The standard designs withstand pressure 16, 25, 40 and 60 bar and temperatures from −160 °C to 538 °C.

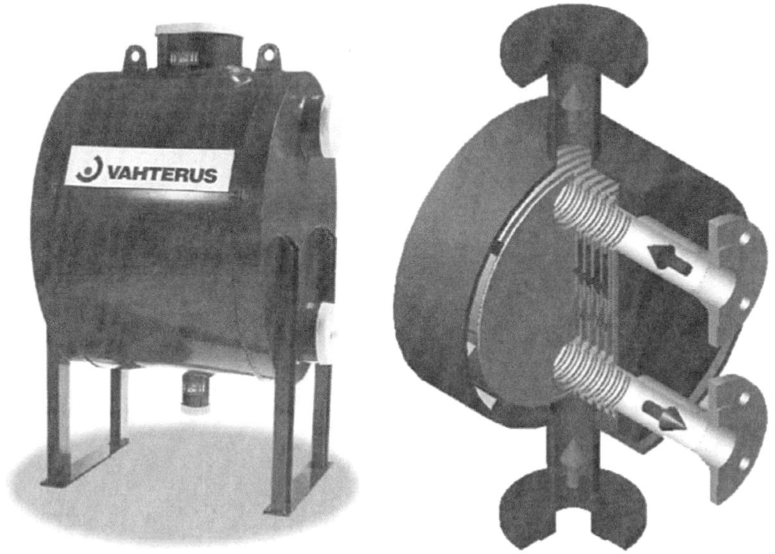

(a) (b)

Figure 3.34 **(See colour insert.)** Plate & Shell® HE: (a) assembled PSHE and (b) flow arrangement in one pass unit (after Vahterus Oy, 2013).

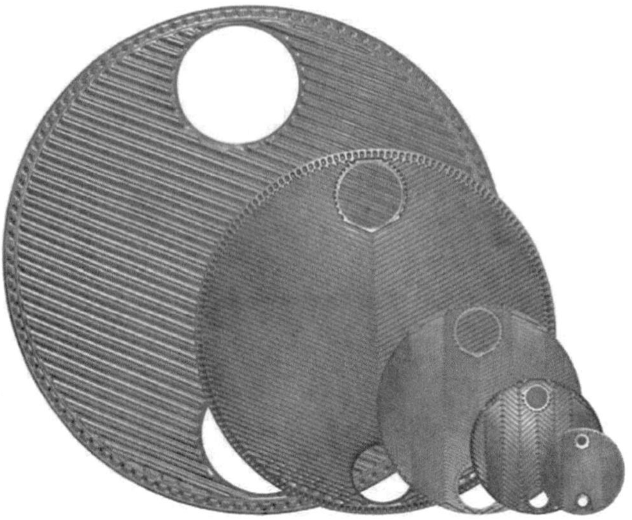

Figure 3.35 Example of the range of plates for HE of plate and shell type (after Sondex, 2012).

HEs with the same construction principles are now produced by a number of PHE manufacturers. Alfa Laval produces AlfaDisc™ with operating pressures from vacuum to 100 bar and temperature from –160 °C to 538 °C at standard design. ViFlow Finland Oy manufactures Laser welded Plate Heat Exchangers Type XPS. Sondex manufactures the SPS (Sondex Plate and Shell) HEs. Tranter SUPERMAX® Shell & Plate Heat Exchanger is designed for pressures up to 200 bar and temperatures up to 900 °C for standard range units. All HEs are available with plates of different diameters (e.g., see range of plates produced by Sondex in Figure 3.35) and total heat transfer area from about 4 to 400 m^2, possible on special design up to 2,000 m^2.

There are different derivatives of Plate and Shell design produced by a number of companies. For example, Figure 3.36 shows two shell welded PHEs produced by Ankor-Teploenergo, the private company founded in 1991 in Kharkiv, Ukraine. Such PHEs have prolonged plates of chevron-type corrugation welded by pairs with resistance seam roller welding and after these pairs are welded by argon arc welding into pack of plates. The pack of plates is placed in shell of prolong form (see Figure 3.36a) or two packs in a round shell (see Figure 3.36b).

Welded PHEs have substantially widened the area of PHE industrial applications. Aside from those mentioned earlier, a number of companies are producing welded PHEs according to developed proprietary designs and manufacturing technologies. With all these advantages, they have two major drawbacks. First, they cannot be disassembled to change plates, and second, due to rigid welded construction and metal fatigue, they cannot withstand frequent vibrations in heat exchanging streams, like streams after reciprocating compressors.

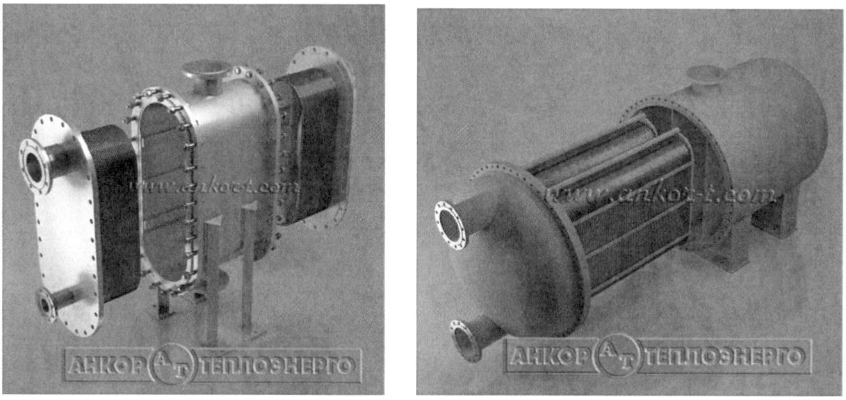

(a) (b)

Figure 3.36 Example of the shell welded PHE: (a) single plate pack in a shell and (b) double plate pack in a round shell (after Ankor-Teploenergo, 2011).

3.7.2.3 Semiwelded PHE with Twin Plates

The semiwelded PHE consists of a series of corrugated twin plates welded in pairs by one side. The other sides of plates are fitted with gaskets. The welded pairs of plates are then compressed together in a rigid frame like in plate-and-frame PHE to create an arrangement of parallel flow channels. Such design is helpful for work with aggressive fluids directed in channels between welded plates and is in contact with ring gaskets only at the nonwelded ports of plates. Figure 3.37 illustrates plates developed by engineers and researchers of NTU KhPI and Ukrniichemmash at Kharkiv in Ukraine and manufactured at Pavlogradchemmash factory starting from 1974. The first of plates (a) and (b) in Figure 3.37 is designed for shell and plate PHE, shown for plate (a) in Figure 3.38a. For plate (b) PHE is looking just as shown in Figure 3.34a. The PHE from plates (c) in Figure 3.37c just as plate-and-frame PHE is shown in Figure 3.38b. The pairs of plates are welded by resistance seam roller welding method.

Nowadays all major PHE manufacturers are producing semiwelded PHEs. The welding of plate pairs is mostly made by laser beam welding with appropriate change of plate construction near the gasket groove, as shown in Figure 3.39a. The PHE is assembled just like plate and frame; see Alfa Laval semiwelded PHE in Figure 3.39b. The design also makes it easy to separate the plates quickly – when, for example, they need to be cleaned. It also makes it easy to adjust capacity by simply adding or removing plates when required. The temperature range is −47 °C to 150 °C; pressure is up to 40 bar, depending on size. Semiwelded PHEs can be used also for refrigeration. The semiwelded HE can handle most refrigerants on the welded side and is particularly suitable for ammonia duties. Standard plates are stamped from stainless

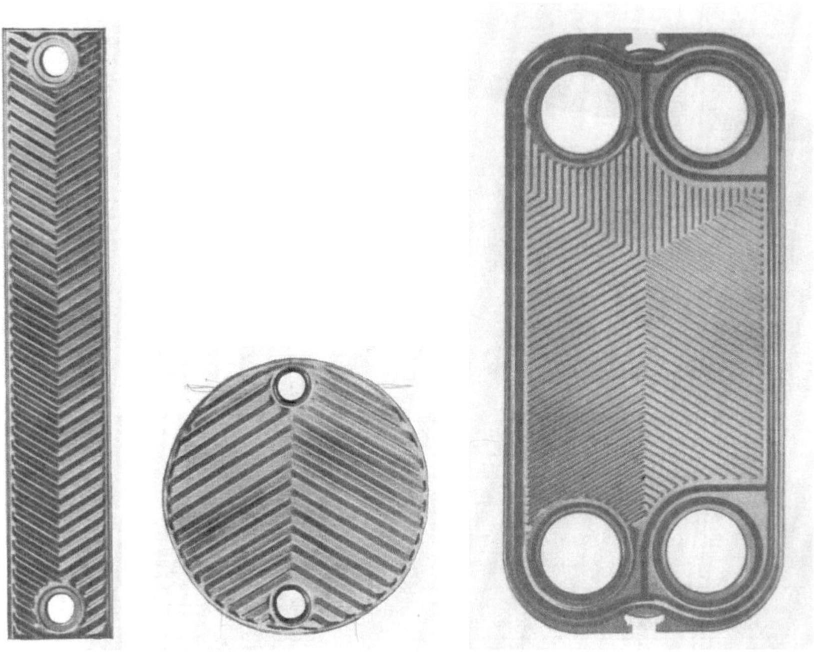

(a) (b) (c)

Figure 3.37 Pairs of plates for semiwelded PHE manufactured at factory Pavlogradchemmash in Ukraine: (a) rectangular plate 0.1 m^2; (b) circular plate 0.2 m^2; and (c) welded pair of twin plates 0.5 m^2.

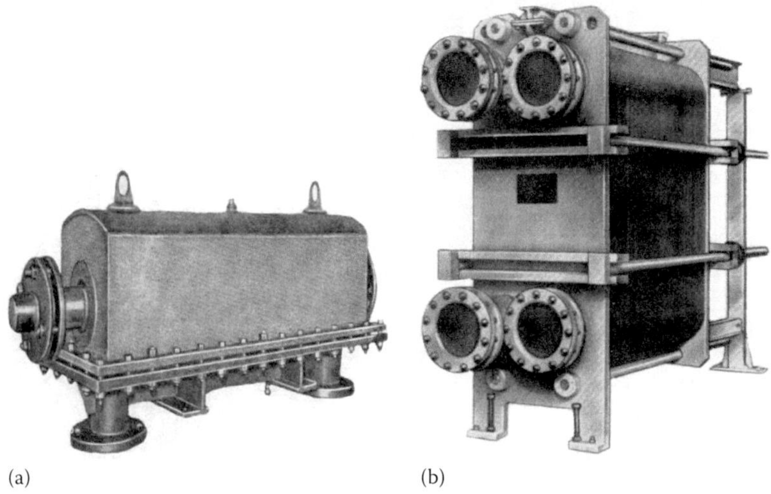

(a) (b)

Figure 3.38 Semiwelded PHE manufactured at factory Pavlogradchemmash in Ukraine: (a) from plates in Figure 3.37a and (b) from plates in Figure 3.37c.

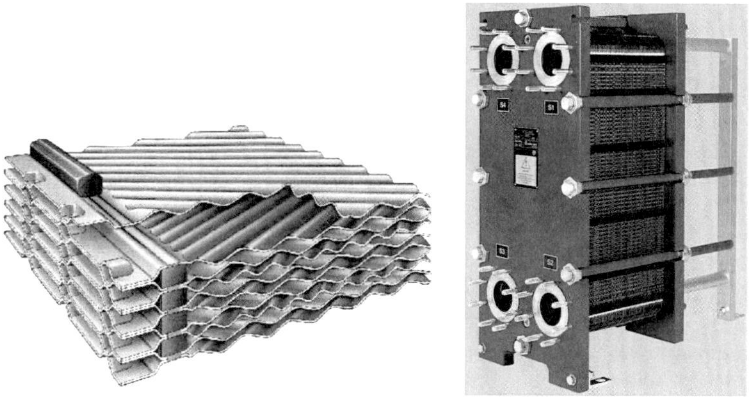

(a) (b)

Figure 3.39 **(See colour insert.)** Semiwelded PHE manufactured by Alfa Laval: (a) plate pack from welded plate pairs and (b) semiwelded PHE (courtesy of OAO Alfa Laval Potok).

steel and titanium. The plate pairs in semiwelded PHE can be removed or changed; due to the damper effect of rubber gaskets, they much better withstand vibrations than all-welded PHEs.

3.7.2.4 Special Design PHEs for Condenser and Evaporator Duties

For specific duties the special type of plate condensers can be produced. In Figure 3.40, plate condenser AlfaCond manufactured by Alfa Laval is presented. This unit was specially designed for condensing vapours at low pressure. The plate condensers are used in applications such as a vacuum condenser in evaporation crystallization systems, as an ethanol condenser in fuel ethanol plants and as a turbine condenser in cogeneration power plants.

The AlfaCond plate condenser is designed for vacuum condensation. It has a large vapour inlet connection, which is placed centrally at the top and the smaller condensate outlets on each side at the bottom. The two medium-sized connections for cooling media are located in the middle of the unit. The plates in this unit are welded in pairs. The vapour is condensed in the welded channel, while the cooling water passes through the corresponding channel with gaskets on periphery. The plate pattern is designed to ensure the most effective condensation possible, using an asymmetric channel configuration with a large gap on the vapour side and a small gap on the cooling water side (Figure 3.23). This makes it possible to maintain a very low pressure drop on the vapour side while still keeping up the velocity and turbulence on the cooling water side, thus maximizing the heat transfer efficiency and minimizing fouling.

The plate evaporators are PHEs with specific design. They are widely used in the following applications: sugar, sweeteners, fruit juice, fish and meat products, stillage,

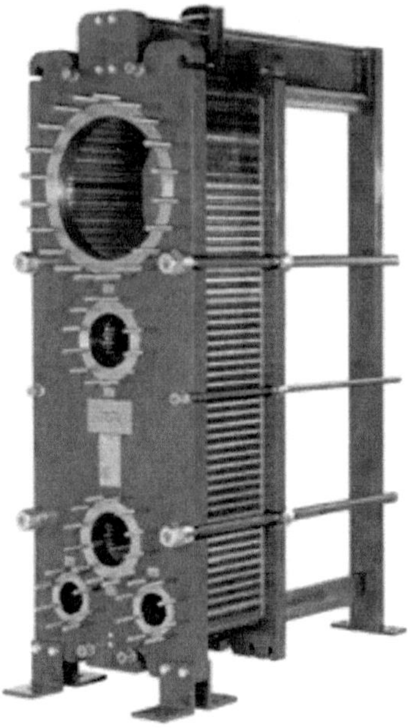

Figure 3.40 (See colour insert.) Semiwelded PHE AlfaCond manufactured by Alfa Laval for condensing duties (courtesy of OAO Alfa Laval Potok).

steep water, waste water, inorganic chemicals and as reboiler. The plate evaporator is presented in Figure 3.41. It consists of the package of plates, welded in pairs and connected with each other using gaskets.

Usually the heating steam is condensed in the welded channels while the evaporated product passes through the channels with gaskets on periphery.

3.7.2.5 Brazed PHE

All PHEs described earlier require a strong frame or shell to create an external pressure for keeping plates close together in a plate pack. In BPHE, the plates with crossing corrugations are brazed together in multiple contact points. Such construction withstands internal pressure inside channels without heavy frame plates or shell. Brazing on the periphery of plates eliminates the gaskets and need in gasket grooves. All this makes BPHE much lighter and more compact. Besides this brazing technology, it proved not much expensive and BPHEs are frequently the cheapest option for many duties.

The brazed HEs presented in Figure 3.42 have plates pack with the brazed seal around the edge of the plates. They form two channels between the plates and corner ports and they are arranged so that the two media flow through alternate channels, always in countercurrent flow. The contact points of the plates are also brazed to withstand the pressure of the media handled. The plates are stamped from thin stainless steel of thickness down to 0.3 mm, depending on operational pressure. Between the

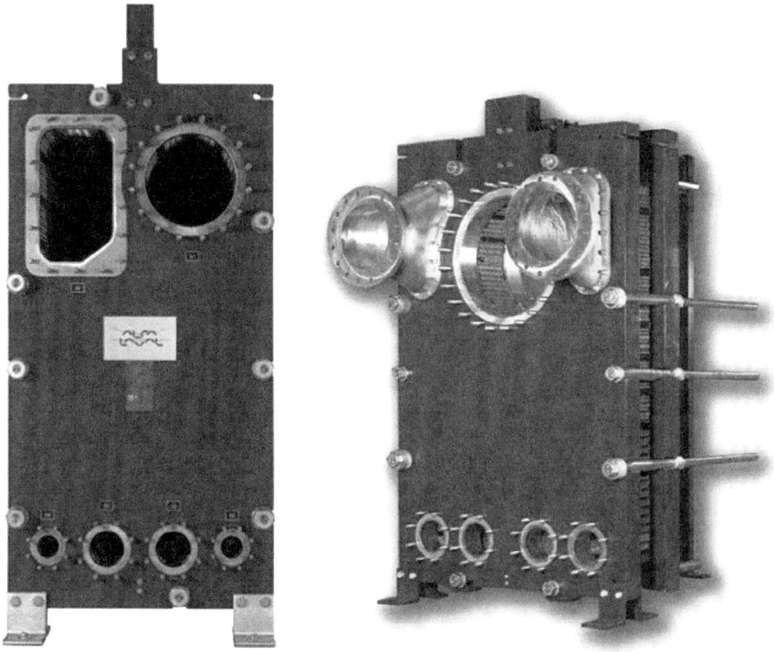

(a) (b)

Figure 3.41 **(See colour insert.)** Evaporator semiwelded PHE AlfaVap: (a) AlfaVap-500 and (b) AlfaVap-700 (courtesy of OAO Alfa Laval Potok).

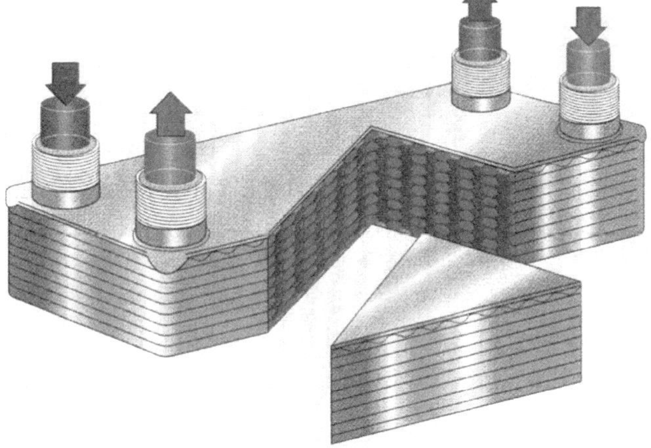

Figure 3.42 **(See colour insert.)** Drawing of BPHE showing the fluid flows in brazed plate pack (courtesy of OAO Alfa Laval Potok).

(a) (b)

Figure 3.43 **(See colour insert.)** Some of BPHEs produced by Alfa Laval: (a) general-purpose BPHEs and (b) special design for oil cooling (courtesy of OAO Alfa Laval Potok).

plates the sheets of brazing material with thickness about 0.1 mm are put. Assembled in such way BPHE is placed in oven under vacuum. After special thermal treatment, it is brazed and ready for testing. The brazing material is usually copper, but for special duties it can be nickel. Figure 3.43 shows some BPHEs of Alfa Laval production.

The operating temperature range for standard BPHE is from –196 °C to 225 °C, depending on type. The operating pressure for different BPHE types can be from 16 bar up to 45 bar, depending on temperature. The heat transfer area in one HE is up to 50 m^2 and more on special request. For refrigeration BPHEs are also made with the possibility to accommodate three streams.

BPHEs are the most efficient and economical solution for field of refrigeration, where they are now dominating all other types of HEs; for District Heating applications and oil cooling, it is also the cheapest option. BPHEs of different construction features and quality are now produced by all PHE manufacturers, the majority of which are SWEP, which is specializing on BPHEs, GEA, APV and others.

3.7.2.6 Fusion-Bonded PHE

The BPHE basically consists of two metals, limiting its performance on temperature, pressure, and for treating fluids aggressive to brazing metal. With invention at Alfa Laval active diffusion bonding technology, it became possible to make PHE from stainless steel only, with no other metal or gasket present. AlfaNova is a PHE made of 100% stainless steel. It is suited in applications that put high demand on cleanliness, applications where ammonia is used, or applications where copper or nickel contamination is not accepted. High resistance to corrosion makes it hygienic and environment-friendly. Moreover, the maximal operational temperature of some standard designs has been increased to 550 °C and limiting pressure up to 70 bar, depending on the type and temperature. Some fusion-bonded AlfaNova PHEs are shown in Figure 3.44.

Figure 3.44 Some of fusion-bonded AlfaNova PHEs produced by Alfa Laval (courtesy of OAO Alfa Laval Potok).

Another development is the tantalum range of PHEs that consists of stainless steel HEs that have undergone a unique treatment where a thin layer of tantalum is metallurgically bonded to all surfaces exposed to corrosive media. Combining a stainless steel core with a tantalum surface maximizes the corrosion resistance and mechanical strength with minimal investment cost.

3.7.2.7 Nonmetallic PHE

The operation with aggressive and high corrosive duties needs special materials. For some duties the metallic plates with low corrosion resistance cannot live up to service life requirements, and the heat transfer efficiency of HEs that use materials such as glass and Teflon is very low. For such conditions the specially designed graphite PHEs are used. In Figure 3.45 DIABON PHEs produced by Alfa Laval are presented. The unit plates are made from DIABON, a dense synthetic resin-impregnated high-quality graphite with a fine and evenly distributed pore structure. It provides performance similar to the material used in SGL's graphite blocks and graphite shell and tube HEs and is suitable for use with corrosive media up to 200 °C (390°F). This type of HEs operates in the following applications: hydrochloric acid and gas in all concentrations, sulfuric acid up to 85%, hydrofluoric acid up to 50%, phosphoric acid, pickling acids in surface treatment plants, electrolytes used in the mining industry, mixed acids, chlorinated hydrocarbons and catalysts such as aluminum chloride.

Figure 3.45 (See colour insert.) Graphite PHE (courtesy of OAO Alfa Laval Potok).

The advantages of graphite PHEs are as follows:

- High resistance to corrosive mediums, even to the most aggressive
- Low scale formation due to special material of plates and enhanced turbulence of the flow
- The high-efficiency heat transfer, which is comparable with heat transfer of metal plates and high strength of the plates' material
- Low thermal enlargement and minimal deformation of plates

3.7.3 Plate-and-Fin Heat Exchanger

PFHE is having secondary heat transfer surfaces in a form of fin structures between plates that are separating the streams (see Figure 3.46). The fins (having triangular and rectangular cross-sectional shapes most commonly), besides being the secondary heat transfer surface with low hydraulic diameter, contain the pressure differential between the streams. Fins are die- or roll-formed and are attached to the plates by brazing, soldering, adhesive bonding, welding, or extrusion. Fins may be used on both sides in gas-to-gas HEs. In cases of gas-to-liquid applications, fins are usually situated only on the gas side, where film heat transfer coefficients are much lower. Employed on the liquid side, they are utilized mainly for structural strength and flow mixing, as also sometimes to withstand differential pressure and rigidity. Plate fins are categorized as (1) plain (i.e., uncut) and straight fins such as plain triangular and rectangular fins, (2) plain but wavy fins (wavy in the main fluid flow direction) and (3) interrupted fins such as offset strip, louver and perforated (e.g., see Figure 3.47). Plate–fin exchangers have

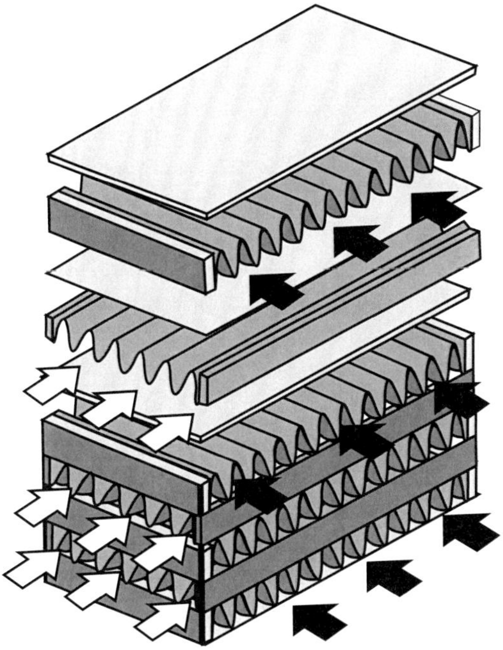

Figure 3.46 Schematic drawing of PFHE.

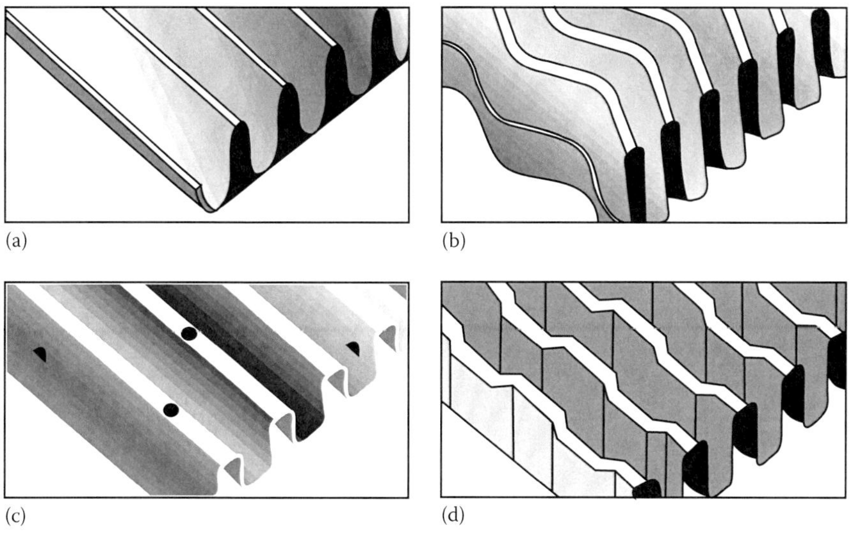

Figure 3.47 PFHE fin types: (a) plain, (b) Herringbone, (c) perforated and (d) offset strip fin (OSF).

been built with a surface area density of up to 5,900 m^2/m^3. There is a total freedom of selecting fin surface area on each fluid side, as required by the design, by varying fin height and fin density. Common fin thicknesses range from 0.05 to 0.25 mm and their heights range from 2 to 25 mm. PFHEs are manufactured in different shapes and sizes from a variety of materials. PFHEs are widely used in many applications like gas turbine, steam, nuclear, fuel cell, automobile, truck, airplane, heat pump, refrigeration, and electronics, cryogenics, gas liquefaction, air-conditioning and waste heat recovery systems.

Initially developed for aircraft and transport vehicles in the 1940s, the PFHEs have evolutionized to a very compact and efficient product in these applications. For example, in the United States, Lytron company is producing aluminum, vacuum-brazed, PFHEs that are custom-designed, as every requirement is unique. PFHE can be designed for use with any combination of gas, liquid and two-phase fluids (see Figure 3.48).

High performance of PFHE and good low-temperature properties of aluminum, as also its high heat conductivity, have stipulated the application of aluminum-brazed

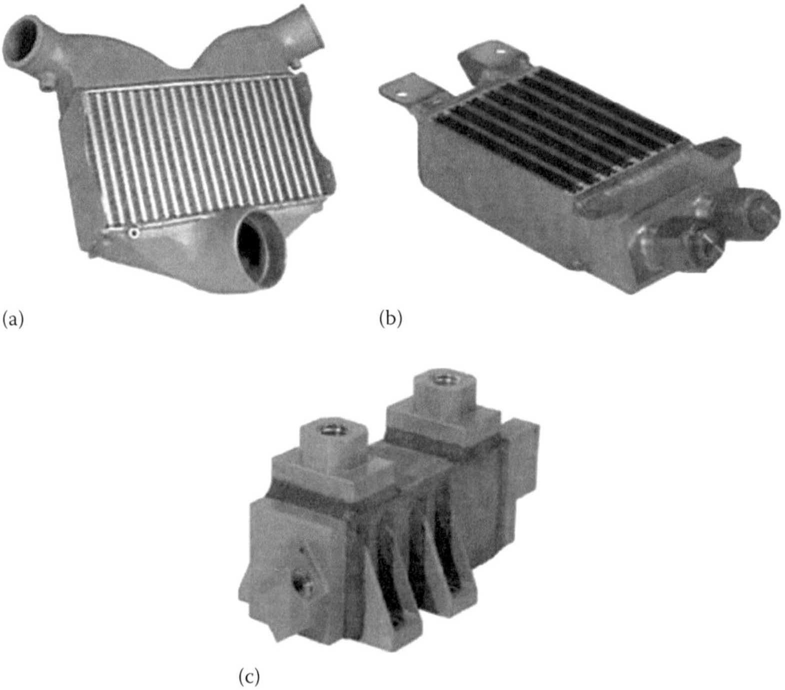

(a) (b) (c)

Figure 3.48 **(See colour insert.)** Custom-designed PFHE produced by Lytron, USA: (a) PFHE for air-to-air cooling, (b) PFHE for liquid cooling by air in an airplane and (c) liquid-to-liquid oil cooler for commercial aircraft (after Lytron, 2012).

Figure 3.49 (See colour insert.) Chart's aluminum-brazed PFHE (after Chart, 2014).

PFHE in the Air Separation, LNG (liquefied natural gas) liquefaction, Nitrogen Rejection, NGL Recovery, ethylene production plants and other low-temperature natural gas and petrochemical processes. The alternating plate–fin construction offers multiple stream capability and simplifies a series of shell and tube units to a single compact structure, saving much weight and cost of the system. A aluminum-brazed PFHE produced by Charts Heat Exchangers, pictured in Figure 3.49, is typically 20% the size of a shell and tube exchanger of comparable performance.

3.7.4 Tube-and-Fin Heat Exchanger

When for one of the fluids exchanging the heat film heat transfer coefficient is much lower, then for other one, it can be compensated by extending the heat transfer surface from that side of the HE. In a gas-to-liquid HE, the heat transfer coefficient on the air side is generally one order of magnitude lower than that on the liquid side. To balance that fins are used on the gas side to increase the surface area there. This is similar to the case of a condensing or evaporating stream on one side and gas on the other. Also if one fluid has a high pressure, it is generally economical to employ tubes for that fluid side. For example, Figure 3.50 shows the finned tube and made on such tubes TFHE.

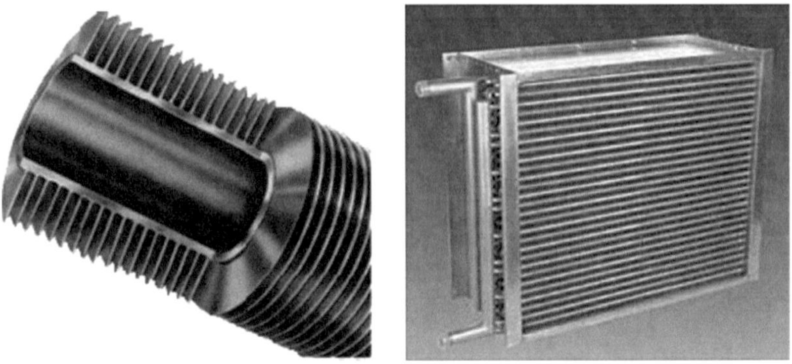

(a) (b)

Figure 3.50 **(See colour insert.)** Example of (a) finned tube and (b) TFHE made of such tubes (manufactured by Defon Heat Exchanger Co., Ltd, in China, see Defon, 2013).

3.7.5 Spiral Heat Exchanger

The SHE, which is presented in Figure 3.51, has a specific design to operate with extremely dirty fluids, on positions where one or both of the fluids may cause fouling, using mediums with high adhesion to wall material, or for the heat carriers with high content of different suspended matters (with nonuniform composition).

The SHE consists of a spiral wound from two sheets of metal strips that form two concentric spiral flow passages (see Figure 3.52). The media flows countercurrently in the two channels. The flow is from the centre to the periphery and vice versa. The sludge channel is open on one side and closed on the other and the water channel is closed on both sides. Each channel has one connection in the centre and one on the periphery. The curved fluid channels provide optimum heat transfer and flow

Figure 3.51 **(See colour insert.)** SHEs (courtesy of OAO Alfa Laval Potok).

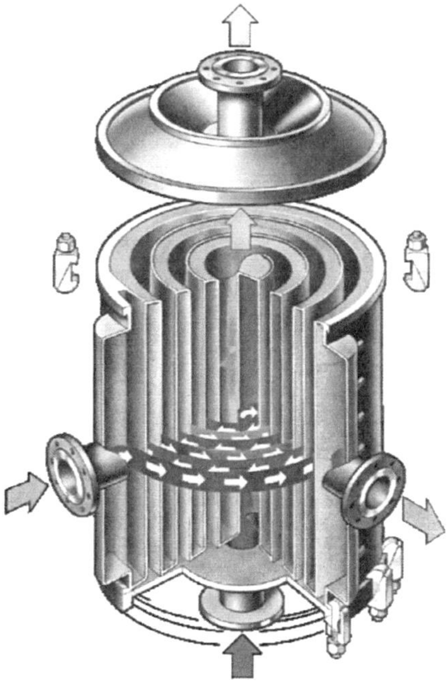

Figure 3.52 **(See colour insert.)** Fluid flow in a liquid–liquid SHE (courtesy of OAO Alfa Laval Potok).

conditions for a wide variety of fluids while keeping the overall size of the unit to a minimum. The result is an HE that provides maximum heat transfer efficiency while only taking up a minimum of installation space.

Single-channel technology means that both fluids occupy a single channel, which allows fully countercurrent flow. One fluid enters the centre of the unit and flows toward the periphery. The other fluid enters the unit at the periphery and moves toward the centre. The channels are curved and have a uniform cross section, which creates a 'spiralling' motion within the fluid. The fluid is fully turbulent at a much lower velocity than in straight-tube HEs, and each fluid travels at constant velocity throughout the whole unit. This removes any likelihood of dead spots and stagnation. Solids are thus kept in suspension, and the heat transfer surfaces are kept clean by the scrubbing action of the spiralling flow (Figure 3.52). The self-cleaning properties of SHEs ensure that the reliable performance of efficient heat transfer is guaranteed, with minimum downtime for maintenance.

In some duties where alternative HEs would need regular cleaning, disassembly, repair and maintenance, an SHE performs for much longer periods and only requires maintenance during routine plant shutdowns. Spiral units provide good access to the heat transfer area – each fluid channel is easily accessed via its cover, exposing the whole of the heat transfer surface.

All SHEs can be drained in position, without any need to disconnect pipework or to open the units. For applications where complete draining of the process fluid is

required on a regular basis, the SHE can be installed in an upright position. This is particularly useful for batch production processes as it makes it possible to empty all fluid from the unit between batches.
The operating conditions are as follows:

Area range per body: from 1 m^2 up to 700 m^2
Design temperature: from −100 °C up to 400 °C
Design pressure: from full vacuum to 40 bar and above

Standard construction materials: stainless steel and carbon steel, and it can also be made from any metal that can be cold formed and welded, including Duplex, titanium, Hastelloy and 904L.

The SHEs can operate in the following industries:

- Petrochemical
- Refinery
- Steel making
- Pulp and paper
- Metal/ore processing
- Wastewater treatment
- Pharmaceutical
- Vegetable oil processing
- Distillery

They can be used as reboilers, condensers and coolers (heaters) for the following duties:

- Liquid/liquid – preheating, heating, cooling, interchanging, heat recovery
- Vapour/liquid – top condensers, reflux condensers, vacuum condensers, vent condensers, reboilers

Type of fluids and gases:

- Fouling liquids – containing solids, fibers, liquors, slurries and sludges
- Gases – pure vapor and mixtures with inert gases

3.7.6 Lamella Heat Exchanger

An LHE has an outer tubular shell with bundle of heat transfer elements called lamella, as shown in Figure 3.53a. These lamellas are flat tubes (pairs of edge welded thin dimpled plates) shown in Figure 3.53b. The inside opening of the lamella channel is in the range from 3 to 10 mm and the wall thickness from 1 to 2 mm. The lamellas are placed close to each other to form narrow channels on the shell side. Lamellas have end fittings with gaskets to prevent the leakage between shell and tube sides. In a large exchanger, lamellas can consist of two or more flat tubes to contain operating pressures. To allow for thermal expansion, one end of the bundle is fixed

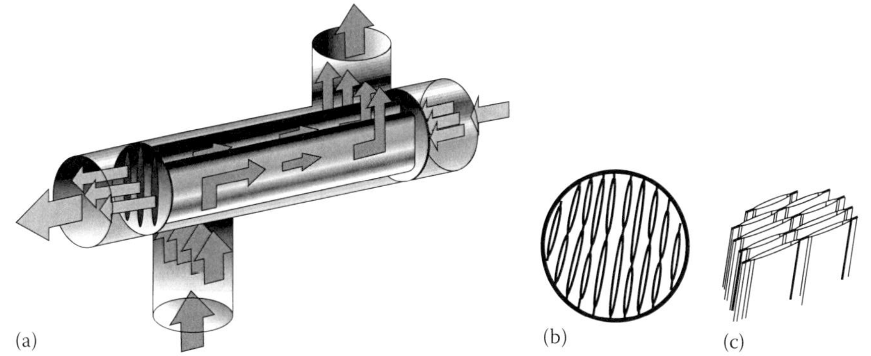

Figure 3.53 LHE: (a) drawing of fluid flow, (b) cross section of lamella bundle, (c) lamella bundle (after Sondex Tapiro OY AB, 2011).

and the other is floating. One fluid flows inside the lamellas and the other fluid moves longitudinally in the spaces between them. The exchanger has a single pass, and the flow arrangement is countercurrent. The tube walls have dimples where they are spot-welded. High heat transfer coefficients are obtained because of small hydraulic diameters and no leakage or bypass streams compare to conventional shell and tube exchanger.

With proper plate spacing, LHE can handle fibrous fluids and slurries. The large units can have surface areas up to 1,000 m^2. LHE is smaller and weighs less than a shell and tube exchanger having the same duty. The operational pressure range is from vacuum up to 35 bar and temperature limits 200 °C for PTFE gaskets and up to 500 °C for asbestos fiber gaskets. LHE is used for heat recovery in the pulp and paper industry, chemical process industry and other industrial applications.

3.7.7 Microchannel Heat Exchanger

As it is discussed in Section 3.5, the decrease of hydraulic diameter of channels can significantly increase not only compactness, but also heat transfer area required for specified temperature programme and allowable pressure drop. These factors have stipulated the interest in developing heat exchangers with hydraulic diameter smaller than in discussed above heat transfer equipment. Such heat exchangers with hydraulic diameter less than 3 mm are called MCHEs, not depending on the other features of construction. These construction features can be different, this is discussed in the following text.

3.7.7.1 Printed Circuit MCHE

The printed circuit heat exchanger (PCHE) is a relatively new concept, originally developed for refrigeration applications, that has been commercially manufactured in United Kingdom by the company Heatric™ Ltd., since 1985. PCHEs are robust heat exchangers with high compactness, effectiveness and the ability to operate with a very large pressure difference between hot and cold sides (Heatric™; see Heatric, 2014). PCHEs are manufactured by the same technique used for producing standard printed circuit boards for electronic equipment. The fluid passages are photochemically etched into the metal plate. Normally, only one side of plate is etched out. The plates are thereafter joined by diffusion bonding. This process creates an exchanger core with no joints, welds or points of failure. Such a unit combines exceptional strength and integrity with high efficiency and performance. PCHE is up to 85 % smaller and lighter than traditional shell and tube heat exchangers.

For the two fluid streams, there are different etching patterns, and cross flow, counter flow or multipass cross-counter flow can be made in heat exchanger. Multiple passes and multiple fluid streams can be made in a single block. For large heat duty applications, several blocks are welded together. Flow plates, as shown in Figure 3.54a, are produced from the material specifically selected according to the duty, process fluids and environment in which the exchanger is operating. The maximum plate size currently available to produce each exchanger core block is 600 × 1,500 mm. If necessary, multiple blocks can then be joined to form the final exchanger core. The plate is typically 1.6 mm thick and the specific design of flow channels is then chemically etched into its surface, creating 2 mm diameter semi-circles, which once are diffusion-bonded create the final solid exchanger core, as shown in Figure 3.54b (Heatric™;

(a) (b)

Figure 3.54 **(See colour insert.)** Example of (a) PCHE Heatric™ plate and (b) block cross section (from Heatric, 2014).

Figure 3.55 **(See colour insert.)** Example of PCHE Heatric™ for oil and gas processing industry (from Heatric, 2014).

see Heatric, 2014). The hydraulic diameter of channels can be anywhere, from < 1 to 3 mm (Hesselgreaves, 2001). The consequence of the surface form is that the porosity of the exchanger is low – of the order of 0.4–0.55, compared with porosity 0.6–0.75 typical for a plate-fin exchanger of similar material. This normally means higher weight and lateral dimensions for similar hydraulic diameters. Operating pressure of such PCHE can be over 1,000 bar, depending on the temperature. Heat exchanger can work at temperatures ranging from –200 °C up to 900 °C, depending on the pressure. The example of PCHE Heatric™ is shown in Figure 3.55. Such PCHEs are used in oil and gas processing, power generation, chemical processing, cryogenic air separation and other applications requiring compactness and reliability.

Nowadays, the constructions of printed circuit MCHEs are undergoing intensive development, leading to application in fields such as Micro-Electro-Mechanical Systems (MEMS), microelectronics, biomedical, fuel processing, aerospace, heat pumps, HVAC, air separation, very-high-temperature gas-cooled nuclear reactors and others. It requires heat exchangers of different sizes, down to centimetres in length for microheat exchangers. The material of heat transfer surface can be copper, nickel, metal alloys, silicon, glass, silicon-carbide and ceramics. Majority of these heat exchangers are one pass counterflow primary heat transfer surface devices. An example of plate stack assembly is shown in Figure 3.56. The range of hydraulic diameters studied by different researchers is from 0.03 to over 2 mm. For metal material of plates, diffusion bonding is usually used and just sometimes brazing. For non-metallic materials, it is ceramic processing. A review of manufacturing processes for PCHE fabrication can be found in a research article written by Ashman and Kandlikar (2006). Besides chemical etching for forming plate channels, they have mentioned micro-machining and some kinds of lithography.

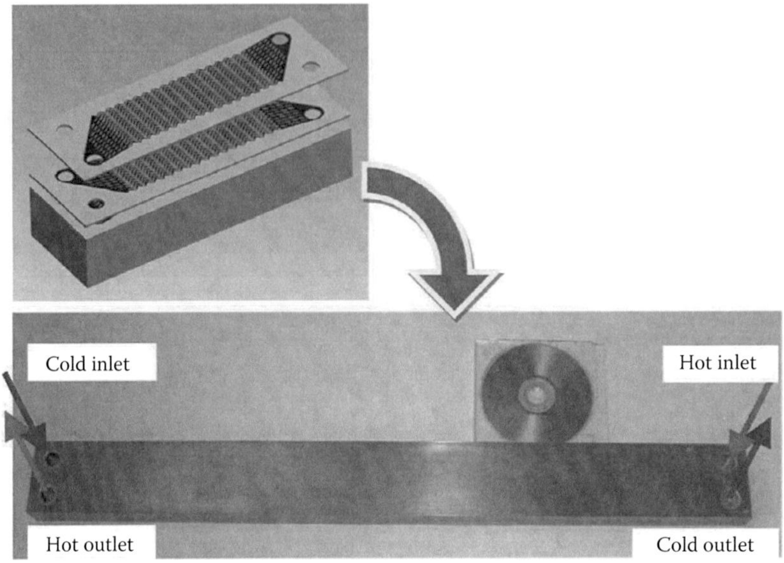

Figure 3.56 (See colour insert.) PCHE for water heating by CO_2 as heat pump working fluid (after Tsuzuki, 2009).

3.7.7.2 Matrix MCHE

The application of matrix heat exchanger in the production of liquid oxygen was introduced in the late 1940s. The main field of its application is cryogenic low power cryocoolers and helium liquefiers. From the beginning, the design consisted of a series of perforated aluminium plates separated by thin neoprene gaskets that together formed a matrix with complex channels for heat exchanging streams. The matrix could be formed in the shape of a box, as shown in Figure 3.57, or be circular and form a tube. With cast aluminium headers at each end, it was held together by means of steel tightening rods. The neoprene spacers prevented axial conduction and provided tight seals, even at liquid air temperatures. The construction was very simple and easy for maintenance and repairs.

The developments in construction of matrix MCHEs and methods of their design are discussed in detail by Moheisen (2009). Nowadays matrix MCHE consists of a stack of high thermal conductivity (copper or aluminium) perforated plates or wire screens made of copper or aluminium. These plates are separated by lower thermal conductivity plastic or stainless steel spacers (plastics, stainless steel). The packet of alternate plates and spacers is bonded together to form leak-free passages for the streams. The bonding techniques can be by soldering, brazing or diffusion. The diffusion bonding of copper plates and stainless steel spacers gives the strong all metal structure that withstand high pressure, up to 115 MPa, as was observed in experiments by Sotnikov et al. (1985) of Refrigeration Technology Institute in Odessa, Ukraine. The gaps between the plates ensure uniform flow distribution and create flow interruptions

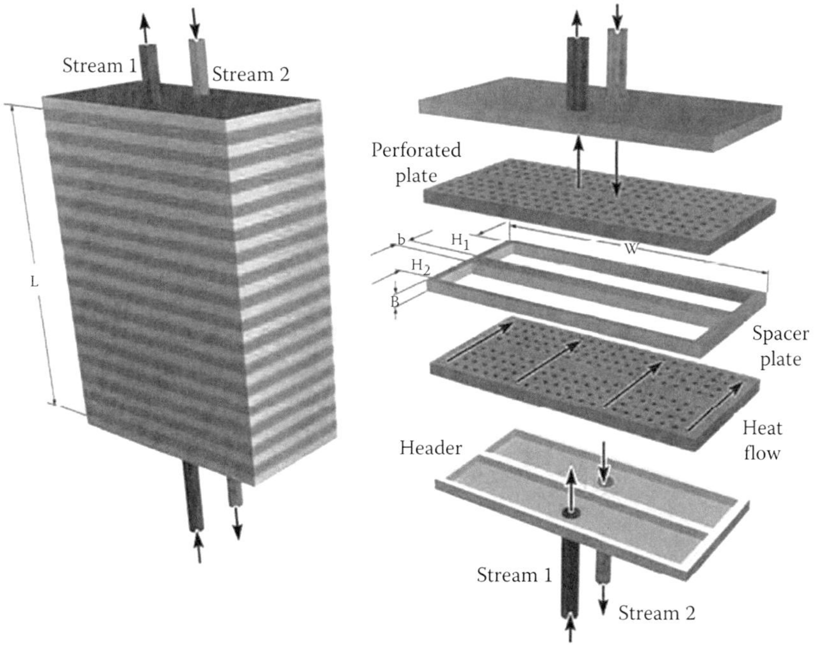

Figure 3.57 (See colour insert.) The construction of matrix MCHE (after Moheisen, 2009).

and turbulence, enhancing heat transfer. The range of Reynolds numbers is usually from 800 to 4,000 and, as it was discussed in Section 3.5, it is important to have smooth transition from laminar to turbulent flow in this region. The construction of matrix creates the counter-current flow arrangement for two streams exchanging heat through the walls of perforated plates with extended heat transfer surface. The spacers with lower heat conductivity are resistant to heat conduction between plates. Small flow passages in plates with diameter 0.3–1.0 mm ensure a high heat transfer coefficient and heat transfer surface area density up to 6,000 m^2/m^3.

3.7.7.3 Miniscale MCHE

As discussed earlier, printed circuit and matrix heat exchangers were initially invented with microchannels as MCHEs. But with the need for compactness and advances in fabrication technologies, any type of heat exchanger can be made smaller by the reduction of scale. Even shell-and-tube heat exchangers can be made with microchannels, see, for example Saji et al. (2001). The best candidates for such miniaturisation are plate-and-fins and tube-and-fins types, which were initially invented as compact heat exchangers.

The need of transport vehicles, refrigeration and air conditioning systems in more compact and efficient heat exchangers have geared the development of

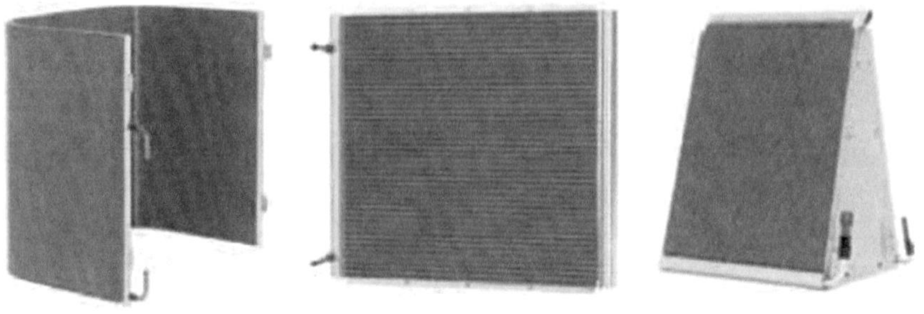

Figure 3.58 Plate-fin MCHEs for refrigeration and air conditioning produced by Danfoss (after Danfoss, 2013).

plate-and-fins and tube-and-fins heat transfer surfaces constructions towards lower hydraulic diameters in the scale of microchannels. Another problem that can be solved with this trend is the decrease of hold-up volume in refrigeration and air conditioning systems. The limitations put on the use of fluorocarbons as refrigerants because of their adverse influence on ozone layer require the usage of natural refrigerants like propane, isobutene, ammonia etc. These refrigerants are more environmentally friendly and have good performance, but the problem with some of them is the fact that they are dangerous in large quantities. Propane is highly flammable, as is isobutene. Ammonia can be toxic, flammable and is corrosive to some metals.

To solve these problems the plate-and-fin heat exchangers principle was used to make microchannels heat exchangers for refrigeration and air conditioning. As an example, in Figure 3.58 are shown some MCHEs produced by Danfoss. It is completely made of aluminium, as illustrated in Figure 3.59. The refrigerant side is made as flat panels with

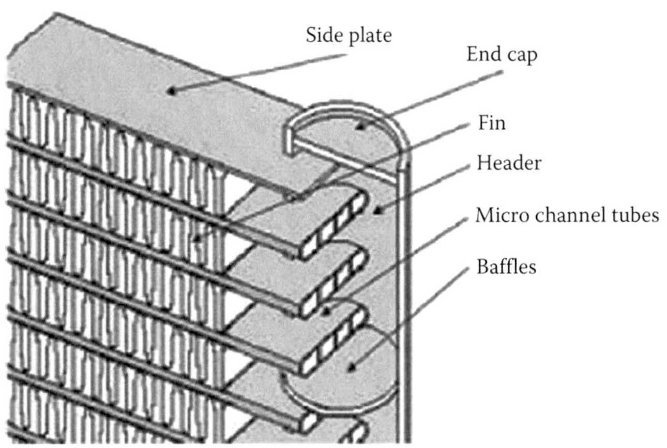

Figure 3.59 The construction principle of plate-fin MCHEs for refrigeration and air conditioning shown in Figure 3.58 (after Popescu et al., 2012).

channels for refrigerant inside with the hydraulic diameter about 0.7 mm. The pitch of fins on the air side is 1.1 mm and hydraulic diameter is about 2 mm. The length of fins is 16 mm that confirms the discussions in Section 3.5, see Figure 3.14.

3.7.8 Non-Metal Compact Heat Exchangers

The developments of compact heat exchangers from metals have significantly widened the field of their application, the operational pressures and temperatures. Nevertheless, the operating limitations of these metallic heat exchangers for some highly aggressive to metals fluids have created the need to develop alternative units from other materials like polymers and ceramics.

3.7.8.1 Polymer Compact Heat Exchangers

Initially the interest in the development of polymer compact heat exchangers was stimulated by their ability to combat corrosion in process functions involving acid solutions. These heat exchangers can handle both liquids and gases, resistant to fouling and corrosion and also can be used in humidification and dehumidification functions. Important point is also that the use of polymers can offer substantial weight, volume, space and cost savings compared to heat exchangers manufactured from exotic metal alloys. The comprehensive review of different polymer compact heat exchangers is published by Zaheed and Jachuck (2004), and more recently, the history of the question reviewed is by Cevallos et al. (2012). T'Joen et al. (2009) in their review have concentrated on HVAC and refrigeration applications of polymer heat exchangers. Summarized main points concerning construction, fabrication and applications of polymer CHEs are as follows.

The main types of high-performance polymers that can be used in polymer compact heat exchangers are as follows:

Polyvinylidene fluoride (PVDF) has high corrosive resistance up to 154 °C. It is suitable for heat recovery processes involving acids, reducing pollution emissions and flue gas cleaning purposes.

Teflon or PTFE (polytetrafluoroethylene) has virtual universal chemical inertness and non-stick properties, except molten alkali metals and fluorine. Teflon can work at temperatures up to 204 °C and can be used in bromine recovery systems, metal pickling, plating solutions and deionised water heating.

Polypropylene (PP) is rigidly constructed, non-toxic, non-staining and has excellent corrosion resistance. It is applicable in mechanical vapour compression (MVR) units.

Polyethylene (PE) is the most widely used plastic. It is transparent to opaque, robust enough to be virtually unbreakable, and, at the same time, quite flexible.

It may be used at temperatures of up to 80 °C. It is suited to a wide range of laboratory apparatus including wash bottles, pipette-washing equipment and tanks.

Polycarbonate (PC) has good chemical resistance to mineral acids, organic acids, greases and oils. It has a service temperature range up to 135 °C.

Polyphenylene sulphide (PPS) performs better in acidic conditions compared to teflon and PVDF.

Polyphenylene oxide (PPO) is similar in chemical composition to *polyphenylene ether* (PPE), and both are generally treated as equivalents. It has poor chemical resistance.

Polyetheretherketone (PEEK) has an estimated continuous working temperature of 250 °C. It has high resistance to chemical attack. The material is fully resistant at aqueous solutions of 50% H_2SO_4 and 50% NaOH at room temperature. The only materials that attack PEEK are supposedly concentrated nitric and sulfuric acids.

Polysulfone (PSU) is an amorphous thermoplastic with a limit of continuous use temperature of 190 °C. It has good creep resistance and thermal stability. It withstands most solvents, oils, acids and alkalis.

Polymers have rather low strength, poor creep resistance, small thermal conductivity and large thermal expansion. But they have high resistance to chemical attack and their relatively low cost are making them attractive option in some applications. Nowadays polymer matrix composite materials offer possibilities to create new materials having properties that cannot be obtained using a single monolithic material. This approach to construction is very promising for future heat exchanger materials which could be developed according to the specific requirements of an application. It bears even more prospects to polymers in heat exchangers. There are three major categories of polymer compact heat exchangers, namely, polymer plate heat exchangers, heat exchanger coils and shell and tube heat exchangers.

Polymer plate heat exchangers are produced by a number of companies, for example AB Segerfrojd producing Monoblock™ heat exchanger that is built up of extruded polypropylene (PP) sheets, as shown in Figure 3.60. The PP sheets are welded together using patented welding method, which enables air-tight joints. It is available in pure cross flow and counterflow arrangements and is available in different arrangements and shapes in the range from 100 × 100 × 100 mm up to 1,200 × 1,200 × 1,000 mm. The temperature range is from −40 °C to 90 °C and pressure drop is from 10 to 400 Pa. The compactness of such heat exchanger is up to 250 m^2/m^3.

Polymer PHEs made of polyphenylene sulphide (PPS) were tested as condensing heat exchangers for large boilers at the Institute Francais du Petrole. An overall heat transfer coefficient value of 40 W/m^2 K was achieved, with condensation on one side and preheating air on the other. Other uses were in boilers for apartment blocks and malt kiln heat recovery, as reported by Zaheed and Jachuck (2004).

Figure 3.60 (See colour insert.) Monoblock™ heat exchanger produced by AB Segerfrojd (www.segerfrojd.com) (after Zaheed and Jachuck, 2004).

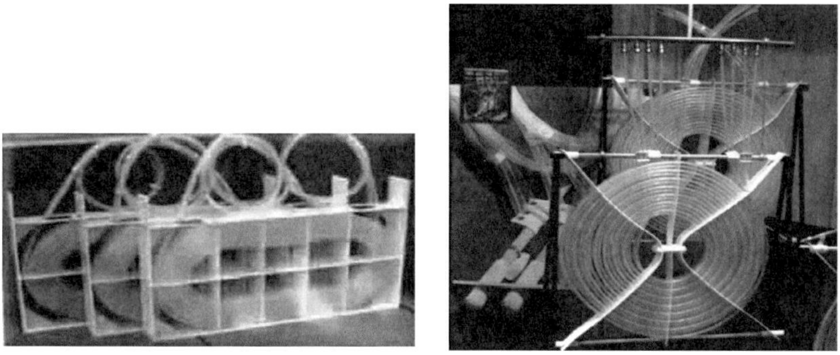

(a) (b)

Figure 3.61 **(See colour insert.)** Fluorotherm immersion heating coils. (a) Polypropylene frame and (b) PTFE frame (after Zaheed and Jachuck, 2004).

For heating and cooling of chemically aggressive fluids immersion coils from teflon (Fluorotherm) are used, as shown in Figure 3.61. They have excellent chemical resistance and reliability.

The polymer shell and tube heat exchanger has a construction similar to the conventional metal shell and tube with tube bundle and shell covered inside with polymer, as can be seen in Figure 3.62. Their applications involve highly aggressive fluid heating or cooling.

After review of the recent developments in experimental new constructions of polymer compact heat exchangers, T'Joen et al. (2009) came to the conclusion that both plate and tubular heat exchangers can be successfully designed, constructed and tested, with their performance being comparable to conventional metal units at lower cost and reduced weight even in HVAC applications.

Figure 3.62 **(See colour insert.)** Ametek shell and tube teflon heat exchanger (after Ametek, 2010).

3.7.8.2 Ceramic CHEs

Main advantages for using ceramic materials in heat exchanger construction over traditional metallic materials are their temperature and corrosion resistance. Ceramic materials can withstand operating temperatures up to 1,400 °C, much exceeding those of conventional metallic alloys. Another major advantage of ceramic-based heat exchangers is their resistance to corrosion and chemical erosion. The major ceramic materials are: silicon carbide, silicon nitride, alumina, zirconia, aluminium titanate, aluminium nitride and ceramic matrix composites (CMCs). Their properties concerning application in heat exchangers are in detail discussed in a review by Sommers et al. (2010).

Introducing ceramic and ceramic matrix composite materials in heat exchangers is possible either by replacing the materials of existing designs or by implementing new configurations. For example printed circuit MCHEs can have the same construction with metals and with ceramics. In Figure 3.63 is illustrated ceramic PCHE. It is one of the wide areas of ceramic application in heat exchangers.

Sommers et al. (2010) identified possible areas of ceramic compact heat exchangers applications as follows: evaporators in evaporative cooling systems for air-conditioning, recuperators and generators in LiBr/H_2O absorption chillers for air-conditioning, primary heat exchangers in gas-fired furnaces for space heating, high temperature recuperators and chemical digesters, open-cell foams for reactive heat exchange processes and filtration.

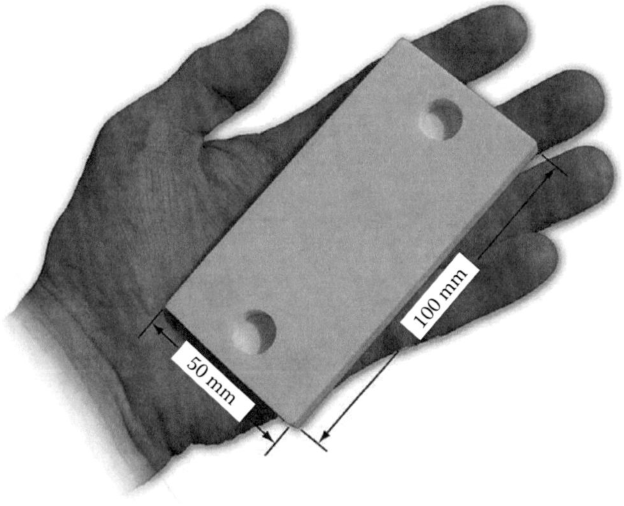

Figure 3.63 **(See colour insert.)** Photograph of a compact ceramic microchannel counterflow heat exchanger. This view shows the hot and cold inlets, with the exhaust flows connected from the bottom side (after Kee et al., 2004, 2011).

REFERENCES

Ametek. 2010. Shell and tube heat exchangers. www.ametekfpp.com/Shell-and-Tube-Heat-Exchangers/Shell-and-Tube-Heat-Exchangers-30-series-models-218-525-and-1000.aspx (accessed 20 February 2014).

Ankor-Teploenergo. 2011. Shell welded heat exchangers. www.ankor-t.com/en/catalog/collapsible-housing.htm (accessed 20 February 2014).

Anxionnaz Z., Cabassud M., Gourdon C., Tochon P. 2008. Heat exchanger/reactors (HEX reactors): Concepts, technologies: State-of-the-art. *Chemical Engineering and Processing: Process Intensification* 47(12): 2029–2050.

API Heat Transfer. 2003. Shmidt all-welded SIGMAWIG plate heat exchangers. Form WPHE-350 5/10, USA, www.apiheattransfer.com/Product/103/SIGMAWIG-All-Welded-Plate-Heat-Exchangers (accessed 18 April 2015).

Arsenyeva O. (2010). The generalized correlation for calculation of hydraulic resistance in channels of PHEs. *Integrated Technologies and Energy Saving* (4): 112–117 (in Russian).

Arsenyeva O., Tovazhnyansky L., Kapustenko P., Khavin G. 2011. The generalized correlation for friction factor in crisscross flow channels of plate heat exchangers. *Chemical Engineering Transactions* 25: 399–404.

Arsenyeva O.P., Tovazhnyanskyy L.L., Kapustenko P.O., Demirskiy O.V. 2012. Heat transfer and friction factor in criss-cross flow channels of plate-and-frame heat exchangers. *Theoretical Foundations of Chemical Engineering* 46(6): 634–641.

Ashman S., Kandlikar S.G. 2006. A review of manufacturing processes for microchannel heat exchanger fabrication. In: *Proceedings of ICNMM2006. Fourth International Conference on Nanochannels, Microchannels and Minichannels*, June 19–21, 2006, Limerick, Ireland, 855–860.

Balasubramanian K., Jagirdar M., Lee P.S., Teo C.J., Chou S.K. 2013. Experimental investigation of flow boiling heat transfer and instabilities in straight microchannels. *International Journal of Heat and Mass Transfer* 66: 655–671.

Calgavin. 2014. How Hitran works. www.calgavin.com/heat-exchanger-solutions/hitran-systems/how-hitran-works/ (accessed 20 February 2014).

Cevallos J.G., Bergles A.E., Bar-Cohen A., Rodgers P., Gupta S.K. 2012. Polymer heat exchangers – History, opportunities, and challenges. *Heat Transfer Engineering* 33(13): 1075–1093.

Chart. (2014) Brazed aluminum heat exchangers. www.chart-ec.com/brazed-aluminum-heat-exchangers.php (accessed 20 February 2014).

Churchill S.W. 1977. Friction-factor equation spans all fluid-flow regimes. *Chemical Engineering* 84(24): 91–92.

Crittenden B.D., Yang M., Dong L., Hanson R., Jones J., Kundu K., Harris J., Klochok O., Arsenyeva O., Kapustenko P. 2013. Crystallization fouling with enhanced heat transfer surfaces. In Malayeri M.R., Muller-Steinhagen H., Watkinson A.P. (Eds.), *Proceedings of International Conference on Heat Exchanger Fouling and Cleaning*, June 9–14, 2013, Budapest, Hungary, 379–385.

Danfoss. 2013. MicroChannel heat exchangers. www.ra.danfoss.com/TechnicalInfo/Literature/Manuals/24/MCHE_520H7163_DKQB.PB.400.A3.22.pdf (accessed 20 February 2014).

Gillham M.W.H., Sennik L. 1984. Selecting elastomers for plate heat exchanger gaskets. *Materials and Design* 5(4): 181–185.

Gnielinski V. 1975. New equations for heat and mass transfer in turbulent pipe and channel flow. *Forschung im Ingenieurwessen* 41(1): 8–16 (in German – Neue Gleichungen für den Wärme- und den Stoffübergang in turbulent durchströmten Rohren und Kanälen).

Gnielinski V. 2013. On heat transfer in tubes. *International Journal of Heat and Mass Transfer* 63: 134–140.

Heatric. 2014. Heatric exchangers. www.heatric.com/diffusion_bonded_exchangers.html (accessed 20 February 2014).

Hesselgreaves J.E. 2001. *Compact Heat Exchangers Selection, Design and Operation.* Pergamon Press, Oxford, UK.

Kalinin E.K., Dreicer G.A., Yarho S.A. 1990. *Intensification of Heat Transfer in Channels.* Mashinostroenie, Moscow, Russian Federation (in Russian).

Kandlikar S., Garimella S., Li D., Colin S., King M.R. 2005. *Heat Transfer and Fluid Flow in Minichannels and Microchannels.* Elsevier, Oxford, UK.

Kays W.M. and London A.L. 1984. *Compact Heat Exchangers*, 3rd edn. McGraw Hill, New York, USA.

Kee R.J., Almand B.B., Blasi J.M., Rosen B.L., Hartmann M., Sullivan N.P., Zhu H. et al. 2011. The design, fabrication, and evaluation of a ceramic counter-flow microchannel heat exchanger. *Applied Thermal Engineering* 31(11): 2004–2012.

Koch Heat Transfer Company. 2014. TWISTED TUBE® heat exchanger. www.kochheat-transfer.com/products/twisted-tube-bundle-technology (accessed 20 February 2014).

Krones. 2014. Compact shell-and-tube heat exchanger. www.krones.com/en/magazine/compact-shell-and-tube-heat-exchanger.php?userLanguage=en&country=United+Kingdom&countryCode=uk (accessed 20 February 2014).

Kukulka D.J., Smith R. 2013. Thermal-hydraulic performance of Vipertex 1EHT enhanced heat transfer tubes. *Applied Thermal Engineering* 61: 60–66.

Li Q., Flamant G., Xigang Yuan X., Neveu P., Luo L. 2011. Compact heat exchangers: A review and future applications for a new generation of high temperature solar receivers. *Renewable and Sustainable Energy Reviews* 15: 4855–4875.

Lytron. 2012. Heat exchangers. www.lytron.com/Heat-Exchangers (accessed 20 February 2014).

Marriott J. 1977. Performance of an Alfaflex plate heat exchanger. *Chemical Engineering Progress* 73(2): 73–78.

Mehendale S.S., Jacobi A.M., Shah R.K. 2000. Fluid flow and heat transfer at micro- and meso-scales with applications to heat exchanger design. *Applied Mechanics Reviews* 53: 175–193.

Moheisen R.M. 2009. Transport phenomena in fluid dynamics: Matrix heat exchangers and their applications in energy systems. Final Technical Report. Air Force Research Laboratory. Tyndall Air Force Base, FL, USA.

Popescu T., Marinescu M., Pop H., Popescu G., Feidt M. 2012. Microchannel heat exchangers – Present and perspectives. *U.P.B. Scientific Bulletin, Series D*, 74(3): 55–70.

Reay D.A., 2002, Compact heat exchangers, enhancement and heat pumps. *International Journal of Refrigeration* 25: 460–470.

Shah R.K., Seculic D.P. 2003. Fundamentals of heat exchanger design. New York, USA: Wiley and Sons.

Saji N., Nagai S., Tsuchiya K., Asakura H., Obata M. 2001. Development of a compact laminar flow heat exchanger with stainless still micro-tubes. *Physica C* 354: 148–151.

Shah R.K. 2005. Compact heat exchangers for microturbines. Micro gas turbines educational notes RTO-EN-AVT-131, Paper 2. RTO, Neuilly-sur-Seine, France, 2-1–2-18.

Sommers A., Wang Q., Han X., T'Joen C., Park Y., Jacobi A. 2010. Ceramics and ceramic matrix composites for heat exchangers in advanced thermal systems –a review. *Applied Thermal Engineering*, 30(11): 1277–1291.

Sondex. 2012. Plate & shell heat exchangers. www.sondex.net/Files/Billeder/PDF/SAW%20SPS%20enkelt%20sidet2.pdf (accessed 20 February 2014).

Sondex Tapiro OY AB. 2011. Heat exchangers. www.tapiro.fi/tuotteet/lammonsiirtimet_e.html (accessed 20 February 2014).

Sotnikov A.A., Vaselev V.A., Bova V.I., Gorenshtein I.V. 1985. Matrix heat exchangers for high pressure systems. *Khim. Neft. Mash.* (4): 27–29 (in Russian); translated: *Chem. Petr. Eng.* (1985) 187–190.

SPX. 2012. APV hybrid – Welded heat exchanger. www.spx.com/en/apv/pd-030_mp-phe-hybrid/ (accessed 20 February 2014).

Teploobmen. 2014. TTAI – High-efficient compact heat exchangers. www.ttai.biz/eng/ (accessed 20 February 2014).

T'Joen C., Park Y., Wang Q., Sommers A., Han X., Jacobi A. 2009. A review on polymer heat exchangers for HVAC&R applications. *International Journal of Refrigeration* 32(5): 763–779.

Tovazhnyanskii L.L., Atroshchenko V.I., Chus M.S., Kapustenko P.A. 1985. Analysis of heat transfer and hydrodynamics in welded plate heat exchangers operating under high pressure and temperature conditions of ammonia synthesis units. *Journal of Heat Recovery Systems* 5(5): 465–466.

Tovazshnyansky L.L., Kapustenko P.O., Khavin G.L., Arsenyeva O.P. 2004. *PHEs in Industry*. NTU KhPI, Kharkiv, Ukraine.

Trelleborg. 2009. Trelleborg develops new materials for PHE gaskets. *Sealing Technology*. www.tss.trelleborg.com. (accessed 20 February 2014).

Tsuzuki N., Utamura M., Tri Lam Ngo. 2009. Nusselt number correlations for a microchannel heat exchanger hot water supplier with S-shaped fins. *Applied Thermal Engineering* 29: 3299–3308.

Vahterus Oy. 2013. Plate & shell heat exchanger. www.vahterus.com/sites/default/files/pshe_eng.pdf (accessed 20 February 2014).

Wang L., Sunden B., Manglik R.M. 2007. *PHEs. Design, Applications and Performance.* WIT Press, Southhampton, UK.

Zaheed L., Jachuck R.J.J. 2004. Review of polymer compact heat exchangers, with special emphasis on a polymer film unit. *Applied Thermal Engineering* 24(16): 2323–2358.

CHAPTER 4

Heat Transfer Intensification

4.1 INTENSIFICATION OF HEAT TRANSFER FOR SINGLE-PHASE FLOWS INSIDE TUBES AND CHANNELS

4.1.1 Artificial Roughness on the Channel Wall

The use of wire coils was the first attempt of enhancing heat transfer in tubes of tubular heat exchangers made in the middle of the nineteenth century by Joule. According to García et al. (2012), those works are considered as the pioneering in the field of heat transfer intensification. Much later, at the end of the 1950s, the researches on heat transfer augmentation using "repeated-rib" roughness in turbulent pipe flow that are analysed in paper by Webb et al. (1971) have started. In the 1970s, the technology to manufacture not expensive deformed tubes was developed and in the 1980s dimpled and corrugated by cold external deformation tubes. These developments were stipulated by a need of more compact and efficient heat exchangers. To improve the performance of heat transfer surfaces with small elements on the wall, which induce artificial turbulence in the flow, a large number of researches were conducted with their results presented in literature of the last decades.

4.1.1.1 Flow Structure and Main Features of Intensification Mechanism

When analysing the enhanced heat transfer in tubes and channels, the film heat transfer coefficients and friction factor obtained with heat transfer intensification are usually compared to their values in smooth tubes and channels without intensifying elements. For heat transfer, it is the ratio of the Nusselt numbers for enhanced and smooth tube with the same diameters and Reynolds number Nu/Nu_S, similar for the friction factors f/f_S. The increase in heat transfer intensity is inevitably leading to increase in friction factor. From a first glance, the desirable result of heat transfer intensification is to obtain highest values of Nu/Nu_S at lowest f/f_S. But as it follows from the discussion in Chapter 3 (see Figure 3.6), the variation in hydraulic diameter and character of thermal and hydraulic characteristics correlations, especially in the transitional and laminar flow regimes, can significantly complicate the situation.

There is a possibility to make more compact and efficient heat exchanger with heat transfer surface having bigger ratio of f/f_S in turbulent flow regime just moving to lower region of hydraulic diameters.

The majority of the researches involving small-scale elements on the channel wall were made for turbulent regime of the fluid flow, as also the analysis of flow structure modifications induced by such elements. Based on results of researches published before 1970, Webb et al. (1971) have proposed the picture of flow structure in tubes and annular channels with periodic ribs inside presented in Figure 4.1. To illustrate the flow structure in a flat channel of solar collector with wire roughness, Prasad and Saini (1988) proposed drawings in Figure 4.2.

To investigate the flow structure in channels with consecutive rib roughness, different experimental techniques were used. The example of flow visualisation in channel having ribs on one side reported by Alamgholilou and Esmaeilzadeh (2012) is presented in Figure 4.3.

Another flow visualisation technique used for channels under consideration is real-time holographic interferometry. Liou et al. (1993a) have used holographic

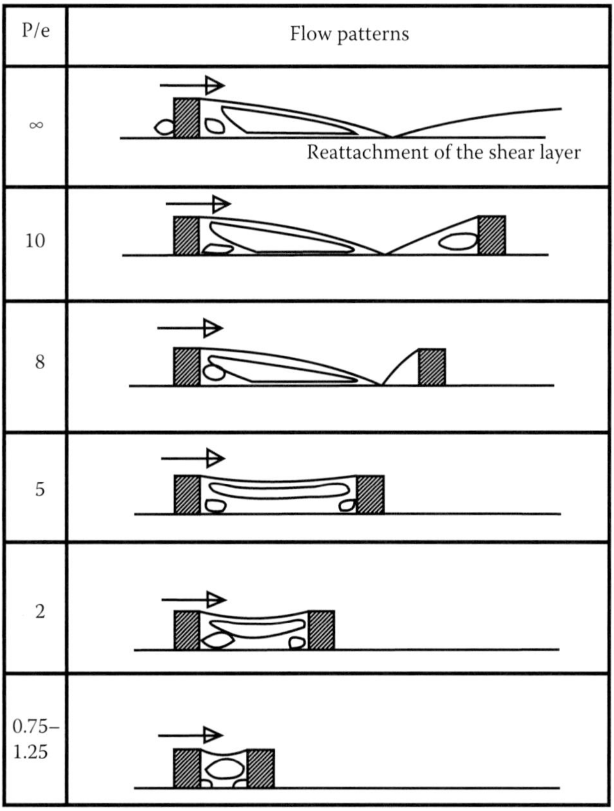

Figure 4.1 Flow structure in the tube with different ratios of a pitch P to a rib height e (after Webb et al., 1971).

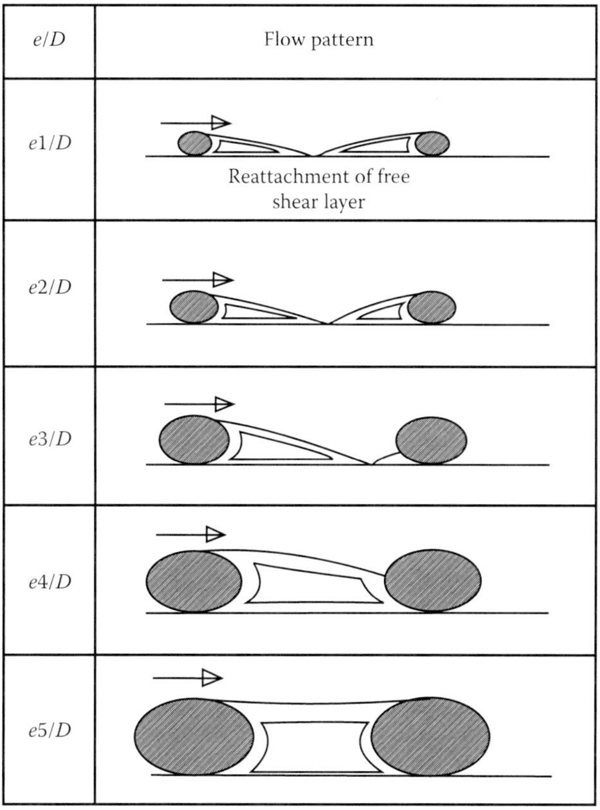

Figure 4.2 Flow structure near the wall of flat channel with transverse periodic wire roughness (*e*1 > *e*2 > *e*3 > *e*4 > *e*5) (after Prasad and Saini, 1988).

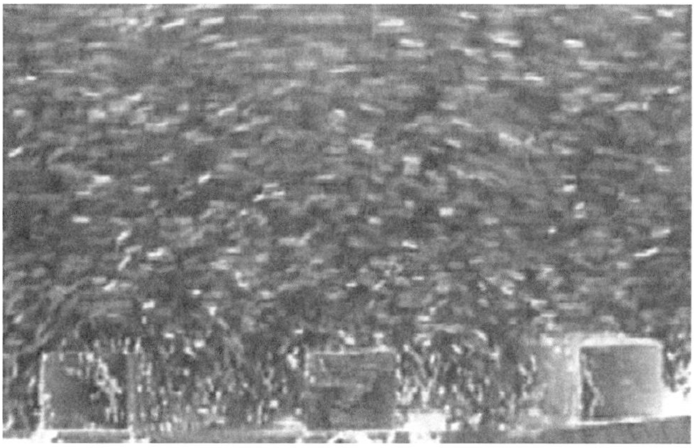

Figure 4.3 Visualisation of flow structure in channel with P/e = 4.5 (after Alamgholilou et al., 2012).

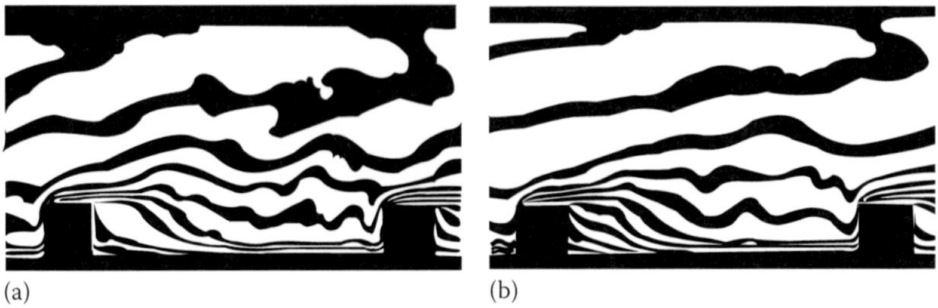

Figure 4.4 Typical examples of the holographic interferometry pictures of the temperature fields: Re = 12,600, e/D_e = 0.1, Pr = 7.2; (a) t = 0 s; (b) after time interval at t = 0.017 s (drawing after Liou et al., 1993a).

interferometry to measure temperature fields in turbulent flow in ribbed channel (see Figure 4.4.) to validate the results of their numerical solution using one of k–ε models of turbulent flow. The results of the researches published in the 1980s concerning measurements of fields with laser Doppler anemometry technique, including the paper by Liou et al. (1993b), and temperature fields with holographic interferometry, as also numerical solutions with different turbulence models, are also discussed in their paper. After the model was validated with experimental results, the detailed picture of the flow structure and temperature distributions in flow and on the wall surface were obtained, which are giving more detailed picture of convective heat transfer process. For example, the distribution of the local Nusselt numbers, presented in Figure 4.5, illustrates the influence of rib pitch on heat transfer enhancement at different Prandtl numbers.

Nowadays, numerical simulation with high-capacity computers is a powerful tool of obtaining insight of internal flow structure for channels of complex geometries. To obtain reliable results on convective heat transfer in turbulent flows, a number of turbulence models and software programs are developed. Most of the researches in this area are concerned with the solution of time averaged Reynolds equations

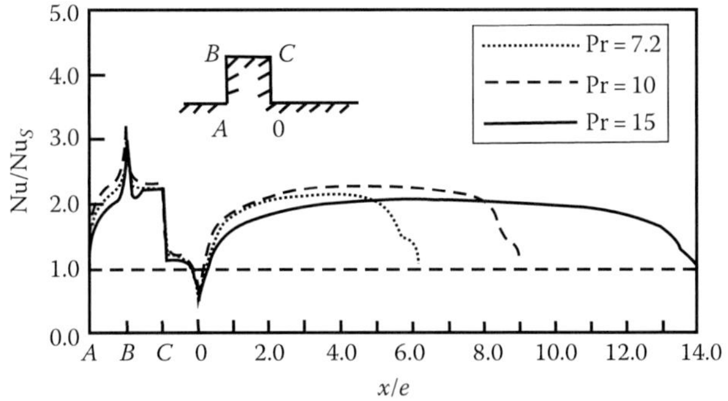

Figure 4.5 Effect of rib spacing on the local Nusselt number distributions (according to Liou et al., 1993a, 1993b).

governing conservation of thermal energy and momentum with different types of the equations system closures. These methods (referred to in the literature as Reynolds-averaged Navier–Stokes, or RANS) remain the primary tool for engineering analysis despite the spectacular advances in computer technology that was expected to make them obsolete. Weihing et al. (2014) have analysed different types of turbulence closures for RANS and made calculations of flow and temperature fields with eight of them for the same conditions and geometry of the channel with transverse ribs. The results have shown significant discrepancies and lead to the conclusion that predictions of the flow and thermal fields in a heated channel with periodic ribs that are obtained using industry-standard turbulence closures fail to reproduce the observed effects of the large-scale separation that occurs between successive ribs.

Other methods of numerical flow simulation involve more general time-dependent Navier–Stokes equations and are called Direct Numerical Simulation (DNS). The rapid progress in high-performance computers has promoted the use of DNS, which provides an accurate solution of the governing equations without introduction of assumptions, as stated by Nagano et al. (2004) who applied DNS for the channel with consecutive rib roughness and studied the effect of rib height y on heat transfer and flow structure in the range of e ratio to channel height H from 0.2 to 0.05. They have also made brief analysis of previously published papers on DNS modelling for considered channel geometry. In one of these papers, Cui et al. (2003) have used DNS with large eddy simulation technique to investigate the effect of the rib pitch. The streamlines for three different cases at Re = 20,000 (calculated on channel height H) are presented in Figure 4.6.

The pictures in Figure 4.6 are looking very similar to the ones in Figure 4.1, which were obtained much earlier just based on flow visualisation and general considerations. But computer modelling, after validation with experimental data, presents considerable information on internal flow structure and heat transfer intensification mechanism. For example, calculated distribution of wall shear stress along the heat transfer surface, which is considerably different from smooth channel and is much affected by rib pitch, is presented in Figure 4.7. The results are also used to identify the pressure and frictional components of resistance in a rib-roughened channel. The noteworthy result is that the averaged in time and space three-dimensional, unsteady and varying in space structure of such complex flow at the same time maintain the classical semi-logarithmic velocity distribution, which is displaced from the law on the smooth walls by the so-called roughness function. Such assumption was used before by some authors (see e.g. Migay, 1981) when analysing heat transfer using heat and momentum analogy.

By the analysis of the streamlines depicted in Figure 4.6, Cui et al. identified heat transfer mechanism and the influence on it on geometrical parameters. For small rib pitch $P = e$ (see Figure 4.6a), a vortex fills the space between the two ribs and the main stream is almost not disturbed. For intermediate rib pitch (see Figure 4.6b), a big vortex is formed between the ribs. This vortex prevents the outer flow from reattaching to the wall between ribs. In the corner behind the rib, a small vortex with opposite circulation appears. Streamlines for P/e = 9 (see Figure 4.6c) show four separation zones between the ribs. A small separation zone is formed on the top of

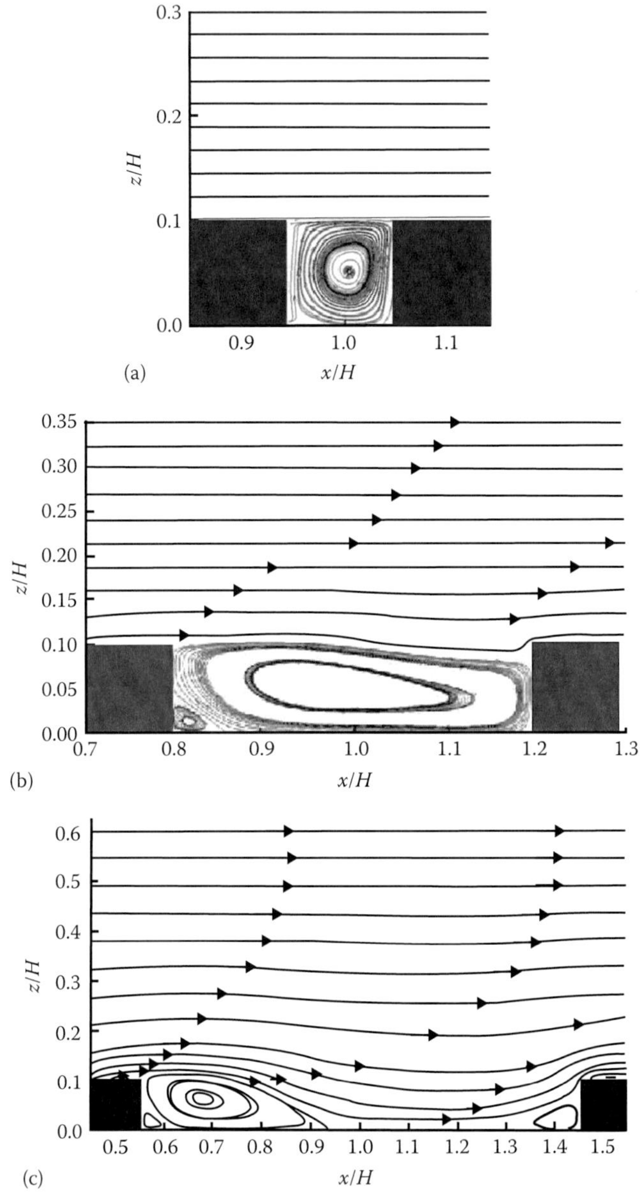

Figure 4.6 Mean streamlines at Re = 20,000 for (a) small roughness pitch P/e, (b) intermediate P/e and (c) large P/e (after Cui et al., 2003).

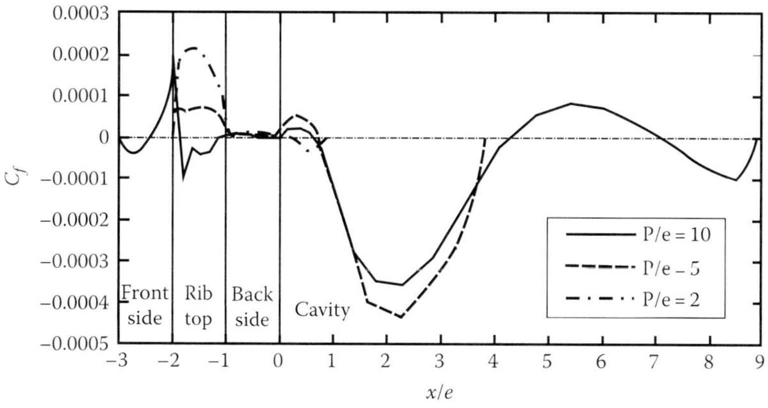

Figure 4.7 Wall shear stress distribution between consecutive rib (after Cui et al., 2003).

the rib at the front edge deflecting the flow. A large separation zone is after the rib and small vortex at the corner behind the rib. Flow reattaches to the channel wall between the ribs and detaches again, creating a vortex before the next rib. Nagano et al. (2004) who studied the effect of rib height on heat transfer and flow structure came to the conclusion that the higher the rib induces the large-scale vortices in the centre of the channel from the spacing between the ribs enhancing the turbulent mixing and the mean heat transfer coefficient, the higher the drag and pressure drop. For lower ribs, the mixing occurs in the region near the wall and the overall heat transfer performance becomes much more effective. It leads to the conclusion made before by some authors based on much more simple analysis of the flow.

Based on analysis of heat transfer in smooth tubes, Migay (1981) estimated that the shares of different flow layers in total resistance to heat transfer at Re = 10,000 in tube, when considering three layers of turbulent flow model for gases, are as follows. For viscous sublayer ($\eta < 6$), it is about 32.3%; for buffer sublayer ($6 \leq \eta \leq 30$), it is 52%; and for flow core ($\eta > 30$), it is 15.7%. Here, η is the non-dimensional distance from the wall surface defined as

$$\eta = \frac{y \cdot (\tau_W/\rho)^{0.5}}{\nu} \tag{4.1}$$

and y is the actual distance from the wall surface, m.

The hydraulic losses of mechanical energy in a channel of heat exchanger are converted to heat energy through dissipation. It is inducing generation of heat, but this effect is usually regarded as important in the total amount of heat transferred only in cases of viscous dissipation at laminar flows in capillary tubes and flows of very viscous fluids; see, for example, Aydin (2005). In turbulent flows, the mechanical energy is predominantly spent for generating big-scale turbulent vortexes, which later are reduced to smaller and smaller sizes dissipating at last in heat energy. Such picture for wall-bounded turbulent shear flows has been confirmed experimentally and also by DNSs provided insights into low-order statistics and coherent structures in these flows; see, for example, Hamlington et al. (2012). In the convective transfer of heat, the major role belongs to big-scale turbulent pulsations promoting interaction of flow core with wall surface by transporting some volumes of fluid in both directions.

As any additional turbulence generation requires more energy input, the crucial role in development of efficient enhanced surface plays the selection of the place and size of turbulence promoters. Accounting for described earlier features of turbulent heat transfer in channels, Kalinin et al. (1990) have proposed to induce additional turbulence only in flow regions close to the wall at $\eta = 30–60$, not in the main stream of the flow. It results in a substantial increase of the Nusselt number Nu/Nu_S with moderate rise of friction factor f/f_S. The schematic drawing of tube with such transverse turbulence promoters in form of consecutive diaphragms is shown in Figure 4.8. The outer grooves on tube are also promoting turbulence on shell side with flow along the tube bundle.

Tubes of that type are used in shell and tube heat exchangers presented in Figure 3.16 of Chapter 3. The heat transfer and hydraulic characteristics for flow inside such tubes are dependent on height of turbulence promoters $h = (D_B - d_B)/2$, their pitch t and geometrical form (rectangular or rounded). Kalinin et al. (1990) have reported about experimental results of varying these parameters and came to the conclusion that the best results can be achieved for gases in the range of t/h from 5 to 20 and d_B/D_B from 0.9 to 0.95. For liquids, it is t/h from 8 to 25 and d_B/D_B from 0.94 to 0.98. It was also confirmed that for some geometries, the increase in friction factor can be equal or even lower than the increase of the Nusselt number. In one case, it was $Nu/Nu_S = f/f_S = 2.0$ and in another $Nu/Nu_S = f/f_S = 2.2$. For most other geometries, it is $Nu/Nu_S < f/f_S$. It was also stated that for the creation of more compact heat exchangers, the increase ratio of Nu/Nu_S is much more important than the decrease of f/f_S and that some approximation ratio of volumes (and masses) of two heat exchangers with turbulent flow is inversely proportional to the ratio of the Nusselt numbers in power 1.4 but directly proportional to the ratio of friction factors in much smaller power 0.4.

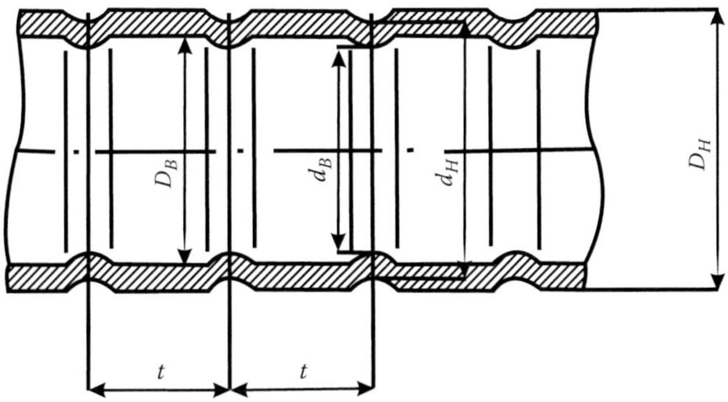

Figure 4.8 Schematic drawing of tubes with transverse turbulence promoters (from Kalinin et al., 1990).

4.1.1.2 Evaluation of Enhanced Heat Transfer Surfaces Performance in Compact Heat Exchanger Design

As it was mentioned by Webb (1981), the analysis reported in 1980 has identified 329 patents on enhanced heat transfer surfaces in US patents only and 1967 literature citations in this field. For nowadays, the amount of such different passive heat transfer augmentation techniques can be estimated in thousands and most of the authors are claiming advantages of proposed heat transfer surface according to certain criteria. The best surface geometry and its efficiency in crucial way depend on specified process conditions and construction features of heat exchanger under consideration. It is illustrated in Chapter 3 that by reducing hydraulic diameter, it is possible to obtain much compact and smaller heat exchanger with flat surfaces even when not using enhancement techniques, like plate–fin microchannel heat exchangers for refrigeration and air conditioning as depicted in Figure 3.57. With the contemporary development of computer technologies and numerical methods, the problem can be formulated as finding the optimal heat transfer surface geometry with the optimal heat exchanger design for specified process conditions, not just the best surface geometry in general.

Nevertheless the objective comparison of the performance and efficiency of different enhanced surfaces for compact heat exchangers is still an important problem. All proposed methods of establishing performance evaluation criteria for enhanced heat transfer surfaces define the reference heat exchanger, presumably one having smooth surfaces, compared to which in reference conditions the benefits of enhanced heat exchanger can be estimated. Starting from the first half of the twentieth century (see, e.g., Guhman, 1938; later Kirpichev, 1942), the main idea was to compare one of the three evaluation criteria based on the ratio to reference heat exchanger of heat transfer surface area r_F, pumping power r_P and heating power r_Q, when two other ratios are equal. This approach was further developed and described by Webb (1981), who based on analysis of previous works have proposed performance evaluation criteria for enhanced heat transfer surfaces in single-phase flows. Defining the heat exchanger performance objectives, he has also introduced design constraints and developed the system of criteria characterising enhanced heat transfer surface performance in certain types of applications. The further development of such system for heat transfer surface performance evaluation and optimisation is described by Webb and Kim (2005).

The exact evaluation of heat transfer surface performance requires accurate enough correlations for heat transfer coefficients and friction factor of enhanced surface. But as was discussed by Webb (1981), approximate estimations with some assumptions can be obtained in turbulent flow regime in internally finned tubes. For example, Sheriff and Gumley (1966) have formulated the condition for the less pumping power at enhanced tube compared to the smooth tube of the same diameter for the same fluid:

$$\frac{f}{f_S} \leq \left(\frac{\mathrm{St}}{\mathrm{St}_S} \right)^3 \tag{4.2}$$

There also attempts to introduce some simple criteria of heat transfer enhancement efficiency when comparing two different tubular heat transfer surfaces. The comparison is made of two heat exchangers having compared heat transfer surfaces inside tubes. The compared enhanced tube is designated with no index, and its diameter is D. The tube with which it is compared is designated with index 1 and its diameter D_S. It requires the introduction of the following assumptions:

1. The tubes have equal diameters $D = D_S$ [m].
2. In both heat exchangers, channels are flowing in the same media with the same flow rate V [m^3/s] and the same physical properties.
3. The temperature programmes are the same.
4. The thermal resistances at the clean wall with no fouling and at another heat transfer media side are much smaller than the thermal resistance to heat transfer at the examined heat transfer surface side, and overall, heat transfer coefficient can be regarded equal to film heat transfer coefficient on this side.

Let us consider the media flowing through n tubes of heat exchanger with flow rate V m^3/s. The pumping power can be expressed as follows:

$$P_P = V \cdot \Delta P = w \cdot \frac{\pi \cdot D^2}{4} \cdot n \cdot f \cdot \frac{L}{D} \cdot \rho \cdot 2 \cdot w^2 \tag{4.3}$$

Here, w is flow velocity, m/s.
For the same fluid properties, the ratio of pumping powers in heat exchanger with enhanced tubes and with compared tubes is

$$r_P = \frac{D \cdot n \cdot f(\mathrm{Re}) \cdot L \cdot w^3}{D_1 \cdot n_1 \cdot f_1(\mathrm{Re}_1) \cdot L_1 \cdot w_1^3} = \frac{(n \cdot L) \cdot f(\mathrm{Re}) \cdot \mathrm{Re}^3}{(n_1 \cdot L_1) \cdot f_1(\mathrm{Re}_1) \cdot \mathrm{Re}_1^3} \tag{4.4}$$

The ratio of heat transfer areas is

$$r_F = \frac{(n \cdot L)}{(n_1 \cdot L_1)} \tag{4.5}$$

The ratio of heating powers is

$$r_Q = \frac{(n \cdot L) \cdot \mathrm{Nu}}{(n_1 \cdot L_1) \cdot \mathrm{Nu}_1} \tag{4.6}$$

For comparison at the same pumping power and heat transfer area, it is assumed that $r_F = 1$ and $r_P = 1$. When comparing with the smooth tube, the following equation must be held:

$$f(\mathrm{Re}) \cdot \mathrm{Re}^3 = f_S(\mathrm{Re}_S) \cdot \mathrm{Re}_S^3 \tag{4.7}$$

When correlations for Nu and f are established, the heating powers on both surfaces can be compared by the ratio of film heat transfer coefficients at equal pumping powers:

$$\eta_Q = \left(\frac{h}{h_S}\right)_{PP} = \left(\frac{\mathrm{Nu}(\mathrm{Re})}{\mathrm{Nu}_S(\mathrm{Re}_S)}\right)_{PP} = \eta \tag{4.8}$$

A number of researchers have used the ratio η to estimate performance of enhanced tubes they had studied, for example, Bilen et al. (2001) called it "performance efficiency" and later Eiamsa-ard et al. (2010a) called it "thermal enhancement index" or Eiamsa-ard et al. (2010b) called it "thermal enhancement factor." Disregarding discrepancies in naming the ratio (4.8) has clear physical meaning as one of performance criteria η_Q, comparing heating powers at equal pumping powers and heat transfer areas of enhanced and smooth tubes in heat exchanger. Meanwhile, the application of this criterion is somewhat complicated, as it requires the calculation of the Nusselt numbers compared to different Reynolds numbers determined by Equation 4.7.

Avoiding this difficulty is possible by introducing the assumption about the exponents at Re in correlations for the Nusselt number and friction factor, as was suggested by Webb (1981). While its exponents can be different from that in correlations for smooth tubes, with some accuracy, it can be assumed that for the range of Reynolds number between values in tube with which comparison is made Re_1 and calculated by Equation 4.7 for enhanced tube under comparison Re, the exponents are the same. Let us assume that these exponents are n in correlation for the Nusselt number and m in correlation for friction factor. For the same properties of fluids, the friction factor in enhanced tube is

$$f(\mathrm{Re}) = B \cdot \mathrm{Re}^m \tag{4.9}$$

In tube to which comparison is made,

$$f_1(\mathrm{Re}) = B_1 \cdot \mathrm{Re}^m \tag{4.10}$$

The ratio of coefficients in these Equations is equal to ratio of friction factors calculated for the same Reynolds numbers inside the range between Re and Re_1. Let us take it equal to Re, then

$$\frac{B}{B_1} = \left(\frac{f(\mathrm{Re})}{f_1(\mathrm{Re})}\right) \tag{4.11}$$

It follows that

$$\frac{f(\mathrm{Re})}{f_1(\mathrm{Re}_1)} = \left(\frac{f(\mathrm{Re})}{f_1(\mathrm{Re})}\right) \cdot \left(\frac{\mathrm{Re}}{\mathrm{Re}_1}\right)^m \tag{4.12}$$

Equation 4.7 can be rewritten as

$$\left(\frac{f(\mathrm{Re})}{f_1(\mathrm{Re})}\right)\cdot\left(\frac{\mathrm{Re}}{\mathrm{Re}_1}\right)^m=\left(\frac{\mathrm{Re}_1}{\mathrm{Re}}\right)^3 \tag{4.13}$$

The ratio of Reynolds numbers in compared tubes under equal pumping power condition is

$$\frac{\mathrm{Re}_1}{\mathrm{Re}}=\left(\frac{f(\mathrm{Re})}{f_1(\mathrm{Re})}\right)^{\frac{1}{3+m}} \tag{4.14}$$

Following the same procedure for the ratio of the Nusselt numbers,

$$\frac{\mathrm{Nu}(\mathrm{Re})}{\mathrm{Nu}_1(\mathrm{Re}_1)}=\left(\frac{\mathrm{Nu}(\mathrm{Re})}{\mathrm{Nu}_1(\mathrm{Re})}\right)\cdot\left(\frac{\mathrm{Re}}{\mathrm{Re}_1}\right)^n \tag{4.15}$$

and substituting Equation 4.14 into Equation 4.15 and to Equation 4.6 with $nL = n_1L_1$,

$$\eta_Q=\left(\frac{\mathrm{Nu}(\mathrm{Re})}{\mathrm{Nu}_1(\mathrm{Re})}\right)\Bigg/\left(\frac{f(\mathrm{Re})}{f_1(\mathrm{Re})}\right)^{\frac{n}{3+m}} \tag{4.16}$$

To estimate the ratio of pumping powers $\eta_P = r_P$ when heating powers and heat transfer surface areas are equal, $r_F = 1$ and $r_Q = 1$ should hold.

From Equations 4.6 and 4.15 at $nL = n_1L_1$, it follows that

$$\left(\frac{\mathrm{Re}}{\mathrm{Re}_1}\right)=\left(\frac{\mathrm{Nu}(\mathrm{Re})}{\mathrm{Nu}_1(\mathrm{Re})}\right)^{-\frac{1}{n}} \tag{4.17}$$

After substituting Equations 4.12 and 4.17 into Equation 4.4,

$$\eta_P=\left(\frac{f(\mathrm{Re})}{f_1(\mathrm{Re})}\right)\cdot\left(\frac{\mathrm{Re}}{\mathrm{Re}_1}\right)^{3+m}=\left(\frac{f(\mathrm{Re})}{f_1(\mathrm{Re})}\right)\Bigg/\left(\frac{\mathrm{Nu}(\mathrm{Re})}{\mathrm{Nu}_1(\mathrm{Re})}\right)^{\frac{3+m}{n}} \tag{4.18}$$

Comparing with Equation 4.16, the relation between η_P and η_Q is as follows:

$$\eta_P=\eta_Q^{-\left(\frac{n}{3+m}\right)} \tag{4.19}$$

To estimate the ratio of heat transfer areas $\eta_F = r_F$ when heating powers and pumping powers are equal, $r_Q = 1$ and $r_P = 1$ should hold. With this assumption from Equations 4.4 and 4.6, substituting Equations 4.12 and 4.15 follows

$$\frac{\mathrm{Nu}(\mathrm{Re})}{\mathrm{Nu}_1(\mathrm{Re})}\cdot\left(\frac{\mathrm{Re}}{\mathrm{Re}_1}\right)^n = \left(\frac{f(\mathrm{Re})}{f_1(\mathrm{Re})}\right)\cdot\left(\frac{\mathrm{Re}}{\mathrm{Re}_1}\right)^{m+3} \tag{4.20}$$

and

$$\frac{\mathrm{Re}}{\mathrm{Re}_1} = \left[\left(\frac{f(\mathrm{Re})}{f_1(\mathrm{Re})}\right)\Bigg/\left(\frac{\mathrm{Nu}(\mathrm{Re})}{\mathrm{Nu}_1(\mathrm{Re})}\right)\right]^{\frac{1}{n-(m+3)}} \tag{4.21}$$

From Equation 4.6 at $r_Q = 1$ and accounting for Equation 4.15, it follows that

$$\frac{n\cdot L}{n_1\cdot L_1} = \left\{\left(\frac{\mathrm{Nu}(\mathrm{Re})}{\mathrm{Nu}_1(\mathrm{Re})}\right)\cdot\left[\left(\frac{f(\mathrm{Re})}{f_1(\mathrm{Re})}\right)\Bigg/\left(\frac{\mathrm{Nu}(\mathrm{Re})}{\mathrm{Nu}_1(\mathrm{Re})}\right)\right]^{\frac{n}{n-(m+3)}}\right\}^{-1} \tag{4.22}$$

Substituting in Equation 4.5 at $\eta_F = r_F$ after rearranging,

$$\eta_F = \left(\frac{\mathrm{Nu}(\mathrm{Re})}{\mathrm{Nu}_1(\mathrm{Re})}\right)^{\frac{n}{n-(m+3)}-1}\Bigg/\left(\frac{f(\mathrm{Re})}{f_1(\mathrm{Re})}\right)^{\frac{n}{n-(m+3)}} = \left[\left(\frac{\mathrm{Nu}(\mathrm{Re})}{\mathrm{Nu}_1(\mathrm{Re})}\right)\Bigg/\left(\frac{f(\mathrm{Re})}{f_1(\mathrm{Re})}\right)^{\frac{n}{m+3}}\right]^{\frac{n}{n-(m+3)}-1} \tag{4.23}$$

Accounting for Equation 4.16, it follows that

$$\eta_F = \eta_Q^{\left(\frac{1}{\frac{n}{m+3}-1}\right)} \tag{4.24}$$

When comparison is made to smooth tube with developed turbulent flow, it can be assumed that $n = 0.8$ in correlation for the Nusselt number (as in Sieder and Tate correlation) and $m = -0.25$ (same as in Blasius correlation) in correlation for friction factor. The criterion for heating powers ratio can be calculated as follows:

$$\eta_Q = \left(\frac{\mathrm{Nu}(\mathrm{Re})}{\mathrm{Nu}_S(\mathrm{Re})}\right)\Bigg/\left(\frac{f(\mathrm{Re})}{f_S(\mathrm{Re})}\right)^{0.291} \tag{4.25}$$

Other two criteria can be expressed through η_Q by Equations 4.19 and 4.24 as follows:

$$\eta_P = \eta_Q^{-0.291} \tag{4.26}$$

$$\eta_F = \eta_Q^{-1.41} \tag{4.27}$$

It is important to note that the criteria η_P and η_F can be expressed through criterion η_Q by simple Equations 4.19 and 4.24. Therefore, the value of this criterion η_Q is characterising the relative performance of enhanced tube on all of these criteria and can be called overall enhancement ratio, following Liu and Sakr (2013) who used in their expression exponent 1/3:

$$\eta = \left(\frac{\mathrm{Nu}(\mathrm{Re})}{\mathrm{Nu}_S(\mathrm{Re})}\right) \Bigg/ \left(\frac{f(\mathrm{Re})}{f_S(\mathrm{Re})}\right)^{1/3} \tag{4.28}$$

Similar relations were proposed by Kalinin et al. (1990), who have taken $m = -0.2$ and obtained exponent 0.286 in Equation 4.25. Counting for approximate character of the proposed equations for η and η_Q, the differences in exponents 0.291, 0286 and 1/3 are not essential, but for correct comparisons of different enhanced surfaces, only one of these exponents must be used. Equation 4.28 was used recently by some researchers on comparisons for performance of different enhanced tubes, as, for example, Hasanpour et al. (2014).

Equations 4.25 through 4.27 are obtained for developed turbulent flow with the assumption that exponents at the Reynolds number in correlations for enhanced tubes are close in value to those in smooth tubes. In practice, the deviations of these exponents are rather frequent. Therefore, these equations can be regarded only as some approximation that allows not only to compare performance of different enhanced surfaces but also to estimate the influence of some parameters on heat transfer performance. The important conclusion is that to enhance performance of heat transfer surface, it is much more important to increase the Nusselt number, even on expense of the considerable rise in friction factor. It is used in the development of many enhanced heat transfer surfaces, as, for example, tubes with inserts inside the flow core or plate heat exchangers (PHEs).

4.1.1.3 Correlations for Heat Transfer and Friction Factor

For the time being, the heat transfer intensification in channels with small transverse ribs is one of the most thoroughly studied subjects in enhanced heat transfer. It was studied by a large number of researchers with experimental and numerical methods. The presented analysis of some of these works gives the insight in the mechanism of heat transfer intensification. Nowadays, the increased interest in the development of more compact and efficient heat exchangers has stipulated researches on heat transfer intensification at different heat transfer surfaces with different forms of artificial roughness inducing turbulence. There are a large number of helically finned tubes and spirally finned, dimpled tubes with big variety of roughness shapes and sizes. The development of solar collectors stipulated appearance of studies on heat transfer in channels with inclined ribs, angle ribs and other forms. There are also researches with flow visualisation and numerical studies for some of such enhanced surfaces, but the main principles of heat transfer intensification are as described earlier. The exact equations for calculation of industrial heat exchangers with enhanced heat transfer are as a rule obtained in experimental studies for specific enhanced tubes and channels and are properties of companies manufacturing this equipment. Hundreds results of different studies with various enhanced tubes and channels are published up to now. In the next section, some published correlations that generalise some of the results available in literature for tubes and channels with different forms of artificial roughness are presented.

Starting from the first experimental studies, most of the researchers of enhanced heat transfer have presented their results in the form of empirical correlations similar to that developed for smooth tubes and channels and strictly valid for the range of Re and Pr numbers at which experiments were conducted and the shape of artificial roughness investigated. To generalise the correlations on heat transfer with accounting for the influence of Pr number, some authors used the heat and momentum transfer analogy based on the law of the wall similarity. For example, Webb et al. (1971) proposed the following correlation accounting for the repeated ribs height e, their pitch P and Pr number influence:

$$\mathrm{St} = \frac{f/2}{1 + \sqrt{(f/2)}\left[4.5 \cdot (e^{+})^{0.28} \cdot \mathrm{Pr}^{0.57} - 0.95 \cdot (P/e)^{0.53}\right]} \tag{4.29}$$

where $e^{+} = (e/D)\mathrm{Re}(f/2)^{0.5}$.

The correlation for friction factor is as follows:

$$\sqrt{(2/f)} = 2.5 \cdot \ln(D/2e) - 3.75 + 0.95 \cdot (P/e)^{0.53} \tag{4.30}$$

The range of parameters for this correlation's validity in turbulent flow regime is $e^{+} > 35$; $10 < P/e < 40$; $0.01 < e/D < 0.04$; $0.7 < \mathrm{Pr} < 80$.

Prasad and Saini (1988) have confirmed the validity of Equation 4.29 also for flat solar air heaters with artificial roughness in the form of transverse ribs on one channel side. Later, Prasad et al. (2014) extended its validity for three-sided artificially roughened solar air heater. In these cases, for the friction factor in Equation 4.29, its average value is used, averaged between values calculated by Equation 4.30 for the roughened part of the channel surface and the value calculated proportionally for smooth channel in the areas of roughened and smooth surfaces. Using also other available literature data, it was experimentally confirmed in the range of the Reynolds numbers from 5,000 to 200,000 and $10 < P/e < 40$; $0.011 < e/D < 0.033$.

With the development of solar thermal systems, considerable numbers of researches were conducted on heat transfer enhancement in flat channels with various roughness geometries. The recent state of the art in this field is analysed by Gawande et al. (2014).

Kalinin et al. (1990) based on extensive experiments with transversely corrugated tubes have proposed the following correlations for friction factor in the range 10,000 < Re < 400,000.

For the range of geometrical parameters of roughness 0.9 < *d*/*D* < 0.97 and 0.5 < *P*/*D* < 10,

$$\frac{f}{f_S} = \left[1 + \frac{100\cdot(\lg \mathrm{Re} - 4.6)\cdot\left(1-\frac{d}{D}\right)^{1.65}}{\exp\left(\frac{P}{D}\right)^{0.3}}\right]\cdot\exp\left[\frac{25\cdot\left(1-\frac{d}{D}\right)^{1.32}}{\left(\frac{P}{D}\right)^{0.75}}\right] \tag{4.31}$$

where f_S is calculated by the following equation:

$$f_S = \frac{0.079}{\mathrm{Re}^{0.254}}\cdot\left(\frac{\mu_{\mathrm{balk}}}{\mu_{\mathrm{wall}}}\right)^n \tag{4.32}$$

The exponent $n = 0.14$ for gas heating, $n = 0$ for gas cooling and $n = 1/3$ for liquid heating.

For the range of geometrical parameters of roughness 0.88 < *d*/*D* < 0.98 and *P*/*D* = 0.5,

$$\frac{f}{f_S} = \left[1 + \frac{(\lg \mathrm{Re} - 4.6)}{3.4\cdot 10^{-5}\cdot \mathrm{Re} + 6}\right]\cdot\exp\left[20.9\cdot\left(1-\frac{d}{D}\right)^{1.05}\right]\cdot\left(1.3-\sqrt{\left|\frac{d}{D}\right|-0.93}\right) \tag{4.33}$$

For the range of geometrical parameters of roughness 0.90 < *d*/*D* < 0.98 and *P*/*D* = 0.25,

$$\frac{f}{f_S} = \left[1 + \frac{(\lg \mathrm{Re} - 4.6)}{6\cdot\left(\mathrm{Re}\cdot 10^{-5}\right)^{0.33}}\right]\cdot\exp\left[17\cdot\left(1-\frac{d}{D}\right)^{0.858}\right]\cdot\left(3\cdot\frac{d}{D}-2\right)\cdot\left(2.5-1.5\cdot\frac{d}{D}\right) \tag{4.34}$$

In the last two Equations 4.33 and 4.34, friction factor for smooth tube is recommended to be calculated by Equation 4.35:

$$f_S = 0.182\cdot \mathrm{Re}^{-0.2} \tag{4.35}$$

To calculate film heat transfer coefficients, Kalinin et al. (1990) have proposed separate correlations for gases with Pr about 0.7 and for liquids.

For gases, there are proposed empirical correlations obtained in the range of the ratio of wall to balk temperatures from 0.13 to 0.16. For the range of geometrical parameters of roughness 0.88 < *d*/*D* < 0.98 and 0.25 < *P*/*D* < 0.8,

$$\frac{\mathrm{Nu}}{\mathrm{Nu}_S} = \left[1 + \frac{(\lg \mathrm{Re} - 4.6)}{35}\right]\cdot\left\{3 - 2\cdot\exp\left[\frac{-18.2\cdot\left(1-\frac{d}{D}\right)^{1.13}}{\left(\frac{P}{D}\right)^{0.326}}\right]\right\} \tag{4.36}$$

For the range of geometrical parameters of roughness $0.88 < d/D < 0.98$ and $0.8 < P/D < 2.5$,

$$\frac{\mathrm{Nu}}{\mathrm{Nu}_S} = \left[1 + \frac{(\lg \mathrm{Re} - 4.6)}{30}\right] \cdot \left[\left(3.33 \cdot \frac{P}{D} - 16.33\right) \cdot \frac{d}{D} + 17.33 - 3.33 \cdot \frac{P}{D}\right] \quad (4.37)$$

In Equations 4.36 and 4.37, Nu_S is calculated at averaged on the tube length wall temperature. For gas heating,

$$\mathrm{Nu}_S = 0.0207 \cdot \mathrm{Re}^{0.8} \cdot \mathrm{Pr}^{0.43} \quad (4.38)$$

For gas cooling,

$$\mathrm{Nu}_S = 0.0192 \cdot \mathrm{Re}^{0.8} \cdot \mathrm{Pr}^{0.43} \quad (4.39)$$

For liquids at $d/D > 0.94$ and $P/D = 0.5$,

$$\frac{\mathrm{Nu}}{\mathrm{Nu}_S} = \left[100 \cdot \left(1 - \frac{d}{D}\right)\right]^{0.445} \quad (4.40)$$

with

$$\mathrm{Nu}_S = 0.0216 \cdot \mathrm{Re}^{0.8} \cdot \mathrm{Pr}^{0.445} \quad (4.41)$$

To calculate heat exchanger with tubes like shown in Figure 4.8, with heat transfer enhancement also on shell side, Kalinin et al. (1990) have also presented equations for heat transfer and friction coefficient in channels along outside surface of such tubes assembled in a tube bundle. The ranges of geometrical parameters for validity of these equations are as follows: the ratio of the distance between tubes to their outer diameter $1.16 < S/D_{\mathrm{out}} < 1.5$, the ratio of the groove depth to the shell side equivalent diameter $0 \le e/d_{eq} \le 0.1$ and the ratio of the protrusion pitch to the shell side equivalent diameter $0.25 \le P/id_{eq} \le 2.0$.

For friction factor, the following equations are proposed:
At $\mathrm{Re} \le 3{,}100$, grooves do not influence friction factor and $f = f_S$.
At $3{,}100 < \mathrm{Re} < 20{,}000$,

$$\frac{f}{f_S} = 1 + \left\{\left(7.55 \cdot \frac{e}{d_{eq}}\right) \cdot (\lg \mathrm{Re} - 3.5) - 0.035 \sin\left[\left(1 - 22.44 \cdot \frac{e}{d_{eq}}\right) \cdot \pi\right]\right\} \cdot \left(1.4 - 0.488 \cdot \frac{e}{d_{eq}}\right) \quad (4.42)$$

At 20,000 ≤ Re < 100,000,

$$\frac{f}{f_S} = 1 + \left\{ \left(3.21 \cdot \frac{e}{d_{eq}} \right) \cdot (\lg \text{Re} - 2.27) + 0.09 \cdot (\lg \text{Re} - 4.3) \cdot \sin \left[\left(1 - 22.44 \cdot \frac{e}{d_{eq}} \right) \cdot \pi \right] \right\}$$
$$\times \left(1.4 - 0.488 \cdot \frac{e}{d_{eq}} \right) \tag{4.43}$$

For calculation of heat transfer, two new parameters are introduced. It is Re_1 below which grooves are not effect heat transfer:

$$\text{Re}_1 = \left(3.6 - 33.8 \cdot \frac{e}{d_{eq}} \right) \cdot 10^4 \tag{4.44}$$

For the Reynolds number higher than Re_2, the heat transfer intensification becomes independent of Re:

$$\text{Re}_2 = \left(4.7 - 18.85 \cdot \frac{e}{d_{eq}} \right) \cdot 10^4 \tag{4.45}$$

To calculate film heat transfer coefficients on the shell side, the following equations are proposed:
At $\text{Re} \leq \text{Re}_1$

$$\text{Nu} = \text{Nu}_S \tag{4.46}$$

At $\text{Re}_1 < \text{Re} < \text{Re}_2$,

$$\frac{\text{Nu}}{\text{Nu}_S} = 1 + 0.6 \cdot \frac{\lg \text{Re} - \lg \text{Re}_1}{\lg \text{Re}_2 - \lg \text{Re}_1} \cdot \left[1 - \exp\left(-35.8 \cdot \frac{e}{d_{eq}} \right) \right] \cdot \left(1 - 0.35 \cdot \frac{e}{d_{eq}} \right) \tag{4.47}$$

At $\text{Re}_2 < \text{Re} < 100{,}000$,

$$\frac{\text{Nu}}{\text{Nu}_S} = 1 + 0.6 \cdot \left[1 - \exp\left(-35.8 \cdot \frac{e}{d_{eq}} \right) \right] \cdot \left(1 - 0.35 \cdot \frac{e}{d_{eq}} \right) \tag{4.48}$$

Further researches in heat transfer intensification inside tubes and channels have stipulated the development of the variety of different forms of artificial roughness, which are not only transverse created by ribs but also inclined to tube axis helical or spiral ribs, schematically depicted in Figure 4.9. Such ribs can be also formed by tube inserts in the form of wire coils. The helical form of artificial roughness deployment on the inner tube surface stipulates the rotational swirl main stream flow inside the tube, in which characteristics are dependent on the rib angle of tube axis α.

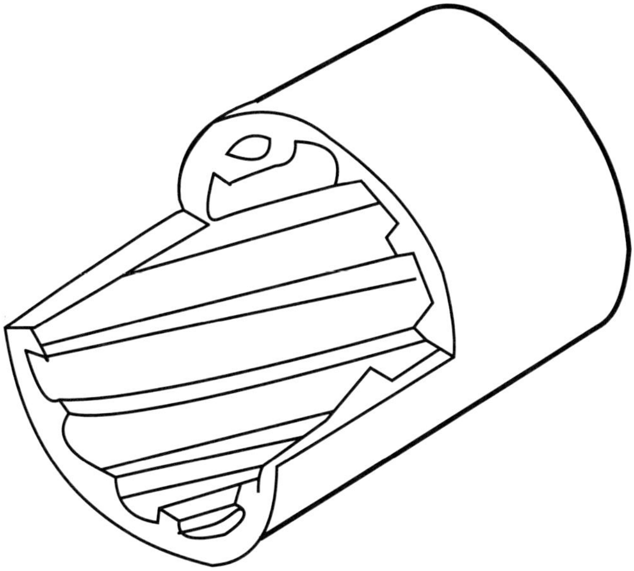

Figure 4.9 Schematic drawing of tubes with internal helical ribs.

The shape of ribs can be different, from rectangular, semicircle and triangular to round in the case of wire coils.

Ravigururajan and Bergles (1996) have collected a database including data published in 17 research papers on friction factor and heat transfer obtained in experiments with enhanced tubes having helical ribs of different geometrical characteristics in the following range: e/D from 0.01 to 0.2, P/D from 0.1 to 7.0 and $\alpha/90$ from 0.3 to 1.0. The range of the Reynolds and Prandtl numbers is $5{,}000 \le Re \le 250{,}000$; $0.66 \le Pr \le 37.6$.

Using statistical approach by applying the databases to an assumed models and minimising the least squares differences in the dependent variables and asymptotic approach proposed by Churhill (1983) to the smooth tube conditions, the following correlations were obtained:

For friction factor

$$\frac{f}{f_S} = \left\{ 1 + \left[\begin{array}{l} 29.1 \cdot Re^{\left(0.67 - 0.06 \cdot \frac{P}{D} - 0.49 \cdot \frac{\alpha}{90}\right)} \cdot \left(\frac{e}{D}\right)^{1.37 - 0.157 \frac{P}{D}} \cdot \left(\frac{P}{D}\right)^{\left(-1.66 \cdot 10^{-6} \cdot Re - 0.33 \cdot \frac{\alpha}{90}\right)} \\ \text{x} \left(\frac{\alpha}{90}\right)^{\left(4.59 + 4.11 \cdot 10^{-6} \cdot Re - 0.15 \frac{P}{D}\right)} \cdot \left(1 + 2.94 \cdot \sin \frac{\beta}{n}\right) \end{array} \right]^{\frac{15}{16}} \right\}^{\frac{16}{15}} \tag{4.49}$$

Friction factor in a smooth tube f_s is determined by Filonenko Equation 4.50:

$$f_S = \left(1.58 \cdot \ln \mathrm{Re} - 3.28\right)^{-2} \tag{4.50}$$

n is the number of sharp corners facing the flow. n is equal to two for triangular and rectangular ribs and infinity for smoother profiles. The contact angle β for semicircular and circular shapes is taken as 90°.
For heat transfer the following correlation was obtained:

$$\frac{\mathrm{Nu}}{\mathrm{Nu}_S} = \left\{1 + \left[2.64 \cdot \mathrm{Re}^{0.036} \cdot \left(\frac{e}{D}\right)^{0.212} \cdot \left(\frac{P}{D}\right)^{-0.21} \cdot \left(\frac{\alpha}{90}\right)^{0.29} \cdot \mathrm{Pr}^{0.024}\right]^{7}\right\}^{\frac{1}{7}} \tag{4.51}$$

In Equation 4.51, the Nusselt number for smooth tube is calculated by the equation proposed by Petukhov and Popov:

$$\mathrm{Nu}_S = \frac{\left(f_s/2\right) \cdot \mathrm{Re} \cdot \mathrm{Pr}}{1 + 12.7 \cdot \left(f_s/2\right)^{0.5} \left(\mathrm{Pr}^{\frac{2}{3}} - 1\right)} \tag{4.52}$$

As have been shown also by experimental results of Ravigururajan and Bergles (1996), the correlations can be applied to a wide range of roughness types and the Prandtl number. The friction factor correlation predicts 96% of the database to within ±50% and 77% of the database to within ±20%. Corresponding prediction figures for the heat-transfer correlation are 99% of the database to within ±50% and 69% of the database to within ±20%.

To improve the accuracy of heat transfer prediction, Wen-Tao Ji et al. (2012) have proposed to use modification of analogy between heat and momentum transfers in a form similar to the Gnielinski equation for smooth tubes of length L:

$$\mathrm{Nu}_S = \frac{\left(f/2\right) \cdot \left(\mathrm{Re} - 1000\right) \cdot \mathrm{Pr}}{1 + 12.7 \cdot \left(f_S/2\right)^{0.5} \left(\mathrm{Pr}^{2/3} - 1\right)} \left[1 + \left(\frac{D}{L}\right)^{\frac{2}{3}}\right] \cdot \left(\frac{\mathrm{Pr}}{\mathrm{Pr}_W}\right)^{0.11} \tag{4.53}$$

The comparison of data of experiments with 16 samples of helically ribbed tubes as also with data available in literature has confirmed the accuracy of Equation 4.27 as follows: the deviations of 89% of experimental data calculated are within ±20% and 99% are within 40%.

Such big scattering of data around generalising correlations is witnessing the existence of the geometrical factors in real manufactured enhanced tubes that are not accounted in proposed equations. These correlations can be used to estimate the general trends of the effects of the main roughness parameters when developing new types of enhanced tubes. For calculations of enhanced heat exchangers in practical applications, the correlations experimentally obtained for specific industrially manufactured and artificially roughened tubes are needed. For example, pictures of ten different types of helically ribbed tubes are presented in a paper by Pethkool et al. (2011) who experimentally investigated nine specific tubes with different pitch and height of ribs at constant helix angle $\alpha = 75°$. In the range of pitch to hydraulic diameter ratio $0.18 \le P/D_H \le 0.27$ and rib height to hydraulic diameter ratio $0.02 \le e/D_H \le 0.06$, the accuracy of their friction factor correlation (4.28) was inside the limits of ±4%:

$$f = 1.15 \cdot \mathrm{Re}^{-0.239} \cdot \left(\frac{e}{D_H}\right)^{0.179} \cdot \left(\frac{P}{D_H}\right)^{0.164} \quad (4.54)$$

The correlation for the Nusselt number has the accuracy estimated as ±9% and is expressed by following equation:

$$\mathrm{Nu} = 1.579 \cdot \mathrm{Re}^{0.639} \cdot \mathrm{Pr}^{0.3} \cdot \left(\frac{e}{D_H}\right)^{0.46} \cdot \left(\frac{P}{D_H}\right)^{0.35} \quad (4.55)$$

The accuracy of these correlations is fairly suitable for practical calculations of heat exchangers with investigated enhanced tubes made with the use of specific manufacturing process.

Most of the experimental researches on enhanced heat transfer in tubes with artificial roughness in form of helical ribs were performed for flows in turbulent or higher end of transitional regimes Re > 3,000. Vicente et al. (2004) have experimentally studied heat transfer and isothermal pressure drop in helically corrugated tubes for laminar and transitional flow regimes. It was observed that roughness accelerates transition from laminar flow to the critical Reynolds numbers even below 1,300. At laminar flow, friction factor was from 5% to 25% higher than in smooth tubes and the Nusselt number can be up to 30% higher. It is much less than for turbulent flow regime. Meyer and Olivier (2011a) got similar results when investigating the adiabatic flow in samples of helically ribbed tubes and determined the critical Reynolds numbers of transition from laminar flow in a form of following correlation valid for $0.022 \le (e/D) \le 0.057$:

$$\mathrm{Re}_{CR} = 2{,}200 \cdot \left[1 + 9.13 \cdot 10^9 \cdot \left(\frac{e}{D}\right)^{5.8}\right]^{-0.1} \quad (4.56)$$

Investigating heat transfer, Meyer and Olivier (2011b) came to the conclusion that the ribs have even negative effect on heat transfer in laminar regime, as the ribs obstruct the flow path for secondary flows. Transition regime occurred at a Reynolds numbers between 2,000 and 3,000. For calculation of the Nusselt number in transition region, some formula averaging the Nusselt number values for laminar flow at Re_{CR} and turbulent flow at Re corresponding to the upper boundary of transitional regime was proposed. Overall, the results have shown that the tubes with artificial roughness have much poorer effect of heat transfer intensification in laminar and transitional flow regimes than in developed turbulent flows.

Similar results were reported by Xiao-Wei et al. (2007) who have conducted experiments on heat transfer intensification in microfin tubes. They concluded that for $Re < Re_{CR}$, the microfin tube behaves like a smooth tube and the heat transfer is not enhanced.

There are hundreds of forms of artificial roughness that were experimentally investigated by different researchers with the aim of heat transfer enhancement. Beside transverse, helical ribs and wire coils with rather systematically classified geometrical parameters of roughness, dimpled tubes with different forms of dimples on their surface are used. One example of such tubes is shown in Figure 3.18 as Vipertex 1EHT stainless steel tubes investigated by Kukulka and Smith (2013).

Wang et al. (2010) have made attempt to investigate the geometrical parameters of dimples on dimpled tubes heat transfer performance, arranging dimples in strict order defined by their pitch and position on tube surface. There were two different forms of dimples, spherical and ellipsoidal. The form of dimples is significantly influencing the heat transfer and friction factor characteristics. For the turbulent flow regime at $Re > 3{,}000$, the characteristics of tubes are changing rather qualitatively similar, with better performance of ellipsoidal dimpled tubes. But in transitional region, the substantial differences were observed in friction factor behaviour, which exhibit the abrupt behaviour in case of spherical dimples and smooth transition in case of ellipsoidal dimples. To calculate heat transfer performance for both tubes, separate correlations for friction factor and the Nusselt number are presented in the range $1{,}500 \leq Re \leq 60{,}000$.

The influence of the form of artificial roughness on heat transfer and hydraulic performance of enhanced tubes in transitional flow regimes was investigated by Garcia et al. (2012). Three enhanced tubes with different types of artificial roughness were studied experimentally in tests with water and oil: helically ribbed tube, dimpled tube and tube with the wire coil insert. In turbulent flow regime at the Reynolds numbers above 2,000, the three roughened tubes produced similar heat transfer coefficients from 2.4 to 2.9 higher than in smooth tube with wire coils that had somewhat higher friction factor coefficients. At $Re = 10{,}000$, the friction factor increase for wire coils was about 5 as compared to 3.7 obtained in corrugated and dimpled tubes. In laminar flow, for $Re \leq 200$, corresponding to pure laminar flow in all investigated tubes, heat transfer in roughened tubes is very similar to that in smooth tubes.

The considerable difference was observed in transitional flow regime. With wire coil inserts, the transition from the fully laminar to the turbulent flow takes place smoothly and the Nusselt number is monotonously changing with Reynolds number. For corrugated and dimpled tubes, the transition starts at $Re \approx 700$ and the Nusselt number abruptly changes to its values at $Re = 2{,}000$ when flow became turbulent. Such behaviour allows to recommend investigated tubes with wire coil for the use in transitional flow regimes, while corrugated and dimpled tubes mostly for turbulent flows. On the other hand, the need for detailed testing of any commercially manufactured enhanced tubes in all flow regimes where they are supposed to be used is emphasised.

4.1.2 Tubes with Inserts

Most of the enhanced tubes described in the previous subsection with different forms of artificial roughness should be produced with special manufacturing technologies. Roughness elements are formed as a part of the tube material. The construction of heat exchanger with such tubes should be specially designed for enhanced tubes optimal usage or modified from traditional shell and tube by substitution of enhanced tubes instead of smooth ones of the same diameter. At the same time, there also exists the problem of increasing efficiency of smooth tubes used in heat exchangers of traditional shell-and-tube design, as also enhancing performance of smooth tubes without modifications on their surface on economy grounds. This problem can be solved with specially designed tube inserts.

One type of such inserts is wire coils that can be neatly fitted inside the smooth tube. In this way, they form a kind of artificial roughness. As it is discussed in Section 4.1.1.3., Garcia et al. (2012) have studied such wire coils in comparison with two other types of artificial roughness as helical ribs and dimples inside the tube. With some differences of performance in transitional flow regime, the principles of heat transfer intensification in turbulent flow in tubes with wire coils are the same as for all types of artificial roughness discussed in Section 4.1.1.1. The main feature is inducing turbulence in layers near the wall to enhance convective heat transfer in this region with smaller disturbances in the flow core and minor increase in friction factor.

Beside wire coils, other tube inserts have elements positioned in a central part of the tube inevitably disturbing flow core and also causing friction and form drug pressure losses. Nevertheless, in turbulent flow, the increase of the Nusselt number can compensate even such higher increase in friction factor, as can be judged from Equations 4.25 and 4.28, enhancing heat transfer performance compared with smooth tubes. The comprehensive survey of different methods for heat transfer enhancements in tubes has been presented by Liu and Sakr (2013). Accounting for this survey, tube inserts for heat transfer enhancement can be classified as follows:

1. Twisted tapes
   a. Typical twisted tape: single tape having length equal to the length of exchanger tube.
   b. *Varying length twisted tape*: tape having the length not equal to the length of the tube (half length, 3/4 length or other part) or short length tapes with different pitches or twist types connecting to each other in some way.
   c. *Multiple twisted tapes*: two or more twisted tapes that are jointly used in one heat exchanger tube.
   d. Twisted tape with rod and spacer.
   e. Twisted tape with attached fins and baffles.
   f. Twisted tapes with slots, holes and cuts.
   g. Helical left–right twisted tape with screw.
   h. *Tapes with different surface modifications*: Some insulation is provided to avoid fin effect. Surface dimpled material can be used for tape.

2. Swirl generators
3. Conical rings
4. Wire coils
5. *Complex wire inserts*: complex wire structures optimised for heat transfer enhancement

The recent researches on different twisted tape inserts were reviewed by Hasanpour et al. (2014). The simple twisted tape is made of a long and thin metal sheet with thickness 0.5–1.5 mm and width on 0.3–0.4 mm smaller than the inner tube diameter. This metal sheet is twisted by any suitable mechanical methods with some pitch of twists P and can be characterised by the twist ratio $Y = P/2D$.

The twisted tapes are commonly made from materials such as

- Carbon steel
- Aluminium
- Stainless Steel AISI 304, 316
- Copper
- Other stainless steel types

The results of the detailed study of flow structure in channels formed by twisted tapes with alternate axis are presented by Eiamsa-ard et al. (2013). The dye injection technique was used, as also numerical solution of the problem with comprehensive analysis of literature data for different types of twisted tapes.

The major effects induced by the twisted tapes are the swirl flow leading to considerable augmentation of fluid mixing and helical fluid motion which ensures longer flow path with a higher flow velocity through smaller cross section of the formed channels. Twisted tape can have also the fin effect with increase of effective heat transfer area. But it has very low influence on the overall heat transfer because of small and uninsured contact area between a tape and a tube.

Overall, the mechanism of heat transfer intensification can be described as augmentation of convection in the main stream area with no emphasis on near-wall region. The convection near the wall is enhanced by increasing velocity gradients in swirling flow through lower cross section and disturbances coming from the main stream agitated by twisted tape with dimples, halls, slots, alternating axis, or other flow-disturbing methods. In turbulent flow of gases and liquids with $Pr \geq 0.7$, such method of heat transfer intensification is less efficient than artificial roughness, which generates turbulence near the wall, where the main part of thermal resistance to heat transfer is concentrated. Nevertheless, the increase of the Nusselt number due to turbulent vortexes intruding to a near-wall region from the main stream renders some efficiency compared to smooth tubes. It is confirmed by comparison of experimental data for different twisted tapes presented by Eiamsa-ard et al. (2013). The data for the Overall Enhancement Ratio calculated by Equation 4.28 are changing from maximum 1.4 at $Re = 5{,}000$ for one kind of twisted tape to 1 and even below to 0.8 for some others. For comparison, the Overall Enhancement Ratio η presented by Pethkool et al. (2011) for helically corrugated tubes can be up to 2.4. To improve efficiency, Promvonge et al. (2012) proposed to insert double twisted tape in a helically ribbed tube and reported increase of η also to 2.4 in some cases at turbulent flow.

The principle of enhancing heat transfer by inducing turbulence not only in the layers near the wall but also in the main flow is implemented in HiTRAN Matrix Element tube insert depicted in Figure 3.19. These inserts are produced by CalGavin company (United Kingdom) and are calculated and designed to improve efficiency of existing shell-and-tube heat exchangers.

Overall, it can be concluded that the tube inserts are mainly useful for enhancing efficiency of existing tubular heat exchangers. Most of experimental data for these devices are obtained for tube about 20 mm in diameter. To project this, results and empirical correlations for diameters less than 10 mm, as for compact heat exchangers, can cause a problem in obtaining accurate calculations. Moreover, the introduction of additional elements in construction can involve additional difficulties in production and increase cost in manufacturing compact heat exchangers on such principle, not giving any better heat transfer efficiency than the tubes with special forms of artificial roughness.

4.1.3 Twisted Tubes

Another way of heat transfer intensification in shell-and-tube heat exchangers is the use of twisted tubes. Typical heat exchanger with such tubes is depicted in Figure 3.20a. The tubes are of elliptical cross section and are twisted by some mechanical method with pitch of twists P. The complex form of flow passages inside tubes induce turbulence and promote high heat transfer coefficients. At the same time, better flow distribution on the shell side is achieved. Tubes are frequently in contact with adjacent tubes, as shown in illustration in Figure 3.20b. It eliminates the problem of tube vibrations and organising complex swirl flow on the shell side with high levels of induced turbulence to enhance heat transfer, which is augmented on both tube and shell sides of the heat exchanger.

The mechanism of developed turbulent flow and heat transfer enhancement in twisted tube was illustrated by Yang et al. (2011) using computational fluid dynamics (CFD) modelling. The changing flow direction in elliptical twisted tube leads to the generation of longitudinal vortex in the tube cross section that results in a fluid flow towards the surface of the tube wall. The isotherms in the elliptical twisted tube are distorted near the two ends of the long axis of the ellipse, and with the increase of the Reynolds number, such distortion becomes more significant, as the temperature gradient normal to the tube wall is also increased. In general, the enhancement of heat convection in regions near the wall is promoted by reorganising the flow structure in the whole stream and penetration nearer the wall of turbulent vortexes generated in other regions at the flow core.

Based on analysis of previous researches of heat transfer inside elliptical twisted tubes and results of their experimental study, Yang et al. (2011) have estimated the influence of the twist pitch and form of tube cross-sectional ellipse on heat transfer and hydraulic resistance. In the range of the Reynolds numbers 600 < Re < 55,000, the following empirical correlations are proposed:

For heat transfer,

$$\mathrm{Nu} = 0.3496 \cdot \mathrm{Re}^{0.615} \cdot \mathrm{Pr}^{\frac{1}{3}} \cdot \left(\frac{A}{B}\right)^{0.49} \cdot \left(\frac{P}{D_H}\right)^{-0.394} \tag{4.57}$$

For Fanning friction factor,

$$f = (1.529/4) \cdot \mathrm{Re}^{0.350} \cdot \left(\frac{A}{B}\right)^{1.686} \cdot \left(\frac{P}{D_H}\right)^{-0.366} \tag{4.58}$$

where

A is the length of the bigger axis of the ellipse, m
B is the size of smaller ellipse axis, m
D_H is the tube hydraulic diameter, m

The correlations given earlier do not reveal any changes corresponding to the transition from between turbulent and laminar flow regimes down to the Reynolds number about 600. It can be explained by artificial turbulence created by vortexes in a flow of such complex structure even at lower Reynolds numbers than 2,300. The highest increase of Nu number compared to smooth tube in developed turbulent flow at Re > 10,000 is about 1.4 times for one of the tested samples, for which the increase in friction is about two times. The overall enhancement ratio η values in this flow regimes are not higher than 1.25. It is much lower than in tubes with artificial roughness and similar to tubes with twisted tape inserts discussed in the previous subsections. At the Reynolds numbers lower than 10,000, the overall enhancement ratio η is increasing taking maximum 3.6 at 1,500 < Re < 2,000 and somewhat decreasing towards Re = 600. The point is that as it is shown earlier in Section 4.1.1.2, the overall enhancement ratio η is strictly justified for developed turbulent flow in pipes and such increase is also connected to deteriorated performance of smooth tubes in transition range of the Reynolds numbers. Nevertheless, twisted tubes have better heat transfer performance at laminar and transitional flow regimes than enhanced tubes with artificial roughness.

The heat exchangers with twisted tubes have another important feature of heat transfer enhancement that is augmentation of heat transfer on the shell side due to better organisation of flow in a space between twisted oval tubes and increase of turbulence in channels of complex geometry on the shell side. Danilov et al. (1986) reported results on detailed research of turbulent flow structure in a tube bundle at the shell side of heat exchangers with twisted tubes in parallel to tube axis flow and cross flow. By their data, the film heat transfer coefficients on the shell side are dependent on the ratio of twist pitch to maximal tube diameter *P/A*. The tested tube bundles have *P/A* equal to 12.45, 8.3 and 4.15. Augmentation of heat transfer compared to tube bundles with straight smooth tubes is 6%, 14% and 42%.

For longitudinal flow in tube bundles consisting of 37 twisted tubes, Ievlev et al. (1982) have reported the following equations:

For heat transfer and friction factor in developed turbulent flow at 50,000 > Re > 2,000 and 2,440> Fr_M > 232,

$$Nu = 0.023 \cdot Re^{0.8} \cdot Pr^{0.4} \cdot \left(1 + \frac{3.6}{Fr_M^{0.357}}\right)^{0.49} \cdot \left(\frac{T_W}{T_F}\right)^{-0.55} \tag{4.59}$$

$$f = \frac{0.3164}{4 \cdot Re^{0.25}} \left(1 + \frac{3.6}{Fr_M^{0.357}}\right) \tag{4.60}$$

where

$Fr_M = P^2/(A \cdot D_e)$ is the modified Froude number which characterises the effect of centrifugal forces on the flow

D_e is equivalent diameter of the channel between tubes, m

T_W is the temperature at the wall, K

T_F is the temperature at the main flow, K

With the decrease of the modified Froude number Fr_M but with the increase of friction factor, heat transfer is becoming more intensive. For $Fr_M = 64$, Dzyubenko and Dreitser (1979) proposed the following correlation for heat transfer at 50,000 > Re > 2,000:

$$Nu = 0.0521 \cdot Re^{0.8} \cdot Pr^{0.4} \cdot \left(\frac{T_W}{T_F}\right)^{-0.55} \tag{4.61}$$

For $Fr_M \leq 100$, Danilov et al. (1986) recommended the following correlation for friction factor:

$$f = \frac{0.3164}{4 \cdot Re^{0.25}} \left(1 + \frac{3.1 \cdot 10^6}{Fr_M^{3.38}}\right) \tag{4.62}$$

The comparison of twisted tube heat exchanger with the traditional rod baffle heat exchanger having smooth tubes is reported by Xiang-hui Tan et al. (2013). The experimental heat exchanger has 37 tubes like that for which correlations (4.59) through (4.62) were obtained. It is observed that the twisted oval tube heat exchanger have much better performance than the traditional one, which can be estimated by proposed comparison coefficient having average value about 2.75. Compared heat exchangers have initial tube diameters about 20 mm, and to make heat exchangers more compact by decreasing the tube diameter, more researches would be needed.

4.1.4 Extended Heat Transfer Surfaces

Heat exchangers with extended surfaces are usually applied for gas–gas or gas–liquid duties. There are plate–fin heat exchangers (schematically shown in Figure 3.45) and tube-and-fin heat exchangers (see Figure 3.49). The fins are situated on primary heat transfer surface and forming channels between adjacent fins bounded by primary surface on one or both sides. The principles of heat transfer intensification in such channels are the same as heat transfer intensification in channels described earlier in this chapter. But the main feature is that the fins are having its own thermal resistance to heat transfer from the flow to the primary surface that is characterised by fin thermal efficiency. Fin efficiency is determined by its geometry, material of construction and the way of connection to primary surface. The heat transfer in such conditions is in detail discussed by Kays and London (1984) and out of the scope of the present book.

4.1.5 Channels of PHEs

The design and operation principle of PHE equipment is described earlier in Section 3.7.2. The heat transfer processes in this heat exchanger take place in the channels of complex geometry formed by plates pressed from thin metal. The plates' corrugation form strongly influences the heat transfer and hydraulic behaviour of inter-plate channels. This effect is similar to that in enhanced tubes.

Plates with straight line corrugations inclined with the certain angle to the plate's vertical axis are generally used in modern PHEs (see Figure 4.10). Assembled together in one unit, they form the channels of criss-cross flow type which are distinguished by complex geometry and by the existence of contact points between the opposite walls in the sites of corrugation crossing. Geometry of plates with different corrugation types (sinusoidal and triangular form) are shown in Figure 4.11. The comprehensive study of hydraulic resistance and mass transfer in the models of such channels with sinusoidal corrugations of height b = 5 mm was reported by Focke et al. (1985). Muley and Manglick (1999) published results for hydraulic resistance and heat transfer in channels formed by commercial plates with b = 2.54 mm.

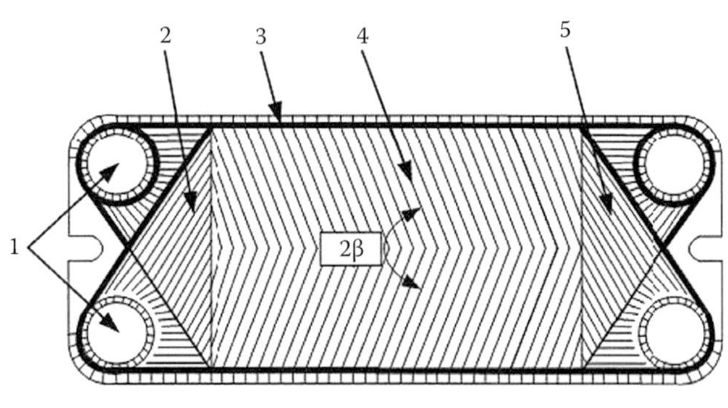

Figure 4.10 Schematic drawing of PHE plate: (1) heat carrier inlet and outlet, (2, 5) zones for flow distribution, (3) rubber gasket and (4) the main corrugated field.

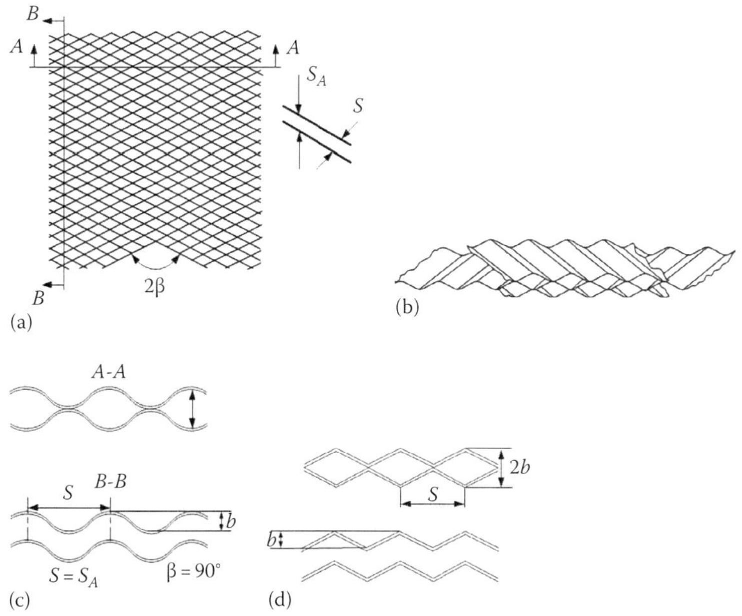

Figure 4.11 Different corrugation forms: (a, b) the intersection of the adjacent plates, (c) channel cross sections for the sinusoidal form of corrugations and (d) channel cross sections for the triangular form of corrugations.

More recently, Dovich et al. (2009) tested the models of channels with corrugations of height $b = 2$ mm. Results of experiments for the models of channels with the triangular shape of corrugations of height 5 and 10 mm were reported by Tovazhnyansky et al. (1980), as also for channels with height down to 1.2 mm by Savostin and Tikhonov (1970). The reviews of these and other works and their results can be found in a book by Wang et al. (2007) and papers of Ayub et al. (2003) and Khan et al. (2010). All of these authors mainly generalise the obtained data in the form of separate empirical correlations for hydraulic resistance calculations, which are valid for the investigated channels only in a limited range of hydrodynamic and thermal parameters.

The plate's surface of industrial plate-and-frame PHE, washed by the fluid (see Figure 4.10), consists of the main corrugated field *4* and zones of flow distribution on the inlet 2 and outlet 5. Most of the heat transfer takes place in the main corrugated field that comes to 80%–85% of the total plate's heat transfer surface area. Despite the fact that on the distribution zones considerably less heat is transferred, their influence on the overall hydraulic resistance of the channel can be significantly higher. First, here is the raised velocity of flow movement, which is increasing in the following range: from the velocity value for the operating field up to the velocity on the outlet from the channel to the gathering collector of PHE due to decrease of channel cross-sectional area (the same is for inlet from the distributing collector of PHE). In addition, the design of distribution parts can considerably vary for different commercial plates, which affects their hydraulic resistance and evenness of flow distribution. As it can be judged from the data of Tovazhnyansky et al. (1980), the share of pressure drop in entrance and exit zones of some PHE channels formed by commercial plates of chevron type can reach 50% and more especially at lower inclination angles of corrugations to the plate vertical axis.

4.1.5.1 Flow Structure and Main Features of Heat Transfer Intensification Mechanism in PHE Channels

The channel of PHE formed by two adjacent corrugated plates can have various geometrical forms determined by the shapes and orientation of corrugations on the plates relative to flow direction. In cases when the directions of corrugations and flow coincide (β = 0°), the plates are forming the straight channel of complex cross section like shown in Figure 4.11(a) or (b). The flow in such channel is along the grooves of corrugations and similar to flow inside tubes with no enhancement of turbulence and heat transfer.

When corrugations are perpendicular to the direction of flow (β = 90°), the plates are forming wavy 2D channels like shown in Figure 4.11(c) or (d). Such channel can be considered as a series of bends and the flow inside it is making series of turnings, which induce complex permanently changing velocity profile along the channel length. The laminar flow structure near the right-angle bend of the channel can be judged from streamlines presented in Figure 4.12. It can be observed that such a channel creates a complex flow pattern involving flow separation, impingement, reattachment, recirculation and flow deflection. In such flow, the velocity and temperature profiles are significantly changing along the channel causing the rise of local heat transfer coefficients and the Nusselt numbers on most parts of the channel wall mostly in zones of flow impingement and deflection, as shown in Figure 4.13. This is leading to enhancement of averaged heat transfer also.

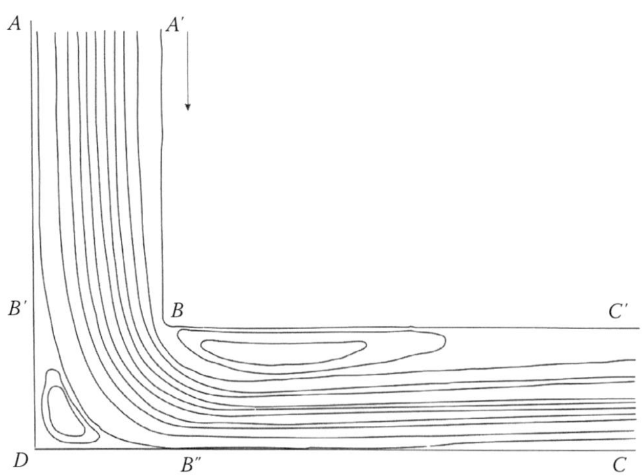

Figure 4.12 The streamlines in flow near the right-angle bend of the channel at Re = 800 according to CFD modelling by Tovazhnyanskyy (1988).

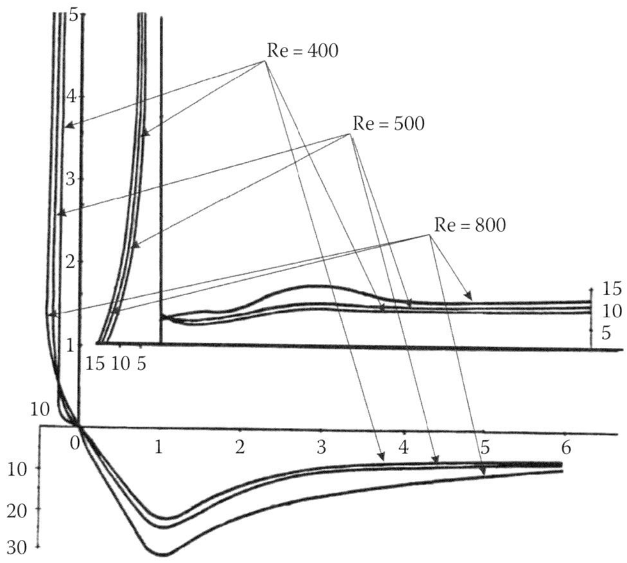

Figure 4.13 The distribution of the local Nusselt numbers at the channel wall near the right-angle bend according to CFD modelling by Tovazhnyanskyy (1988).

Qualitatively similar picture was observed by Amano (1985) for turbulent flow based on CFD modelling using k–ε model of turbulence. The results of detailed experimental study of turbulent flow structure in channels with bends at 90° and 120° by using laser Doppler anemometry technique were reported by Tovazhnyanskyy (1988). The substantial increase of the intensity of turbulent pulsations in all flow regions is observed. At the same time, in regions nearby the wall, which are corresponding to viscous and buffer sublayers, the velocity profile is very close to that for flows inside straight tubes and channels.

For criss-cross flow channels formed by plates with corrugations inclined to overall flow direction at certain angle $0° < \beta < 90°$, the form of the channel is having complex 3D character. The flow structure is also 3D but having some features that are inherent to flows in channels with $\beta = 0°$ and $\beta = 90°$. At low values of the corrugation inclination angle β, the main part of the flow is directed along the corrugation grooves and the features corresponding to strait tubes and channels are predominant, with smaller influence of flow separation, impingement, reattachment, recirculation and flow deflection. With higher β values, the flow becomes mainly directed across the corrugations and features inherent to channel with $\beta = 90°$ becoming predominant. It was observed in experimental study with flow visualisation technique reported by Dović et al. (2009).

More detailed picture of 3D flow in criss-cross channels of PHE was obtained with CFD modelling by Kanaris et al. (2009). The complex flow structure contains all the elements like flow separation, impingement, reattachment, recirculation and flow deflection which lead to heat transfer enhancement and significant redistribution

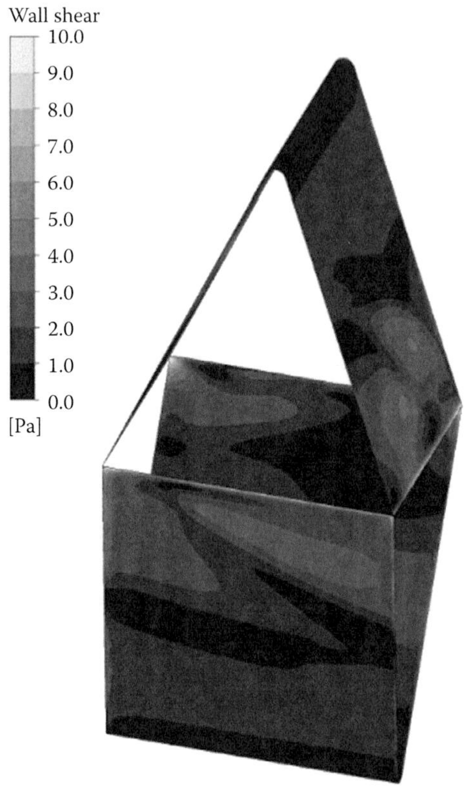

Figure 4.14 The distribution of local shear stresses at the walls of PHE channel in its representative elementary unit at Re = 8,900 according to CFD modelling by Stogiannis et al. (2013).

of local heat transfer and shear stresses on the wall. For example, the shear stress distribution on the channel walls obtained by CFD modelling by Stogiannis et al. (2013) is presented in Figure 4.14.

Overall analysis of the flow structure and heat transfer intensification mechanism in PHE channels reveals its significant difference from that one in tubes and channels with artificial roughness, where the most action on the stream is applied in the flow regions close to the wall. In turbulent flows of fluids with the Prandtl numbers close to one and bigger in these regions near the wall is located main part of the thermal resistance to heat transfer. It enables to obtain heat transfer intensification with moderate rise of friction factor. In PHE channels, the disturbing action is applied mostly in the main flow stream with heat transfer augmentation caused mostly by flow separation, impingement, reattachment, recirculation and flow deflection. The turbulent pulsations are mainly induced in the main flow and after that transferred to the wall augmenting convective turbulent heat transfer there. With such heat transfer enhancement mechanism, which is more similar to that in tubes with inserts and twisted tubes, the rise of friction factor is much higher than the increase of the Nusselt number.

As it is discussed in Section 4.1.1.2, in developed turbulent flows, the influence on enhanced heat transfer surfaces performance of the Nusselt number and friction factor increase compared to smooth tubes can be estimated by Equations 4.25 through 4.27. If to estimate the beneficial for heat transfer relative increase of Nusselt number (Nu/Nu_S) by exponent equal to 1, the adverse influence of increase in friction factor (f/f_S) can be estimated by exponent 0.291, as follows from Equation (4.25). This exponent is 3.43 times smaller than 1. Therefore, to have performance better than in smooth tubes, the enhanced surface with $Nu/Nu_S = 2$ should have the increase in friction factor smaller than 2 in power 3.43 or $f/f_S \leq 10.8$ and with $Nu/Nu_S = 5$ smaller than $f/f_S \leq 250$. The PHE channels have usually much smaller increase in friction factor at such Nusselt number increases and compared to smooth tubes can offer significantly better performance that can match or even overcome performance of enhanced tubes.

The comparisons by criteria determined by Equations 4.25 through 4.27 are valid for developed turbulent flows at channels of the same hydraulic diameters. As it is discussed in the previous chapter, the abrupt transition to laminar flow at Re ≤ 2,300 in smooth tubes can cause deterioration of heat transfer performance and increase of required heat transfer area for the same duty. The same feature is observed for enhanced tubes with artificial roughness that also exhibits such behaviour at laminar flow, as shown, for example, in Figure 3.6. The flow structure in PHE channels is leading to promotion of flow separation and other effects enhancing convective heat transfer even at the Reynolds numbers below 2,300 down to about 100. This effect can be called artificial turbulence promotion and is leading to smooth transition from laminar to turbulent flow regime, which is characterised by smooth curves of correlations for the Nusselt number and friction factor like shown in Figures 3.1 and 3.2. It enables to get good heat transfer performance at relatively low Reynolds numbers and to decrease hydraulic diameter of PHE channels that is leading to heat transfer enhancement also.

The influence on PHE thermal and hydraulic performance of geometrical parameters of corrugations allows optimisation of geometrical form of corrugations and plate sizes. It requires detailed information about heat transfer intensification in such channels and reliable correlations for calculation of heat transfer and friction factor in channels of different geometrical forms.

4.1.5.2 Hydraulic Resistance of PHE Channels

The generalisation of data on hydraulic resistance in PHE channels without accounting for the existence of distribution zones was proposed by Martin (1996) based on semi-empirical model of flow inside PHE channels. The relation expressing the dependence of the hydraulic resistance coefficient on the Reynolds number and geometrical parameters of plates' corrugation is presented. But the calculation deviation for this relation from experimental data of other authors in some cases runs up to 50% and more. In other works, Dovic et al. (2009) have obtained similar results in terms of accuracy. The low accuracy of generalisation in the mentioned works can be explained, first of all, by significant differences in data for experimental PHE models and industrial plates, which were considered all together.

To eliminate the influence of distribution zones on hydraulic resistance of PHE channels, the results of researches on the models of corrugated field at which the effects on entrance and exit zones of the channel can be negligible compared to pressure drop at corrugated field should be considered. Focke et al. (1985) have reported the results of comprehensive research on friction factor at the models of PHE channels with sine-shaped corrugations in a wide range of corrugation inclination angle β to flow direction. Plates with β equal to 0°, 30°, 45°, 60°, 72°, 80° and 90° were tested. The height of corrugations was 5 mm and the corrugation doubled height to pitch ratio $\gamma = 2 \cdot b/S$ was $\gamma = 1$. For different corrugation inclination angles, the following set of correlations to determine friction factor ζ is reported:

$$\left.\begin{array}{lcc}
\beta = 0^\circ & \text{Laminar (theory)} & \zeta = 114.4 \cdot \mathrm{Re}^{-1} \\
 & 8{,}000 < \mathrm{Re} < 56{,}000 & \zeta = 0.552 \cdot \mathrm{Re}^{-0.263} \\
\beta = 30^\circ & 260 < \mathrm{Re} < 3{,}000 & \zeta = 0.37 + 230 \cdot \mathrm{Re}^{-1} \\
 & 3{,}000 < \mathrm{Re} < 50{,}000 & \zeta = 3.59 \cdot \mathrm{Re}^{-0.263} \\
\beta = 45^\circ & 150 < \mathrm{Re} < 1{,}800 & \zeta = 1.21 + 367 \cdot \mathrm{Re}^{-1} \\
 & 1{,}800 < \mathrm{Re} < 30{,}000 & \zeta = 5.5849 \cdot \mathrm{Re}^{-0.177} \\
\beta = 60^\circ & 90 < \mathrm{Re} < 400 & \zeta = 5.03 + 755 \cdot \mathrm{Re}^{-1} \\
 & 400 < \mathrm{Re} < 16{,}000 & \zeta = 26.8 \cdot \mathrm{Re}^{-0.209} \\
\beta = 72^\circ & 110 < \mathrm{Re} < 500 & \zeta = 19.0 + 764 \cdot \mathrm{Re}^{-1} \\
 & 500 < \mathrm{Re} < 12{,}000 & \zeta = 132 \cdot \mathrm{Re}^{-0.296} \\
\beta = 80^\circ & 130 < \mathrm{Re} < 3{,}700 & \zeta = 140 \cdot \mathrm{Re}^{-0.28} \\
\beta = 90^\circ & 200 < \mathrm{Re} < 3{,}000 & \zeta = 5.63 + 1{,}280 \cdot \mathrm{Re}^{-1} \\
 & 3{,}000 < \mathrm{Re} < 16{,}000 & \zeta = 63.8 \cdot \mathrm{Re}^{-0.289}
\end{array}\right\} \tag{4.63}$$

Results of another experimental research on the influence of corrugation inclination angle β on friction factor were reported by Tovazhnyansky et al. (1980). The experiments were made on the models of the criss-cross flow channels formed by plates with inclined corrugations of triangular shape of rounded on the edges. The height of corrugations at different samples was varied from 5 to 10 mm with corrugation doubled height to pitch ratio equal to $\gamma = 0.556$ and β equal to 30°, 45° and 60°. The following correlation for friction factor is presented:

$$\zeta = \frac{0.34 \cdot \exp(1.52 \cdot tg\beta)}{\mathrm{Re}^{0.25 - 0.06 \cdot \mathrm{tg}\,\beta}} \tag{4.64}$$

For accurate comparison of the results obtained by correlations (4.63) and (4.64), the difference in definition of characteristic linear dimensions should be accounted. In obtaining correlations (4.63), Focke et al. (1985) have used the actual channel length L and equivalent diameter equal to double plate spacing $d_e = 2b$. Tovazhnyansky et al. (1980) in obtaining correlation (4.64) employed effective channel length equal to the length of channel wall in flow direction and hydraulic diameter of the channel.

Arsenyeva et al. (2011) have recalculated data of paper by Tovazhnyansky et al. (1980) according to convention $d_e = 2b$ and actual channel length L. The resulting data together with correlations (4.63) were generalised using the form of equation proposed by Churchill (1977) for straight tubes. This equation accounts for limiting cases at laminar and turbulent flow regimes and ensures the smooth description of friction factor in area of transitional flow regime.
The general correlation is as follows:

$$\zeta_s = 8\cdot\left[\left(\frac{12+p2}{\mathrm{Re}}\right)^{12}+\frac{1}{(A+B)^{\frac{3}{2}}}\right]^{\frac{1}{12}} \tag{4.65}$$

$$A=\left[p4\cdot\ln\left(\frac{p5}{\left(\frac{7\cdot p3}{\mathrm{Re}}\right)^{0.9}+0.27\cdot 10^{-5}}\right)\right]^{16};\quad B=\left(\frac{37530\cdot p1}{\mathrm{Re}}\right)^{16}$$

where $p1$, $p2$, $p3$, $p4$ and $p5$ are the parameters defined by channel corrugation form.

In Figure 4.15 is shown the comparison of calculations by empirical formulas presented by Focke et al. (1985) with calculations according to the Equation 4.65 for the following values of its parameters:

$$p1=\exp(-0.15705\cdot\beta);\quad p2=\frac{\pi\cdot\beta\cdot\gamma^2}{3};\quad p3=\exp\left(-\pi\cdot\frac{\beta}{180}\cdot\frac{1}{\gamma^2}\right) \tag{4.66}$$

$$p4=\left(0.061+\left(0.69+\mathrm{tg}\left(\beta\cdot\frac{\pi}{180}\right)\right)^{-2.63}\right)\cdot(1+(1-\gamma)\cdot 0.9\cdot\beta^{0.01});\quad p5=1+\frac{\beta}{10}$$

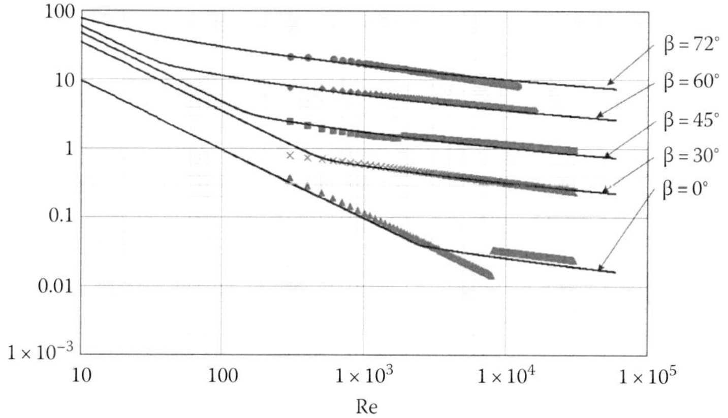

Figure 4.15 The correlation by Equation 4.65 of the data presented by Focke et al. (1985).

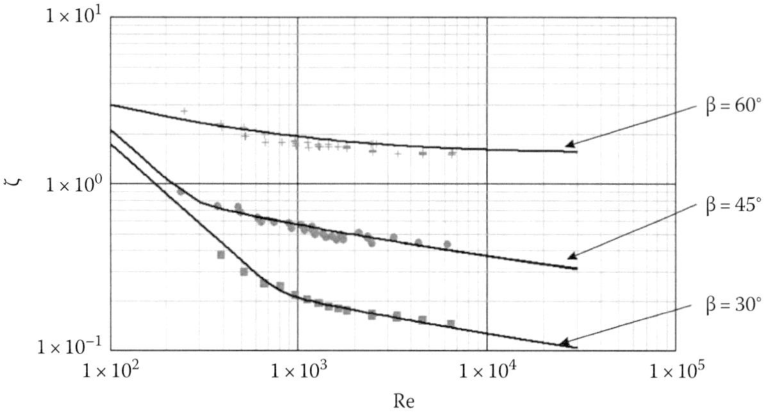

Figure 4.16 Correlation by Equation 4.65 of data from paper by Tovazhnyansky et al. (1980).

Here, $\gamma = 2 \cdot b/S$ is the corrugation doubled height to pitch ratio. For the corrugations of plates from the paper of Focke et al. (1985), its value is $\gamma = 1$. The average mean-square error of the calculation on Equation 4.65 for corrugation forms mentioned in cited paper at $\beta \leq 72°$ constitutes ±9%.

For the proper comparison of hydraulic parameters of channels formed by plates with triangular corrugation patterns, the processing of data from the paper of Tovazhnyansky et al. (1980) was carried out at $d_e = 2b$ and characteristic length L. The processing was done using the initial data of tests by Tovazhnyansky et al. (1980). The results of tests are also well predicted by Equation 4.65 at the value of $\gamma = 5/9 = 0.556$ for the triangular corrugations (with rounded edges) of the prototypes used in that work.

Calculated (solid lines) and experimental data (dots) are presented in Figure 4.16. The mean-square error of the experimental data generalisation comes to ±6.5%. As for the plates with sinusoidal corrugation, the influence of the corrugation inclination angle is the same and can be defined by the same correlation.

One of the earliest studies on heat transfer and hydraulic resistance for models of channels with triangular form of corrugations was reported by Savostin and Tikhonov (1970). The agreement with Equation 4.65 is rather good for β in the range 14°–72° (the error mostly less than 20%). It increased at β = 10° to 35% (see Figure 4.17). The parameter γ varied from 0.872 to 1.02. Data for bigger γ are not considered and γ = 1.02 is regarded as the upper limit of Equation 4.5 application.

Figure 4.18 shows the comparison of calculations according to Equation 4.65 at the parameter values determined by Equation 4.66 with the experimental data presented in the paper of Dović et al. (2009). These data are obtained for two test models of channels with sinusoidal corrugations with inclination angels of 28° and 65°. The value of parameter γ was equal to 0.52. Data for ζ and Re were corrected to account for difference in d_E and d_h. The data agree with calculation rather good. The discrepancy for β = 65° is less than 20% (the calculation according to the formula presented by Dović et al. (2009) gives the overestimation of up to 50%). For the β = 28°,

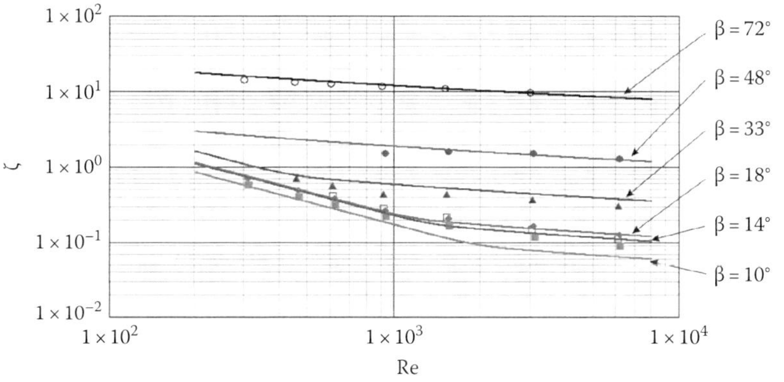

Figure 4.17 The comparison of calculations by Equation 4.65 with experimental data of Savustin and Tikhonov (1970).

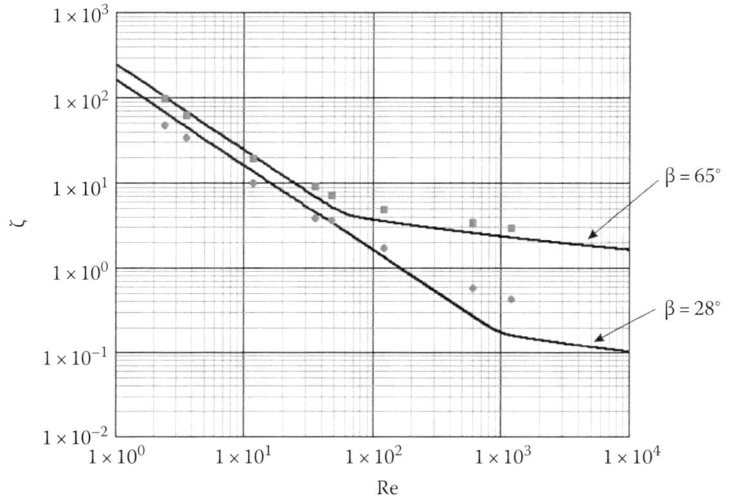

Figure 4.18 The comparison of calculations by Equation 4.65 with experimental data of Dović et al. (2009).

the discrepancy is the same for Re up to 300 but increases at Re = 1,200 up to 50%. It should be noticed that data were estimated from the graphs given in a cited paper and they give rather qualitative picture. But it allows us to conclude that the accuracy of data generalising is higher than it was for the model presented in that paper.

The experimental study of the influence of corrugation inclination angle β on heat transfer and friction factor was reported by Zimmerer et al. (2002). The experiments were made on the models of channels with corrugated walls, so the influence of inlet and outlet zones is minimal. The data for friction factor of Figure 4.19 are in good agreement with the calculations obtained from Equation 4.65, which are

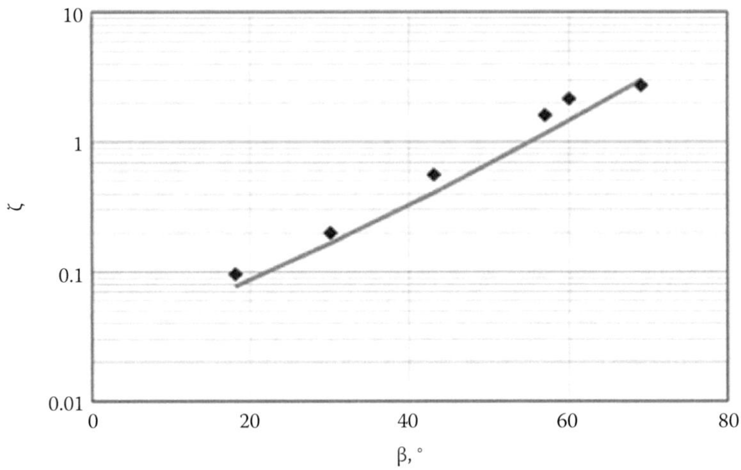

Figure 4.19 The influence of corrugation inclination angle β on friction factor. Comparison of data from Zimmerer et al. (2002) with calculations by Equation 4.65 at Re = 2,000.

presented by the solid line (biggest discrepancy is 20%). The change of β from 18° to 68° results in 30 times increase on the friction factor.

The correlation expressed by Equation 4.65 enables to predict friction factors at corrugated field of criss-cross flow channels, formed by plates with inclined corrugations in a wide range of corrugation parameters. It was confirmed for corrugation inclination angle from 14° to 72° and double height to pitch ratio from 0.52 to 1.02 in the range of the Reynolds numbers from 5 to 25,000. The difference between sinusoidal and triangular (with rounded edges) shapes is practically not significant for friction factor.

Pressure drops on other elements in the channels formed by the adjacent plates should be considered also to calculate the pressure drop at PHE formed by commercially manufactured plates beside the main corrugated field as will be discussed in the next chapter.

4.1.5.3 Heat Transfer in PHE Channels

The link between heat and momentum transfer in channels with enhanced heat transfer is very important for the selection of the optimal geometry and correct design of heat exchangers. Dović et al. (2009) and Martin (1996) used the Lévêque equation to generalise heat transfer data for PHEs published by different researchers. Contrary to the fact that Leveque equation was initially developed for the heat transfer through developing thermal boundary layers in a hydrodynamically developed *laminar* duct flow, the accuracy of prediction is rather good for some cases, but the error in estimation by cited authors sometimes reach up to 40%–52%. This can be partly explained by an attempt to generalise together the data for models of corrugated fields of PHE channels and data for commercial plates that have rather higher pressure losses at inlet and outlet, as it is discussed in the previous subsection. The other reason lays on the accuracy of the employed relation between heat transfer and hydraulic resistance, which has theoretical background for laminar flows.

Arsenyeva et al. (2012a) have proposed heat transfer data generalisation based on the data of hydraulic resistance for PHE channel main corrugated field using turbulent flow model and analogy between heat and momentum transfer. For flow in channels, such analogy is a well-established fact. The simplest Reynolds analogy and its different modifications are sufficiently well described in the literature (see, e.g., books by Shah and Seculic 2003, or previous book by Kutateladze and Leont'ev (1990). Semi-empirical models, developed based on this analogy, proved useful for theoretical analysis of different transport phenomena. For PHE channels of complex geometry, the existence of the link between heat and momentum transfer was shown by Tovazhnyansky and Kapustenko (1984); it was employed also for the prediction of heat transfer in condensation and fouling conditions. The way to establish the link between heat transfer and friction factor in PHE channels based on a modification to the Reynolds analogy using experimental results is described below in this section.

Let's assume that the relationship between heat transfer and frictional shear stress on the channel wall of PHE channels is the same as in tubes. In other words, for equal shear stress at the wall, the film heat transfer coefficients in PHE channel and straight tube will be equal, supposing that equivalent diameter, temperature conditions and physical properties of both flows are the same. In a dimensionless form, it can be written

$$\mathrm{Nu} = \frac{h \times d_e}{\lambda} = \mathrm{Nu}_0 = \frac{h_0 \times d_e}{\lambda} \tag{4.67}$$

The shear stress on the wall can be calculated by the following formulas:
For the straight tube,

$$\tau_{w0} = \frac{1}{8} \times \xi_0 \times \rho \times W_0^2 \tag{4.68}$$

For the PHE channel,

$$\tau_w = \frac{1}{8} \times \xi_\tau \times \rho \times W^2 \tag{4.69}$$

where

- W and W_0 are the flow velocities in PHE channel and in tube, m/s
- ρ is the fluid density, kg/m^3
- ζ_0 is the friction factor in straight tube
- ζ_τ is the friction factor in PHE channel, which is accounting for pressure losses due to friction on the wall

When shear stresses (4.68) and (4.69) are equal, the following equation can be written:

$$\mathrm{Re}^2 \times \xi_\tau = \mathrm{Re}_0^2 \times \xi_0 \tag{4.70}$$

where $\mathrm{Re} = \dfrac{\rho W d_e}{\mu}$ and $\mathrm{Re}_0 = \dfrac{\rho W_0 d_e}{\mu}$ are the Reynolds numbers for the tube and the PHE channel, and μ is the dynamic viscosity, Pa·s.

For the turbulent flow of fluid with Pr = 1 in a straight smooth tube, the Reynolds analogy holds:

$$\mathrm{Nu}_0 = \mathrm{Re}_0 \times \frac{\xi_0}{8} \tag{4.71}$$

where the friction factor ζ_0 can be calculated from the Blasius equation:

$$\xi_0 = 0.3164 \times \mathrm{Re}_0^{\frac{-1}{4}} \tag{4.72}$$

By substituting Equation 4.72 into Equations 4.67, 4.70 and 4.71 and after rearranging, it obtained

$$\mathrm{Nu} = 0.065 \times \mathrm{Re}^{\frac{6}{7}} \times \left(\xi_\tau\right)^{\frac{3}{7}} \tag{4.73}$$

Let's assume that the effect of the Prandtl number on heat transfer in the range of $0.5 \le \mathrm{Pr} \le 20$ can be taken into account by introducing the factor Pr raised to the power 0.4. To account for temperature change across the channel cross section, we introduce the term (μ/μ_w) raised to the power 0.14. Then, the modified Reynolds analogy for PHE channel can be expressed by the following equation:

$$\mathrm{Nu} = 0.065 \cdot \mathrm{Re}^{\frac{6}{7}} \cdot \left(\psi \cdot \frac{\xi_s}{Fx}\right)^{\frac{3}{7}} \cdot \mathrm{Pr}^c \cdot \left(\frac{v}{v_w}\right)^{0.14} \tag{4.74}$$

where

- ζ_s is the friction factor for the overall hydraulic resistance which accounts for all pressure losses (due to friction and a form drag) that can be calculated by Equation 4.65
- ψ is the share of friction losses in the total pressure loss in the channel ($\psi = Fx\,\zeta_\tau/\zeta_s$)
- Fx is the ratio of actual to projected heat transfer area that is introduced as ζ_s in Equation 4.65 is determined for projected channel length

The main parameter, which is not determined in Equation 4.74 is the share of pressure loss due to friction on the wall in total loss of pressure ψ. Its value can be

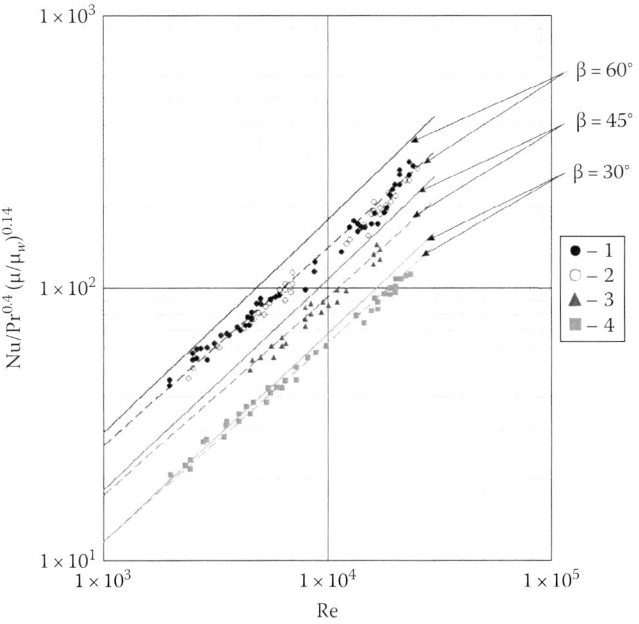

Figure 4.20 Heat transfer in samples of PHE channels (data from Tovazhnyansky et al., 1980 and Tovazhnyansky and Kapustenko, 1984) at $\gamma = 0.556$ (lines are calculated by Equation 4.74: ______: – $\psi = 1$; _ _ _ – ψ by Equation 4.75). 1, 2, 3, 4 – samples numbers; see Table 4.1.

estimated from experimental data on heat transfer in models of PHE channels. In Figure 4.20 are shown the experimental data of Tovazhnyansky et al. (1980) and Tovazhnyansky and Kapustenko (1984) for four experimental samples of PHE channel. The parameters of the samples are given in Table 4.1. The Nusselt and Reynolds numbers are recalculated for $de = 2b$.

The analysis of the data presented in Figure 4.20 leads to the following conclusion: the discrepancies between the calculations using Equation 4.74 for $\psi = 1$ with experimental data increase with the increase of the corrugation angle β and Reynolds number. The analysis of flow patterns in PHE channels by Dović et al. (2009) has shown a stronger mixing at higher β and Reynolds numbers. The mixing

Table 4.1 Geometrical Parameters of the Experimental Plates by Tovazhnyansky et al. (1980)

Sample No.	Pitch *S*, mm	Height *b*, mm	β, °	Length *L*, m	Width *W*, mm	d_h, mm	*Fx*
1	18	5	60	1.0	225	9.6	1.15
2	36	10	60	1.0	225	19.3	1.15
3	18	5	45	1.0	225	9.3	1.15
4	18	5	30	1.0	225	9.0	1.15

is associated with flow disruptions which contribute to the rise of form drag and consequent decrease in the share ψ of pressure loss due to friction on the wall. Correlating the values of ψ calculated from Equation 4.74 using experimental values of Nu, the expression describing its dependence on β and the Reynolds number is

$$A_1 = 380/\left[\mathrm{tg}(\beta)\right]^{1.75};$$
$$\psi = \left(\mathrm{Re}/A_1\right)^{-0.15\sin(\beta)} \text{ at Re} > \text{A}; \quad \psi = 1 \text{ at Re} \leq A \qquad (4.75)$$

Equation 4.74 with ψ determined by Equation 4.75 correlates experimental data of Figure 4.20 with a mean-square error of 6.2%. The accuracy of Equation 4.75 was also confirmed by direct calculation using CFD modelling of average shear stress on the wall of PHE channel formed by plates with $\beta = 45°$ reported by Stogiannis et al. (2013).

To validate the Equation 4.74 and establish the limits of its application, the results of calculations are compared below in this section with different experimental data available in literature.

The experimental study of the influence of corrugation inclination angle β on heat transfer and friction factor was reported by Zimmerer et al. (2002). The experiments were made on the models of channels with corrugated walls, so the influence of inlet and outlet zones is minimal. The data for friction factor of Figure 4.19 are in good agreement with the calculations obtained from Equation (4.65), which are presented by the solid line (biggest discrepancy is 20%). The change of β from 18° to 68° results in 30 times increase on the friction factor. At the same time, the film heat transfer coefficient, characterised by Nu in Figure 4.21, increases about five times. The accuracy of calculations by the Equation 4.74 is fairly good for β up to 60° (deviation less than 2%), but for $\beta = 68°$, discrepancy increases up to 20%. It can be concluded that $\beta = 68°$ is outside the upper limit of model application. In any

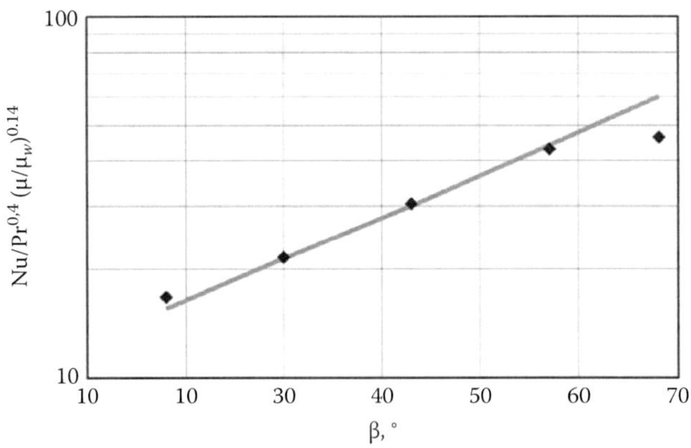

Figure 4.21 The influence of corrugation inclination angle β on heat transfer. Comparison of data from Zimmerer et al. (2002) with calculations by Equation 4.74.

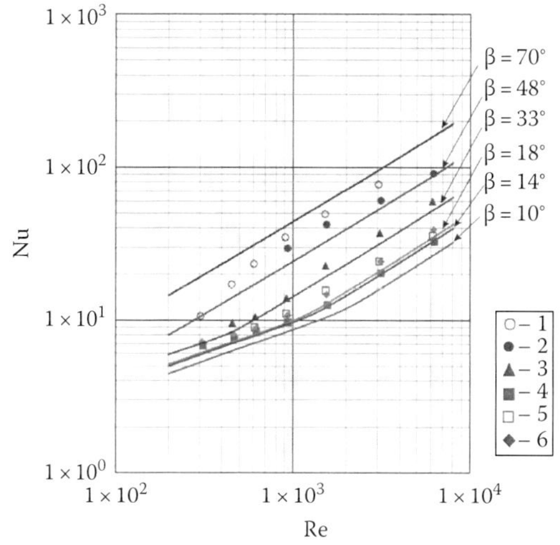

Figure 4.22 Comparison of the Nusselt numbers calculated by Equation 4.74 with data published by Savostin and Tikhonov (1970): 1 – β = 72°, γ = 0.872; 2 – β = 48°, γ = 0.926; 3 – β = 33°, γ = 0.911; 4 – β = 18°, γ = 0.926; 5 – β = 14°, γ = 1.460; 6 – β = 10°, γ = 0.934.

way, the corrugation inclination angle affects friction factor and heat transfer to a considerable extent and can be used as one of the main variables when optimising the geometry of heat exchanger plates.

The data for models of criss-cross flow channels were published by Savostin and Tikhonov (1970). These data are presented in Figure 4.22. The experiments were conducted with air at 368 K and 463 K. For these conditions, Pr = 0.69 and $(\mu/\mu_w)^{0.14} = 1$. The corrugations had triangular shape with radiuses (about 0.6 mm) at the edges. The height of the corrugations for the cases presented in Figure 4.22 was in the range of 1.12–1.22 mm. The experimental Nu and Re were multiplied on $2b/d_h$, for a comparison to be made based on the same definition of equivalent diameter. The discrepancies of calculations by Equation (4.78) for β = 14°, 18°, 33° and 48° do not exceed 15%. For β = 10°, the error is up to 25% and it can be concluded that the lower limit for the application of the model is β = 14°. The upper limit is certainly below 72°, as discrepancies at β = 72° go up to 50% at Re < 800.

Another comprehensive study of heat transfer with extended set of mixed commercial plates (produced by APV) of different corrugations was reported by Heavner et al. (1993). These data are presented in Figure 4.23 by points calculated from correlations presented by Wang et al. (2007). The equations were obtained for the exponent of Pr equal to 1/3. To account for differences with exponent in Equation 4.74, which is equal to 0.4, the data for Nu were calculated at average Pr number equal to 4.6. The estimation for geometrical parameters was taken from the paper of Dović et al. (2009): $\gamma = 0.7$, $Fx = 1.26$. The results of the calculation with Equation 4.74 are presented by solid lines in Figure 4.23. The agreement with experiments is rather good for β from 23 to 23 + 90 degrees. The discrepancies do not exceed 10%. For the

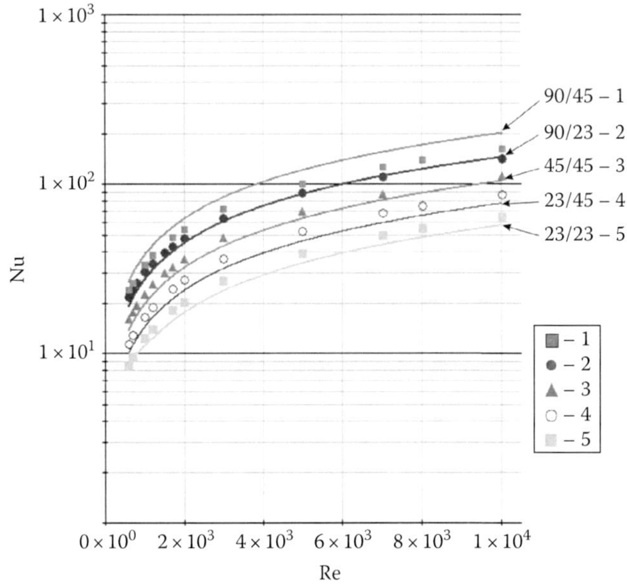

Figure 4.23 Comparison of the Nusselt numbers calculated by Equation 4.74 with data published by Heavner et al. (1993): 1 – β = (90° + 45°)/2, 2 – β = (90° + 23°)/2, 3 – β = 45°, 4 – β = (23° + 45°)/2, 5 – β = 23°.

highest β at combination of plates with β = 90° and β = 45° (average β = 67.5°), the experimental data are 14%–23% lower than predicted.

The research for β = 65° was reported by Dović et al. (2009). The authors have performed tests with two models of PHE channels having corrugations with angles at β = 65° and β = 28°. The tests were made with water and water–glycerol solutions. The water data corresponds to Re numbers higher than 200. The heat transfer results of these authors are presented in Figure 4.24. As the hydraulic diameter they used is $d_h = 2b/Fx$, the values of Nu and Re were corrected by $Fx = 1.19$. The data for Nu were calculated at Pr = 3.54 (water 50 °C) to account for the difference in the exponent of Pr. The data were taken from a graph, so the accuracy is limited, but for Re >100, the prediction (solid lines on Figure 4.24) is fairly good. The error for β = 65° does not exceed 10%. So we define the upper β limit of the model as 65°. The lower limit for the Reynolds number can be considered as 100.

The heat transfer and friction factor data for different arrangements of two plates with β = 60° and β = 30° are reported by Muley and Manglik (1999). The authors studied the cooling of hot water (2 < Pr < 6) and used an equivalent diameter of $de = 2b$. The exponent of the Pr number was taken to be 1/3, so we performed calculations for Pr = 4 to exclude the influence of Pr. The predictions by Equation 4.74 (solid lines in Figure 4.25) are rather good for β = 60° and combination of plates with β = 60° and β = 30° (average β = 45°); the error is not bigger than 10%. But for a low angle β = 30°, the model underestimates data by up to 30%. This can be explained based on the influence of the inlet and outlet distribution zones of the short tested

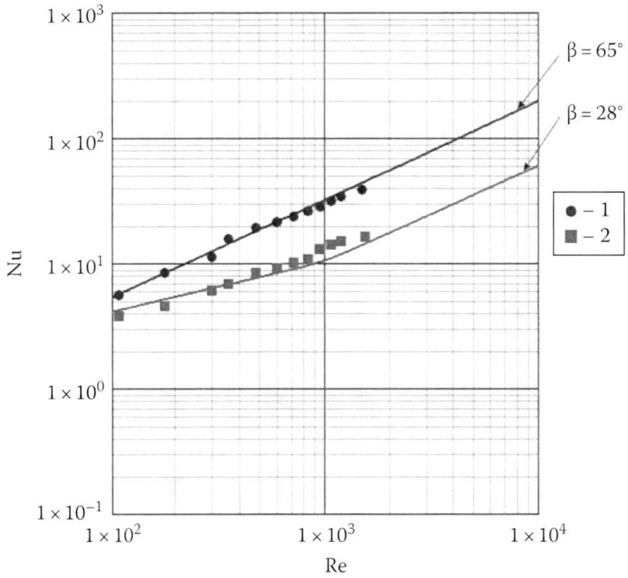

Figure 4.24 Comparison of the Nusselt numbers calculated by Equation 4.74 with heat transfer data by Dović et al. (2009): 1 – β = 65°, 2 – β = 28° (γ = 0.52, *Fx* = 1.19).

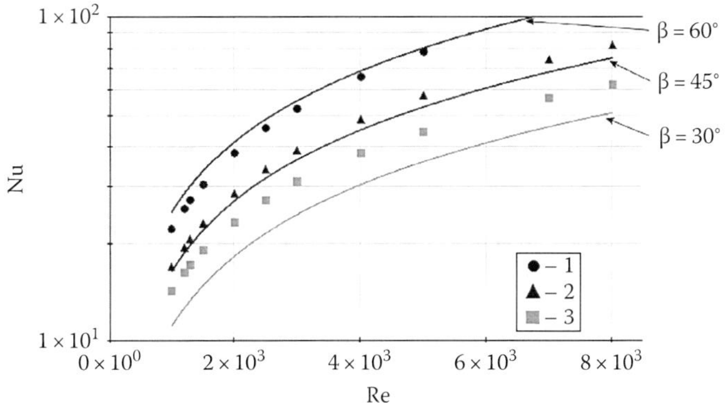

Figure 4.25 Comparison of the Nusselt numbers calculated by Equation 4.74 with data reported by Muley and Manglik (1999): 1 – β = 60°, 2 – β = (60° + 30°)/2, 3 – β = 30° (γ = 0.556, *Fx* = 1.29).

plates (the distance between the ports is 392 mm, the length of the main corrugated field is approximately $L_F = 278$ mm). While the area of the distribution zones is about 20% of the total heat transfer area, the higher level of turbulence generated at the entrance has an influence on the heat transfer of all plate, especially at low β, when there is about 15 cells formed by the corrugations along the length of a channel. For β = 60°, the number of such cells is 27 and 23 cells for β = 45°. For this plate, the

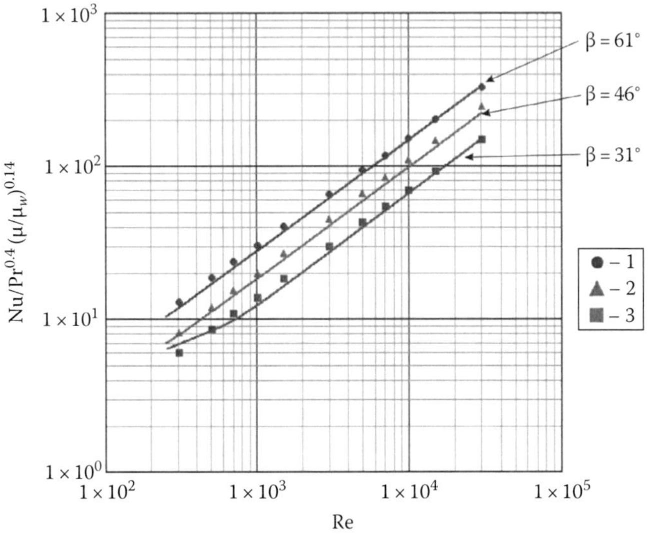

Figure 4.26 Comparison of the Nusselt numbers calculated by Equation 4.74 with data for commercial plate M10B from Arsenyeva et al. (2011): 1 – $\beta = 61°$, 2 – $\beta = (61° + 31°)/2$, 3 – $\beta = 31°$ ($\gamma = 0.588$, $Fx = 1.21$).

ratio $Lp/de = 55$, but we can expect that such effect will be eliminated at $Lp/de > 100$. The data for heat transfer of longer plates were reported by Arsenyeva et al. (2011a). These data for the commercial plates M10B of Alfa Laval are presented in Figure 4.26. The length Lp is 720 mm and Lp/de is about 120. The error of the prediction by Equation 4.74 does not exceed 10% for all the β analysed: (1) $\beta = 61°$, (2) mixed $\beta = 61°$ and $\beta = 31°$ (average 46°) and (3) $\beta = 31°$.

Another study of PHE consisting of commercial plates with $\beta = 60°$ was reported by Gherasim et al. (2011). The corrugations on the plates are of triangular cross section, but with small, about 1 mm flat zones at the edges. The pitch is 9 mm and the corrugation height is $b = 2.5$ mm, with an aspect ratio of $\gamma = 0.556$. The comparison of data calculated from the empirical correlation of the cited paper (represented by the dotted line) with Equation 4.74 results (solid line) at Pr = 3.54 are presented in Figure 4.27. The error is not more than 5%. This confirms the applicability of the model for corrugations of different shapes: sinusoidal, triangular and even triangular with flat zones about 12% of the corrugation pitch.

The comparison of experimental results for heat transfer in PHE channels of various geometrical forms is shown, as reported in 8 papers of different authors. The natures of heat and momentum transfer phenomena in flows at all these channels are similar. The average values of friction factors and film heat transfer coefficients can be calculated by the same equations of proposed dimensionless form. These equations account for the main parameters that characterise the geometrical form of the channel, but they are not sensitive to the size of the channel in the range of channel spacing from 1.12 to 10 mm. The range of applicability for these equations includes

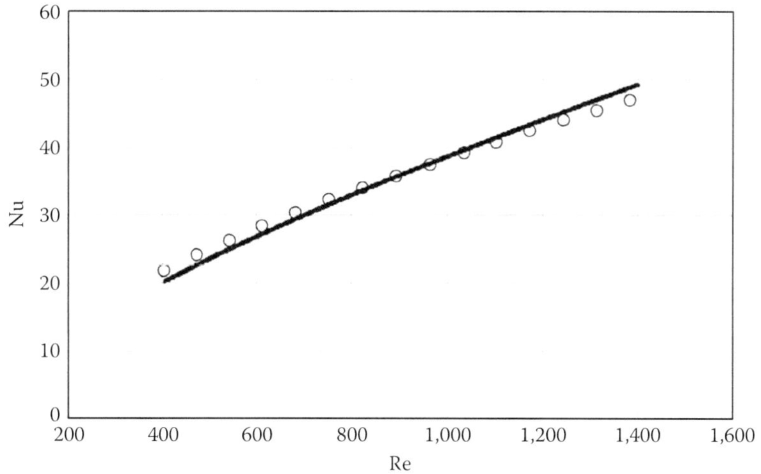

Figure 4.27 Comparison of the Nusselt numbers calculated by Equation 4.74 with data reported by Gherasim et al. (2011) with $\beta = 60°$, $\gamma = 0.556$, $Fx = 1.21$.

the possible range of geometrical parameters for commercial plates. The error on the heat transfer prediction by Equation (4.65) is not larger than 15% in the following range of corrugation parameters: β from 14° to 65°, γ from 0.5 to 1.5 and Fx from 1.14 to 1.5. This applies for the Reynolds numbers from 100 to 25,000 for both shapes of individual corrugations: sinusoidal and triangular. Equation 4.74 with Equation 4.75 can be used for the optimisation of PHEs' geometrical parameters that meet the required process conditions.

4.1.5.4 Analogy of Heat and Momentum Transfer in PHE Channels and Accounting for the Prandtl Number Influence on Heat Transfer

The relation between heat transfer and hydraulic characteristics of inter-plate channels is very important for the selection of the optimal geometry and correct design of PHEs. The semi-empirical analysis of the link between heat and momentum transfer for turbulent flows in straight round tubes was started by Reynolds in 1874 and modified later by Prandtl in 1928 and von Kármán in 1939. It was utilised and developed further by many researchers and proved itself useful in obtaining more accurate picture and physical background of turbulent heat transfer processes. In a number of practical cases, it enabled to correlate experimental data and to extrapolate correlations for a wider range of their application. For example, the correlation for heat transfer in straight tubes and channels presented by Gnielinski (1975), based on the Prandtl analogy, has been proved accurate for turbulent and also transitional flow regimes in a wide range of the Prandtl numbers and is recommended by authors of *Perry's Chemical Engineers Handbook* (2008) for usage in practical applications.

To predict heat transfer based on the data of friction factor in PHE channels, Martin (1996) have utilised equation of Leveque analogy, initially proposed for laminar flow. The same approach was employed later by Dović et al. (2009) and gave reasonable accuracy in predicting heat transfer data of different authors. Modification of the Reynolds analogy for PHE channels was proposed by Tovazhnyansky and Kapustenko (1984). It gave a good agreement with their experimental data for models of PHE channel corrugated field. Similar modification of the Reynolds analogy that is described in the previous subsection in this book is proposed by Arsenyeva et al. (2012) and has proved fairly accurate compared with experimental data of different heat transfer studies for PHE channels. All these generalisation attempts are having fixed power at the Prandtl number in correlating equations (0.33 in the Leveque equation and 0.4 taken in the modified Reynolds analogy). According to empirical correlations for different PHE channels published by different authors, the powers at the Prandtl number vary in fairly wide range, mostly from 0.3 to 0.5. Attempting to use PHE channels, Gnielinski equation (1975) gives discrepancies with experimental results up to 300%.

The modification of von Kármán analogy for PHE channels is proposed in the paper by Arsenyeva et al. (2014). One of the developments for von Kármán analogy in pipes presented in the paper by Lyon (1951) is used. The following equation is derived there:

$$\mathrm{Nu}^{-1} = 2\cdot\int_0^1 \frac{\left(\int_0^{\xi}\omega\cdot\xi\cdot d\xi\right)^2}{\left(1+\varepsilon\cdot\mathrm{Pr}\cdot\frac{\nu_T}{\nu}\right)\cdot\xi}d\xi \tag{4.76}$$

where

- $\xi = R/R_0$ is the relative distance from centre of the tube
- $\omega = w/W$ is the relative velocity and w is the local velocity, m/s
- W is the average flow velocity, m/s
- ν is the kinematic viscosity, m^2/s
- ν_T is the turbulent viscosity (momentum eddy diffusivity), m^2/s
- $\varepsilon = \lambda_T/(c\rho)/\nu_T$ is the ratio of the eddy diffusivities for heat and momentum
- $\mathrm{Pr} = c\rho\nu/\lambda$ is the Prandtl number
- c is the fluid specific heat, $J/(kg\cdot K)$
- ρ is the fluid density, kg/m^3
- λ is the fluid heat conductivity, $W/(m\cdot K)$

Equation 4.76 can be used regardless of the flow regime, with the proper estimation of velocity, ν_T and ε distributions, as it was shown by Lyon (1951). In a book by Kutateladze (1979), the method to use Equation 4.76 for turbulent heat transfer in pipes was proposed. The main assumptions of that method are adopted to turbulent flow in PHE channels as it is shown below in this section.

The turbulent flow in direction perpendicular to PHE channel wall is assumed divided on viscous sublayer, buffer layer and turbulent main stream. It is important to estimate the thickness of those layers compared to the channel equivalent diameter D_e. Near smooth wall surface, the distribution of velocity in uncompressible flows is closely related to shear stress on that wall τ_W. Using the friction factor for the total hydraulic resistance of PHE channel ζ_S, it can be written for average shear stress on the wall:

$$\tau_W = \frac{\xi_S \cdot \psi}{Fx} \frac{\rho \cdot W^2}{8} = \xi_\tau \cdot \frac{\rho \cdot W^2}{8} \tag{4.77}$$

where

ψ is the relative share of pressure loss due to friction in total pressure loss at main corrugated field of the channel

Fx is the surface area enlargement factor

ζ_τ is the friction factor accounting for shear stress only

As it is discussed earlier in a wide range of corrugation geometrical parameters, the friction factor ζ_S can be estimated by Equation 4.65 and the relative share of pressure loss due to friction in total pressure loss ψ by Equation 4.75. It was shown in the paper by Arsenyeva et al. (2012b) that for water flowing in PHE channels for effectively working PHEs, the shear stress on the wall should be in the range from 10 to 100 Pa and even higher. Let's introduce dimensionless distance from the wall η. Then the thickness of buffer layer (including viscous sublayer) can be estimated from the following relation (assuming $\eta = \eta_2 = 30$):

$$y = \frac{\eta \cdot \nu}{(\tau_W/\rho)^{0.5}} \tag{4.78}$$

For properties of water at 50 °C and $\tau_W = 10$ Pa, the upper boundary of buffer sublayer is $y_2 = 0.17$ mm. It diminishes to $y_2 = 0.05$ mm for $\tau_W = 100$ Pa. That constitutes from about 4% to 1% of equivalent diameter D_e. The viscous sublayer is about 5 times even thinner, as by different published data, its dimensionless upper boundary is estimated from $\eta_1 = 5$ to $\eta_1 = 7$. With such small thickness of viscous sublayer, it can be concluded that in this region,

a. Variable ξ is very close to 1, changing maximally from 0.95 to 1.
b. Counting on to such dimensions, the surface of the plates forming PHE channels can be regarded as smooth. To maintain metal quality during stamping, it should have curvature radius at least 1 – 1.5 mm to be pressed from sheet metal of even 0.4–0.5 mm without cracks.
c. On integration of the inner integral of Equation 4.76, most of its parts are outside of the buffer and viscous sublayers. It permits to assume $\omega = 1$ for all flow region.

The right side of Equation 4.76 can be regarded as a sum of integrals, which are corresponding to division of flow on turbulent main stream I_T, buffer I_B and viscous I_L sublayers. These integrals are representing the influence on heat transfer of the specific region. For Pr > 1, the main change of temperature is happening in buffer and viscous sublayers, as can be judged from literature (see e.g. Lyon, 1951). Increasing the Prandtl number shifts it closer to the wall, and the part of thermal resistance in the main stream compared to thermal resistance of all flow is diminishing. In such cases, the integral I_T corresponding to turbulent main stream can be estimated in assumption; its value is approximately equal to that one in a flow core of the smooth tube of diameter equal to equivalent diameter of the channel and with the same shear stress on the wall. Assuming also that eddy diffusivities of heat and momentum are equal ($\varepsilon = 1$) and that there $\nu < \nu_T$, the Equation 4.76 can be rewritten as follows:

$$\mathrm{Nu}^{-1} \approx \frac{1}{2}\cdot(I_T + I_B + I_L) = \frac{1}{2}\cdot\left(\int_0^{\xi_2}\frac{\xi^3 d\xi}{\mathrm{Pr}\cdot\frac{\nu_T}{\nu}} + \int_{\xi_2}^{\xi_1}\frac{\xi^3 d\xi}{1+\mathrm{Pr}\cdot\frac{\nu_T}{\nu}} + \int_{\xi_1}^{1}\frac{d\xi}{1+\mathrm{Pr}\cdot\frac{\nu_T}{\nu}}\right) \tag{4.79}$$

At the central part of the tube ($0 \le \xi \le \xi_2$) more accurate than $\omega = 1$, distribution of velocity is needed. The logarithmic velocity profile can be used:

$$w = w^*\left(C_* + \frac{1}{\chi}\cdot\ln\frac{v^*\cdot y}{\nu}\right) \tag{4.80}$$

where

$w^* = (\tau_W/\rho)^{1/2}$, m/s

C^* is the constant

χ is the constant determined according to experimental data for turbulent flows in tubes

Assuming the local shear stress is equal to turbulent shear stress and is proportional to the distance from the centre $\tau = \tau_W \cdot \xi$, the turbulent viscosity is estimated as

$$\nu_T = \frac{\tau_T}{\rho\cdot dw/dy} = \chi\cdot w^*\cdot y\cdot\xi$$

In a dimensionless form, accounting for Equations 4.77 and 4.78,

$$\frac{\nu_T}{\nu} = \chi\cdot\mathrm{Re}\cdot\sqrt{\frac{\xi_\tau}{32}}\cdot(1-\xi)\cdot\xi \tag{4.81}$$

For heat transfer in the main turbulent stream, after substituting Equation 4.81 in the first integral I_T of Equation 4.79 and integration,

$$I_T = \int_0^{\xi_2}\frac{\xi^3 d\xi}{\mathrm{Pr}\cdot\frac{\nu_T}{\nu}} = \frac{\sqrt{32}}{\mathrm{Pr}\,\chi\,\mathrm{Re}\sqrt{\zeta_\tau}}\left[\ln\left(\frac{\mathrm{Re}}{\eta_2}\sqrt{\frac{\zeta_\tau}{32}}\right) - \frac{1}{2}\left(1-\frac{\eta_2}{\mathrm{Re}}\sqrt{\frac{32}{\zeta_\tau}}\right)^2 - 1 + \frac{\eta_2}{\mathrm{Re}}\sqrt{\frac{32}{\zeta_\tau}}\right] \tag{4.82}$$

In the buffer layer of PHE channel ($\xi_2 \le \xi \le \xi_1$), the logarithmic velocity profile correlation can be used, as established in experiments with smooth tubes and for turbulent flow near flat surface:

$$w = w^* \left(C'_* + \frac{1}{\chi'} \cdot \ln \frac{v^* \cdot y}{\nu} \right) \tag{4.83}$$

where $C^{*\prime}$ and χ' are the empirical constants for buffer layer.

Taking that the local shear stress τ in buffer sublayer is the sum of viscous and turbulent τ_T shear stresses and is approximately equal to wall shear stress, the turbulent viscosity can be determined as follows:

$$\nu_T = \frac{\tau_W}{\rho \cdot dw/dy} - \nu \approx \chi' \cdot w^* \cdot y - \nu \tag{4.84}$$

In a dimensionless form, with accounting for Equations 4.77 and 4.78,

$$\frac{\nu_T}{\nu} \approx \chi' \cdot \mathrm{Re} \cdot \sqrt{\frac{\zeta_\tau}{32}} \cdot (1-\xi) - 1 \tag{4.85}$$

For heat transfer in the buffer sublayer, after substituting Equation 4.85 into the second integral I_B of Equation 4.79 and integration,

$$I_B = \int_{\xi_2}^{\xi_1} \frac{\xi^3 d\xi}{1 + \mathrm{Pr} \cdot \frac{\nu_T}{\nu}} = \int_{\xi_2}^{\xi_1} \frac{\xi^3 d\xi}{\mathrm{Pr} \cdot \chi' \mathrm{Re} \sqrt{\frac{\zeta_\tau}{32}} (1-\xi) - (Pr-1)}$$
$$= \frac{\sqrt{32}}{\mathrm{Pr}\,\chi' \mathrm{Re} \sqrt{\zeta_\tau}} \left[Z^3 \ln\left(\frac{1 + \mathrm{Pr}(\chi' \cdot \eta_2 - 1)}{1 + \mathrm{Pr}(\chi' \cdot \eta_1 - 1)} \right) + \frac{\eta_2 - \eta_1}{\mathrm{Re}\sqrt{\frac{\zeta_\tau}{32}}} (1 - Z + Z^2) \right] \tag{4.86}$$

where $Z = 1 - \dfrac{\mathrm{Pr} - 1}{\mathrm{Pr}\,\chi' \mathrm{Re} \sqrt{\frac{\zeta_\tau}{32}}}$

In Equation 4.79, the third integral I_L characterises the heat transfer in viscous sublayer. The biggest temperature gradient at the Prandtl numbers much higher than unity is observed in this area and its parts closest to the wall. It requires accounting for the influence of turbulent pulsations intruding from outer layers. Following Kutateladze (1979), it can be done by the following relation:

$$\frac{\nu_T}{\nu} \approx \beta_M \cdot \eta^3 \tag{4.87}$$

Here, β_M is the proportionality coefficient, whose value is estimated empirically as $\beta_M \approx 0.03$. According to Kutateladze (1979), in integral I_L for heat transfer, another empirical coefficient $\beta_T = \beta_M/\eta_1^2$ should be used, which gives the following:

$$I_L = \int_{\xi_1}^{1} \frac{d\xi}{1+\Pr\cdot\frac{\nu_T}{\nu}} = \frac{1}{\mathrm{Re}}\sqrt{\frac{\xi_\tau}{32}}\int_0^{\eta_1}\frac{d\eta}{1+\Pr\cdot\beta_T\cdot\eta^3} \tag{4.88}$$

The analytical expression for the integral of the function type $1/(a^3 + x^3)$ is rather cumbersome and long. It is easily estimated by numerical integration on a computer. For the turbulent velocity profile parameters and empirical coefficients in equations, the following is assumed: $\eta_2 = 30$, $\eta_l = 6.8$, $\chi = 0.37$, $\chi' = 0.2$, $\beta_T = \beta_M/\eta_1^2$. As a result of calculation of the Nusselt number, the following is obtained:

$$\mathrm{Nu} = \frac{0.131\cdot R_\xi\cdot\Pr}{\ln\left(\frac{R_\xi}{760}\right) - \frac{14{,}450}{R_\xi^2} + \frac{340}{R_\xi} + B_z + 2.52\Pr\cdot\varphi(\Pr)} \tag{4.89}$$

where

$$B_z = 1.85\left[Z^3\ln\left(\frac{1+5\Pr}{1+0.36\Pr}\right) + \frac{131.24}{R_\xi}\left(1 - Z + Z^2\right)\right];\quad R_\xi = \mathrm{Re}\cdot\sqrt{\frac{\zeta_S\cdot\psi}{Fx}}$$

$$Z = 1 - \frac{(\Pr-1)\sqrt{32}}{0.2\Pr \mathrm{R}_\zeta} \tag{4.90}$$

$$\varphi(\Pr) = \frac{1}{\eta_1}\cdot\int_0^{\eta_1}\frac{d\eta}{1+\Pr\cdot\beta_T\cdot\eta^3}$$

The numerical solution of integral (4.90) can be approximated for Pr > 1 by the following equation:

$$\varphi(\Pr) \approx \frac{1.14}{\eta_1}\cdot\Pr^{-0.04}\cdot\mathrm{arctg}\left(\eta_1\cdot\sqrt[3]{\Pr\cdot\beta_T}\right) \tag{4.91}$$

This approximation of the integral I_L solution deviate from numerical one not more than ±2% for $3 \le \Pr \le 10^4$ and not more than −8% for Pr as low as 0.69. Accounting that for such low Prandtl numbers the share of the viscous sublayer in total resistance to heat transfer becomes relatively smaller compared to other parts, this approximation can be used to save computing time.

To check the limits of Equation 4.89 application and its accuracy, the results of calculations by it are compared with experimental data on heat transfer in channels of different geometries. The experimental results for models of PHE channels of

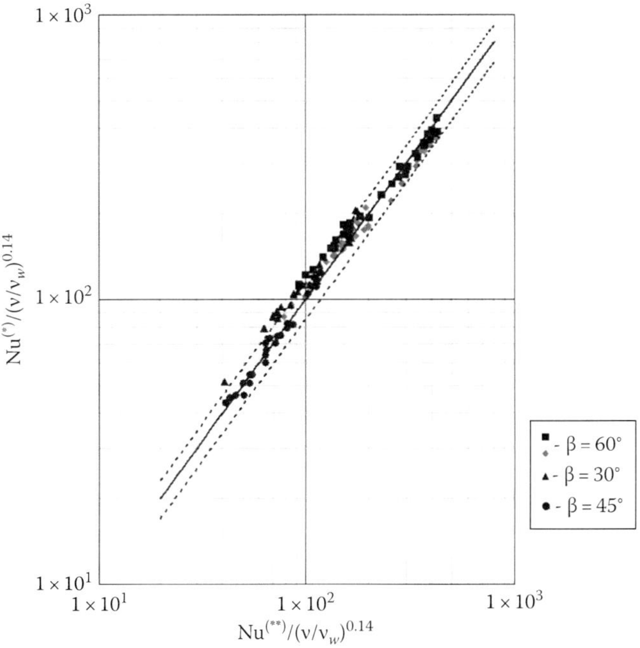

Figure 4.28 The comparison of Nucalc (solid line) calculated by Equation 4.89 with experimental Nuexp: dashed lines correspond to error ±15%.

different corrugation size and inclination angle presented in paper by Tovazhnyansky et al. (1980) are compared with prediction by Equation 4.89, as shown in Figure 4.28. The values of the Prandtl numbers are taken from data of experiments. The discrepancies between Nu numbers calculated by Equation 4.89 and experimental ones have not exceeded 15% with mean-square error 6.5%.

The results of experiments presented in Figure 4.28 were obtained for water as tested fluid in a limited range of the Prandtl numbers (from 1.9 to 7). Most of the available literature data on heat transfer in PHE channels are reported as empirical correlations for specific range of the Prandtl numbers, with no data on Pr in individual experimental runs. In Equation 4.74 that was derived based on the Reynolds analogy, as presented in the previous subsection, to account for the Prandtl number influence on heat transfer, the exponent at the Prandtl number is fixed to $c = 0.4$. The comparison with quite a number of literature data was made by adjusting results accounting for the value of Pr exponent in correlation under comparison and the range of Pr presented in those papers. The deviations did not exceeded ±15%. The comparison between calculations by Equations 4.74 and 4.89 at $(\nu/\nu_W) = 1$ and $\text{Pr} = 1$ is presented in Figure 4.29. The deviations did not exceed 5%. Therefore, the conclusion can be taken that in the region $0.7 \leq \text{Pr} \leq 7$, the Equation 4.89 predicts the influence of corrugation geometry on heat transfer practically with the same accuracy as Equation 4.74.

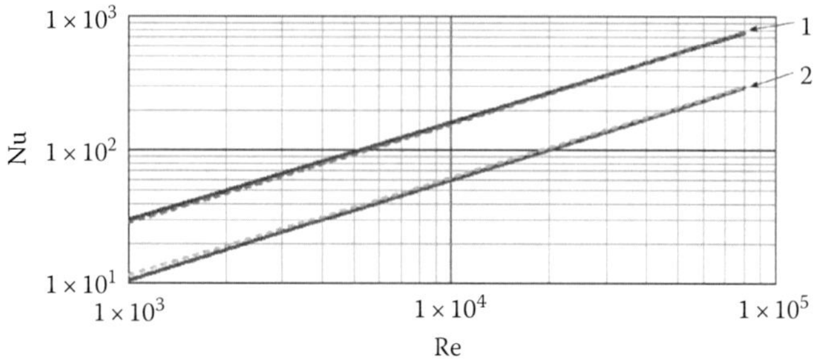

Figure 4.29 The comparison of Nu(89) (solid lines) calculated by Equation 4.89 with Nu(74) calculated by Equation 4.74 (dashed lines) for $\nu/\nu_W = 1$ and Pr = 1: 1 – β = 65°; 2 – β = 30°.

The accuracy of Equation 4.89 in accounting the Prandtl number influence on heat transfer is estimated by comparison of results of calculations by Gnielinski (1975) equation for flows inside pipes and straight smooth channels:

$$\mathrm{Nu} = \frac{\zeta_\tau \cdot \mathrm{Pr} \cdot (\mathrm{Re} - 1{,}000)}{8 \cdot \left[1 + 12.7 \cdot \sqrt{\zeta/8} \cdot \left(\mathrm{Pr}^{\frac{2}{3}} - 1\right)\right]} \tag{4.92}$$

Here, ζ_τ is the friction factor in a smooth tube determined by correlation

$$\zeta_\tau = (0.79 \cdot \ln \mathrm{Re} - 1.64)^{-2} \tag{4.93}$$

For comparison in Equation 4.89, it was assumed that $\zeta_s = \zeta_\tau$; $F_x = 1$; $\psi = 1$. The discrepancies of the results, which are presented in Figure 4.30, do not exceed 6%. It is confirmed in papers of Gnielinski (2009) and by authors of *Perry's Chemical*

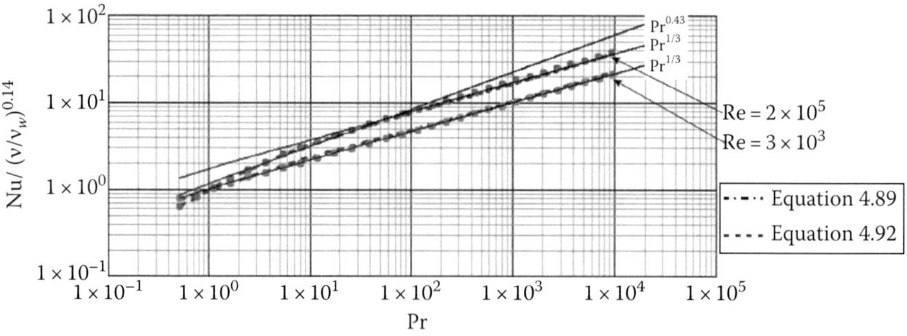

Figure 4.30 The comparison of Equation 4.89 with calculations by Gnielinski Equation 4.92 and correlations with fixed Pr powers.

Engineers Handbook (2008) that the accuracy of Equation (4.92) is suitable for practical applications in a wide range of the Prandtl numbers (0.5 – 100,000). Therefore, the good agreement of the results calculated by both equations recommend Equation 4.89 for the same range of the Prandtl numbers not only for smooth tubes but also for PHE channels.

The influence of the Prandtl number on heat transfer can be analysed with graphs in Figure 4.30. For fully developed turbulent flow (the curves for Re = 200,000 in Figure 4.30) for $1.5 \le Pr \le 12$, the curves can be approximated by using Pr exponent 0.4, as it is made in the Nusselt equation (see Kutateladze, 1979). In known equation of Mikheyev for this range of Pr, the exponent 0.43 is used. When the Prandtl number is bigger than 20, the exponent 0.33 can be used, which is usually used for heat transfer in laminar flows. This effect can be explained by shifting for higher Prandtl numbers the main part of thermal resistance to viscous sublayer (3rd integral I_L in Equation 4.79).

For lower Reynolds numbers (see the curves for Re = 3,000 in Figure 4.30) that are corresponding to transitional flow regime in smooth tube, the exponent at the Prandtl number can be taken as 0.33 for all range of the Prandtl numbers. It is explained by lower intensity of turbulence in the main stream, causing smaller influence on regions with effects of laminar heat transfer (buffer and viscous sub layers). Therefore, the influence of the Prandtl number on heat transfer is determined not only by its value but also by the Reynolds number, which is characterising the flow hydrodynamics in channel. The accurate prediction of the Prandtl number influence in a wide range of the Prandtl and Reynolds numbers is not possible by introducing it as multiplier with some fixed exponent. The calculations has to made with equations like (4.89) or (4.92), which are accounting for flow conditions and the Reynolds number influence.

The equation with variable exponent at the Prandtl number for calculation of film heat transfer coefficients in PHE channels was used by Bogaert and Bölcs (1995). They have reported the results of experimental study on heat transfer and pressure drop in brazed PHE. The mineral oil NUTO H5, provided by ESSO, was used as a hot fluid. It was cooled at experimental set-up by water. In their calculations and presentation of results, Bogaert and Bölcs (1995) have used exponent at the Prandtl number expressed by Equation 4.94. It was extracted from the SWEP calculation procedure. Its accuracy is confirmed for all investigated range of the Reynolds ($1 < Re < 1{,}000$) and the Prandtl ($2 < Pr < 100$) numbers:

$$c_1(\mathrm{Pr}) = \frac{1}{3} \cdot e^{\frac{6.4}{\mathrm{Pr}+30}} \tag{4.94}$$

By this equation, the exponent at the Prandtl number is changing from 0.41 at $Pr \approx 1$ to ≈ 0.333 at $Pr > 800$ (see Figure 4.31).

Bogaert and Bölcs (1995) estimated that in their experiments, the developed turbulent flow occurred at Re > 85 and up to higher of investigated Re = 1,000. The influence of Re and Pr on heat transfer for corrugation parameters corresponding to experimental PHE of cited paper is estimated by calculations with Equation 4.89. The corrugation inclination angle is $\beta = 68°$, the corrugation aspect ratio $\gamma = 0.6$ and

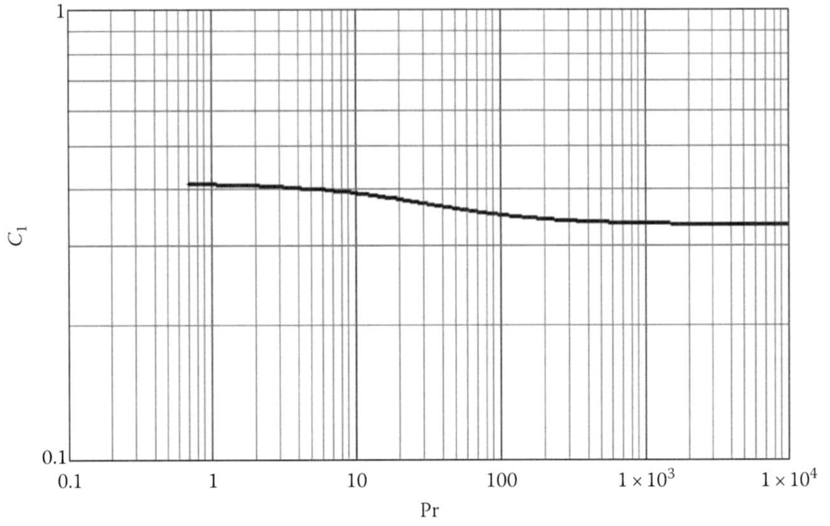

Figure 4.31 The change of the Prandtl number exponent as is calculated by Equation 4.94.

heat transfer area enlargement factor $Fx = 1.2$. For the range of Reynolds numbers corresponding to turbulent flow, the influence of the Prandtl number calculated by Equation 4.89 is predicted to be in a good agreement with the Prandtl power c_1 determined by Equation 4.94, as it is shown in Figure 4.32 for Re = 200. The discrepancies for Pr < 100 do not exceed 4.2%, but become bigger at higher Pr up to 9.9% at Pr = 1,000. The discrepancies at Re = 1,000 do not exceed 5% up to Pr = 30, but become 9.5% at Pr = 100 and 14.7% at Pr = 1,000. Much bigger discrepancies are observed for Re = 10,000, as shown in Figure 4.33. To account for the Reynolds number influence on the Prandtl number exponent, the correction factor is introduced into Equation 4.94, which becomes as follows:

$$c_2(\mathrm{Re}, \mathrm{Pr}) = \frac{1}{3} \cdot \frac{e^{\frac{6.4}{\mathrm{Pr}+30}}}{1 - 0.012 \cdot \mathrm{Re}^{0.27}} \tag{4.95}$$

The comparison of the Prandtl number influence on heat transfer calculated by Equation 4.89 and predicted by introduction of the Prandtl exponent on Equation 4.95 is presented in Figure 4.33. The discrepancies do not exceed 6% in the range of the Reynolds numbers from 100 to 20,000 and the Prandtl numbers from 0.7 to 1,000. The Prandtl number exponent calculated by Equation 4.95 can be used in simplified Equation 4.74 for prediction of film heat transfer coefficients in PHE channels with Equation 4.96 as the following:

$$\mathrm{Nu} = 0.065 \cdot \mathrm{Re}^{\frac{6}{7}} \cdot \left(\psi \cdot \zeta_s / Fx\right)^{\frac{3}{7}} \cdot \mathrm{Pr}^{\frac{1}{3} \cdot \frac{e^{\frac{6.4}{\mathrm{Pr}+30}}}{1-0.012 \cdot \mathrm{Re}^{0.27}}} \cdot \left(\frac{\nu}{\nu_W}\right)^{0.14} \tag{4.96}$$

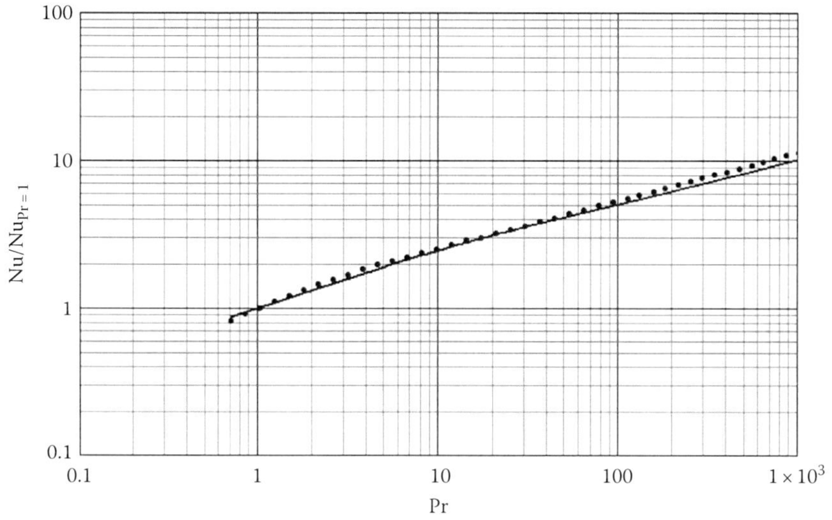

Figure 4.32 The influence of the Prandtl number on heat transfer $Nu/Nu_{Pr=1}$ at Re = 200: dotted lines calculated by Equation 4.89; solid lines calculated as Prc^1.

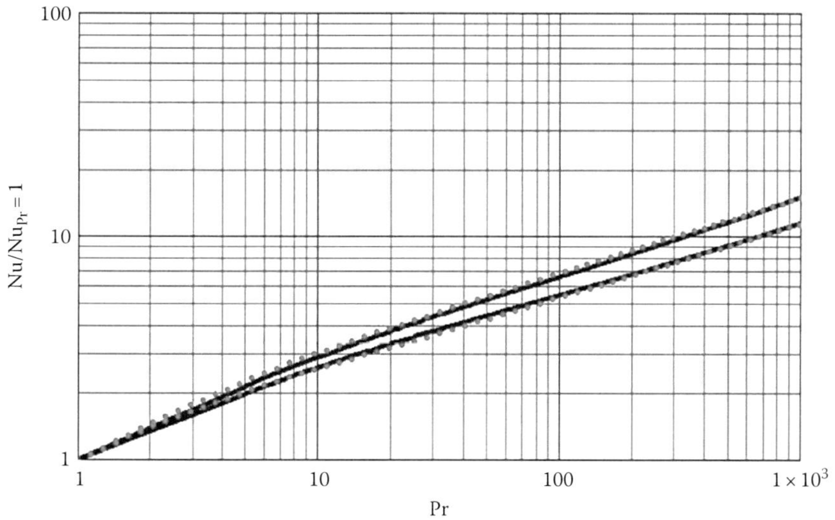

Figure 4.33 The influence of the Prandtl number on heat transfer $Nu/Nu_{Pr=1}$ at Re = 200 (lower lines) and Re = 10,000 (upper lines): dotted lines calculated by Equation 4.89; solid lines calculated as Prc^2 with Equation 4.95.

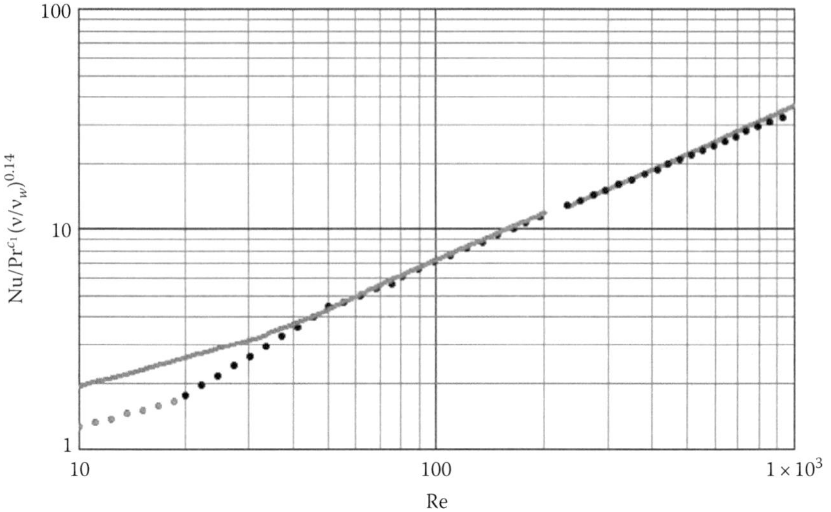

Figure 4.34 The comparison of calculation by Equations 4.74 and 4.95 (solid lines) with experimental results of Bogaert and Bölcs (1995) (dotted lines).

The comparison of the calculated by Equation 4.96 results with data on heat transfer obtained by Bogaert and Bölcs (1995) has demonstrated a good accuracy of prediction, as can be seen in Figure 4.34. In the range of the Reynolds numbers 200 > Re > 40, the deviation from their correlations did not exceed 2.1% at average Pr = 50 for oil, and for 1,000 > Re > 230, maximal discrepancy is 6.5% at Re = 1000 at average Pr = 4 for water. According to Bogaert and Bölcs, this region of the Reynolds numbers corresponds to fully developed turbulent flow (Re > 85) and transitional quasi-turbulent for the Reynolds number down to 40 ÷ 50. For lower Reynolds numbers, the flow becomes predominantly laminar and experimental values of film heat transfer coefficients are much lower than predicted by Equation 4.96.

The similar trend is observed when comparing calculation results with experiments of Dovic et al. (2009), as shown in Figure 4.35. The results of tests with two models of PHE channels having corrugations with angles at $\beta = 65°$and $\beta = 28°$ were reported in that research. The study was made by using water and water–glycerol solutions. The water data corresponds to Re numbers higher than 200. As the hydraulic diameter they used is $d_h = 2b/Fx$, the values of Nu and Re were corrected by $Fx = 1.19$. The data for Nu at Re > 200 were calculated at Pr = 3.54 (water 50 °C) to account for the difference in the exponent of Pr. For lower Reynolds numbers, we considered average Pr = 100. The data were taken from a graph, so the accuracy is limited, but for Re > 40 at $\beta = 65°$, the model prediction (upper solid line in Figure 4.35) is fairly good. The error does not exceed 10%. For the Reynolds numbers lower than 40, the experimental results become much lower than the calculated ones, same like results of a study by Bogaert and Bölcs (1995); see Figure 4.34. For the channel with smaller angle of corrugations $\beta = 28°$, the higher discrepancies are observed at Re < 80. It is explained by lower level of turbulence in PHE

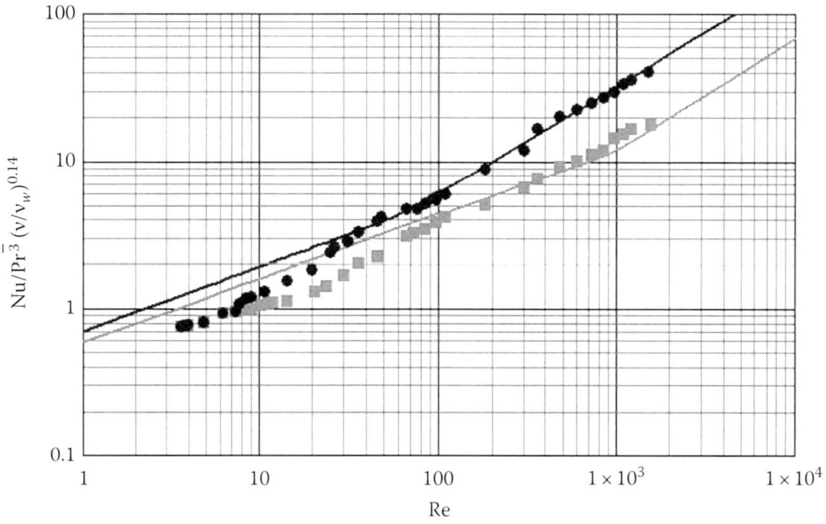

Figure 4.35 The comparison of calculation with experimental results of Dovic et al. (2009): solid lines are calculated by Equation (4.96); dotted lines are calculated with correlations presented in the paper of Dovic et al. (2009); round dots for β = 65° and squires for β = 28°.

channels with smaller angle of corrugations, which also lead to smaller heat transfer coefficients at flow region with predominantly turbulent heat transfer Re > 80. The predominantly laminar flow mechanism is taking place in such channels to higher Reynolds numbers, up to 80, compared to channels with β = 65°and β = 68°.

Another two studies of heat transfer in PHE channels that have been made for the same channels in a wide range of the Prandtl numbers were reported by Muley et al. (1999) for 130 < Pr < 290 and 30 < Re < 400. Muley and Manglik (1999) reported also data for the same channels at 2 < Pr < 6 and 1,000 < Re < 9,000.

The results calculated using Equation 4.96 are in good agreement with experimental data on heat transfer from the paper by Muley et al. (1999), as it is shown in Figure 4.36. The experimental data are presented by empirical correlations of that paper for average Pr = 210, as there are no data of the Prandtl numbers for individual experiments. As reported in the cited paper, the accuracy of these empirical correlations is ±10%. The maximal discrepancy of calculated results with those by correlations is not bigger than 12% that confirms the accuracy of the proposed calculation method. The experimental data for channel with β = 30° also show deviation to lower values at Re < 80, which is more clearly seen in a graph with experimental points in the cited paper. The good agreement of calculation by Equation 4.74 with data from the paper of Muley and Manglik (1999) was illustrated earlier in the previous subsection. The experiments of both cited papers were carried out with the same plate geometry and altogether in the range of the Reynolds numbers 20 < Re < 10,000 and the Prandtl numbers 1 < Pr < 290. It confirms the applicability of Equation 4.96 for calculation of heat transfer in PHE channels in a wide range of the Reynolds and the Prandtl numbers.

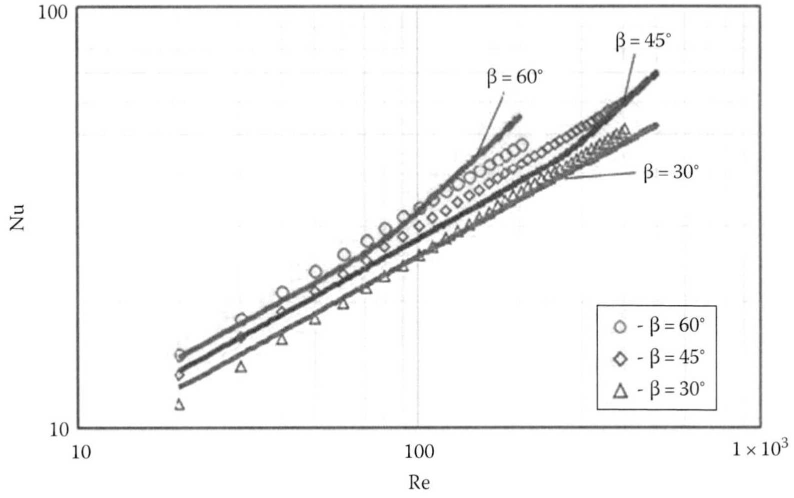

Figure 4.36 The comparison of calculation with experimental results of Muley et al. (1999): solid lines are calculated by Equation 4.96.

The introduction of Equation 4.96 allows to widen the range of application of proposed calculation procedure compared to fixed Prandtl number exponent in correlation (4.74). It is estimated as follows: the Prandtl number Pr from 0.7 to 10^3, the Reynolds number Re from 80 to 25,000, corrugation inclination angle β from 14° to 68°, corrugation aspect ratio γ from 0.5 to 1.5 and area enlargement factor Fx from 1.14 to 1.5. This range covers most of the possible conditions for PHE applications in process industries and for low-grade heat utilisation, as well as feasible options of corrugations on the PHE plates. Together with equations for estimation of local pressure drop in channel distribution zones and ports, the presented equations allow to develop thermohydraulic mathematical model of PHE. Such model can be used for optimisation of PHEs channel geometry for different conditions in the process industries. It can be employed for developing software for optimisation of individual PHEs and for optimising PHEs heat exchange networks.

4.2 INTENSIFICATION OF HEAT TRANSFER FOR TWO-PHASE FLOWS

4.2.1 Condensation Enhancement

Condensation is a process of fluid conversion from gaseous to liquid state with release of latent heat corresponding to phase change. It can take place in a volume of gaseous phase or on cooled below saturation temperature surfaces. In recuperative heat exchangers, the released latent heat of condensation together with the sensible heat of liquid and gaseous phase cooling is discharged through heat transfer surface to cooling media.

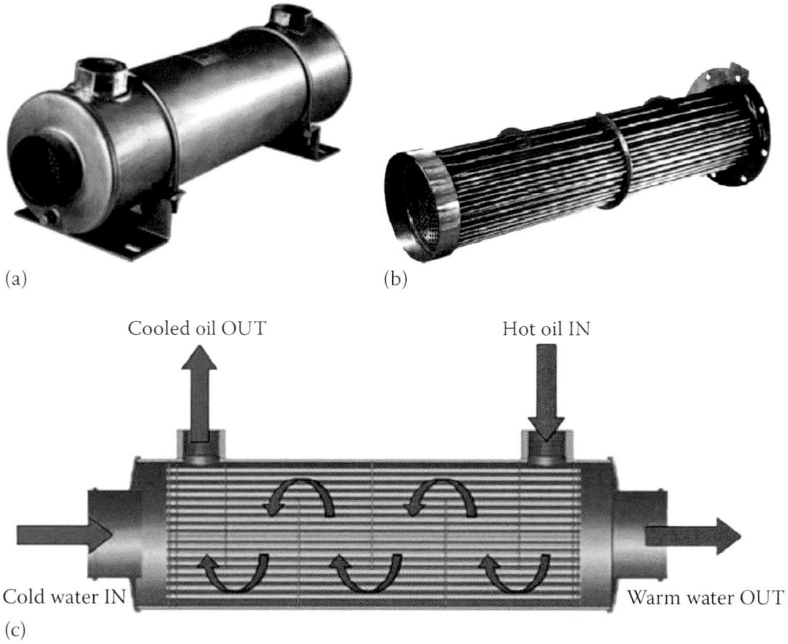

Figure 3.15 Compact brazed shell and tube HE: (a) assembled HE, (b) tube bundle and (c) the movement of heat exchanging streams in HE (courtesy of OAO Alfa Laval Potok, Korolev, Russian Federation).

Figure 3.17 Side view of the Krones shell and tube HE module (after Krones, 2014).

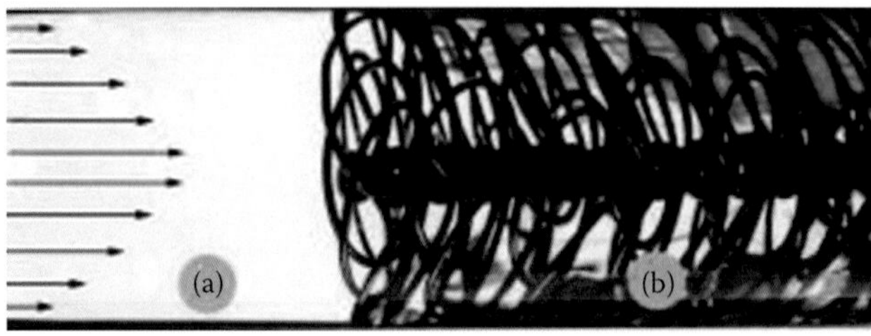
(a)
(b)

Figure 3.19 HiTRAN Matrix Element tube insert promoting turbulence and enhancement of heat transfer: (a) smooth tube; (b) part of tube with insert (after Calgavin, 2014).

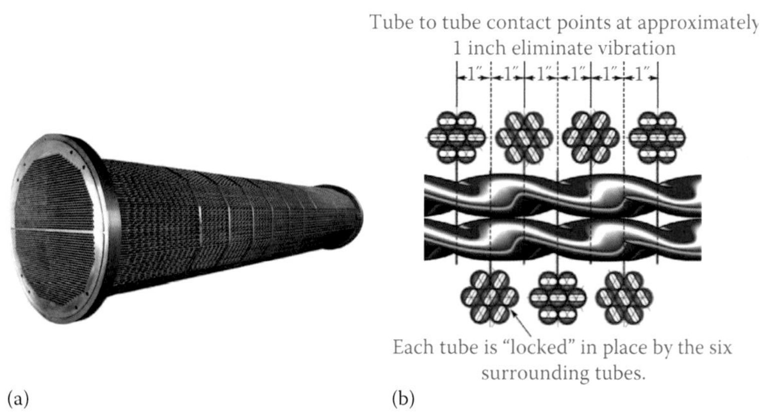
Tube to tube contact points at approximately 1 inch eliminate vibration
1″ 1″ 1″ 1″ 1″ 1″
Each tube is "locked" in place by the six surrounding tubes.
(a)
(b)

Figure 3.20 (a) Koch Heat Transfer Company's TWISTED TUBE® bundle and (b) scheme of tube arrangement (see Koch Heat Transfer Company, 2014).

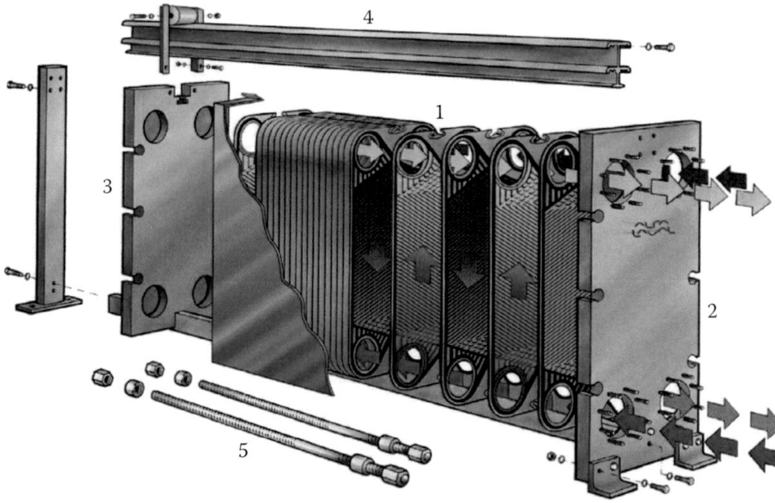
4
1
3
2
5

Figure 3.21 Schematic drawing of one-pass frame-and-plate PHE: 1, pack of plates with gaskets; 2, fixed frame plate; 3, movable frame plate; 4, carrying bar and 5, tightening bolts (courtesy of OAO Alfa Laval Potok).

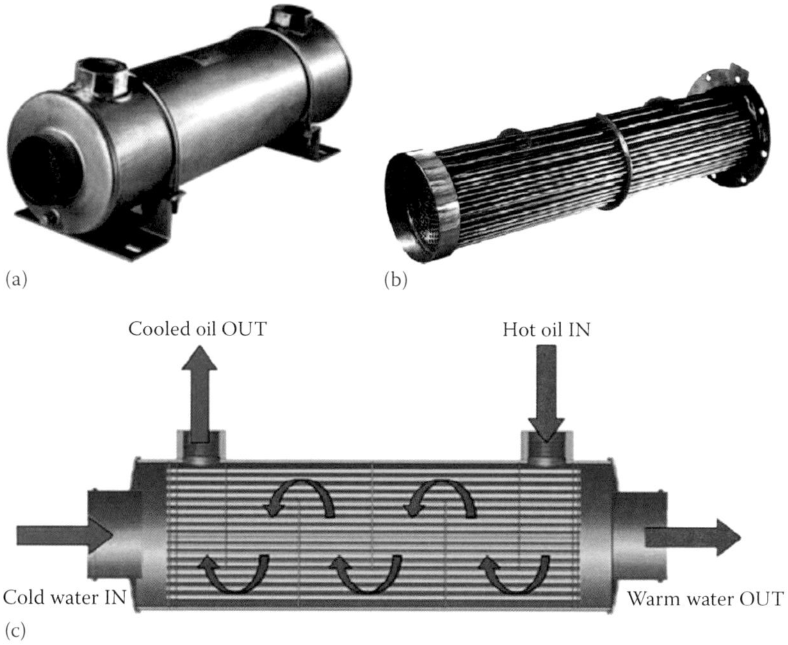

Figure 3.15 Compact brazed shell and tube HE: (a) assembled HE, (b) tube bundle and (c) the movement of heat exchanging streams in HE (courtesy of OAO Alfa Laval Potok, Korolev, Russian Federation).

Figure 3.17 Side view of the Krones shell and tube HE module (after Krones, 2014).

(a)
(b)

Figure 3.19 HiTRAN Matrix Element tube insert promoting turbulence and enhancement of heat transfer: (a) smooth tube; (b) part of tube with insert (after Calgavin, 2014).

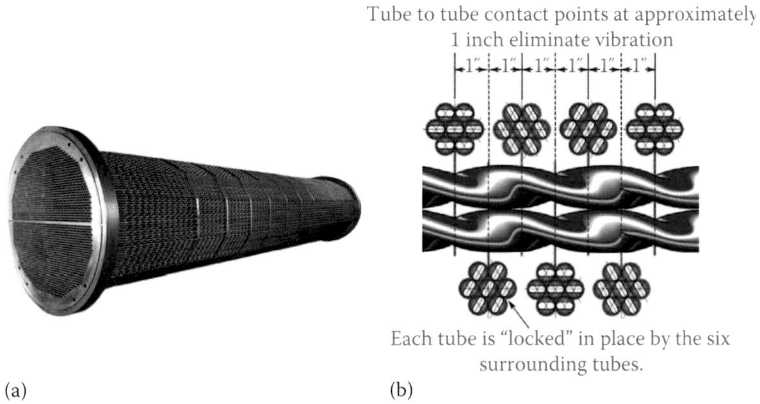
Tube to tube contact points at approximately 1 inch eliminate vibration
1″ 1″ 1″ 1″ 1″ 1″
Each tube is "locked" in place by the six surrounding tubes.
(a)
(b)

Figure 3.20 (a) Koch Heat Transfer Company's TWISTED TUBE® bundle and (b) scheme of tube arrangement (see Koch Heat Transfer Company, 2014).

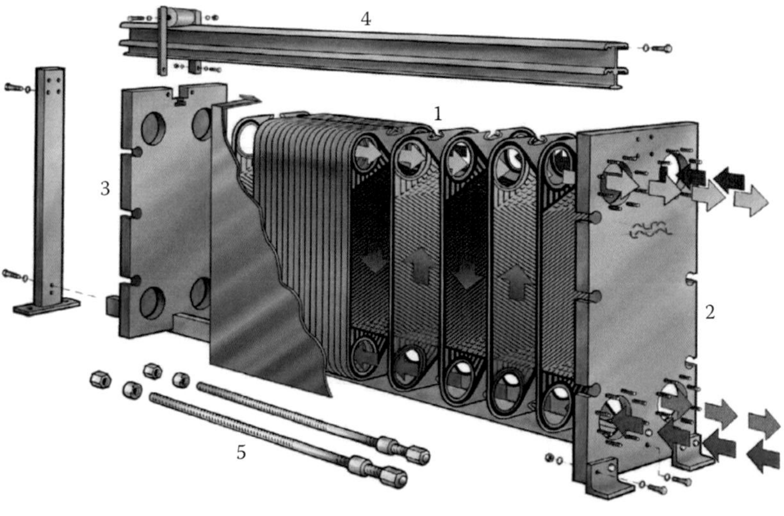
4
1
3
2
5

Figure 3.21 Schematic drawing of one-pass frame-and-plate PHE: 1, pack of plates with gaskets; 2, fixed frame plate; 3, movable frame plate; 4, carrying bar and 5, tightening bolts (courtesy of OAO Alfa Laval Potok).

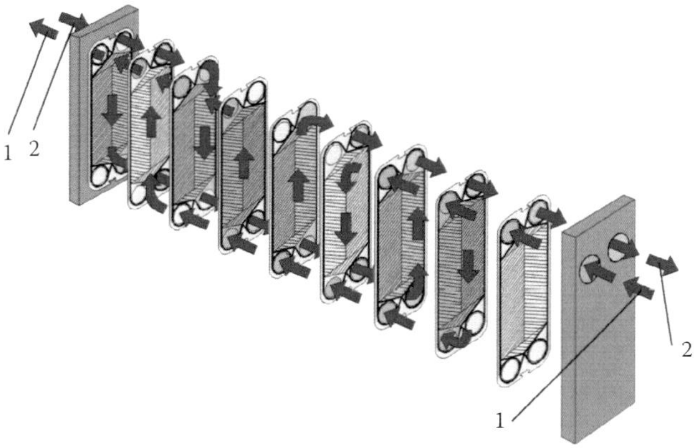

Figure 3.22 Schematic drawing of two-pass frame-and-plate PHE: 1, hot stream and 2, cold stream (courtesy of OAO Alfa Laval Potok).

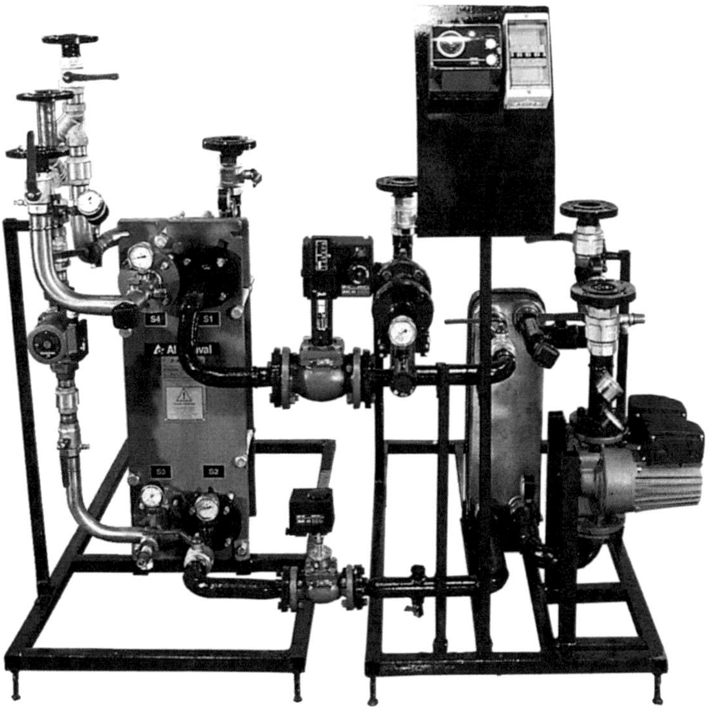

Figure 3.25 District heating substation module with plate-and-frame PHE produced with Alfa Laval plates by AO Spivdruzhnist-T LLC in Kharkiv, Ukraine.

Figure 3.28 Drawing of Alfa Laval Packinox welded PHE (courtesy of OAO Alfa Laval Potok).

Figure 3.29 Alfa Laval AlfaRex welded PHE with cut of welded plate pack (courtesy of OAO Alfa Laval Potok).

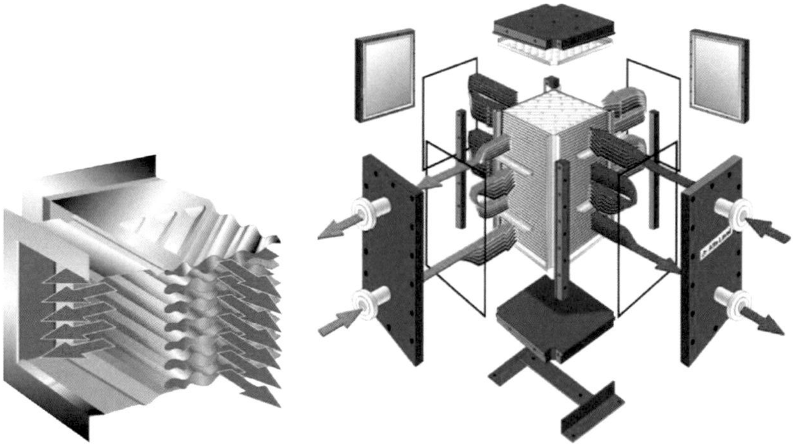

(a) (b)

Figure 3.31 Compabloc™ PHE of Alfa Laval production: (a) schematic drawing of a plate stack and (b) exploded view showing cross-countercurrent arrangement (courtesy of OAO Alfa Laval Potok).

Figure 3.32 Compabloc™ PHE of Alfa Laval production installed by AO Spivdruzhnist-T LLC at CHP unit in Sochi, Russian Federation.

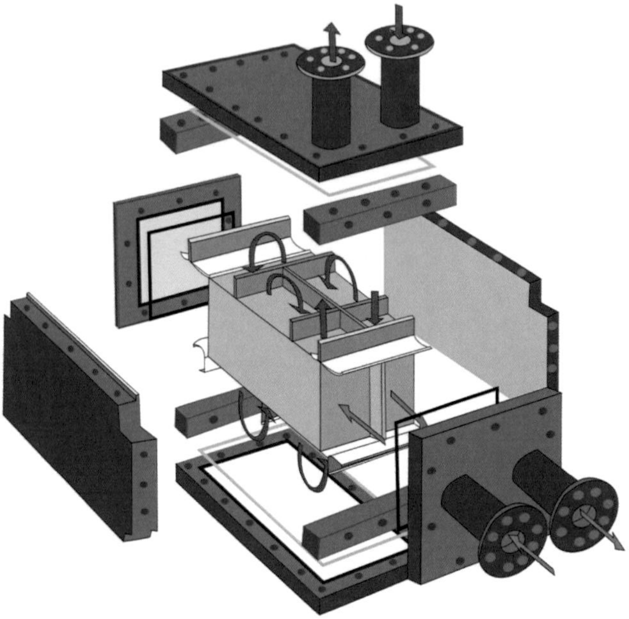

Figure 3.33 APV Hybrid PHE: exploded view showing cross-countercurrent arrangement (after SPX, 2012).

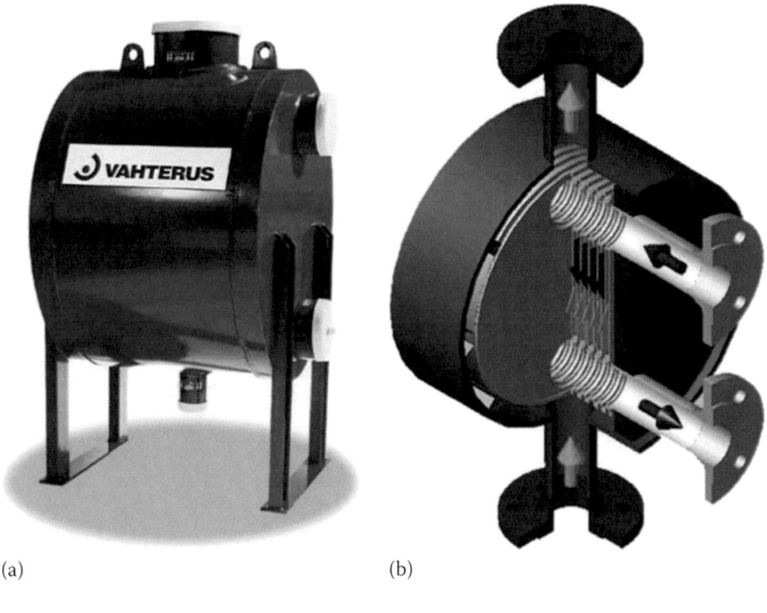

(a) (b)

Figure 3.34 Plate & Shell® HE: (a) assembled PSHE and (b) flow arrangement in one pass unit (after Vahterus Oy, 2013).

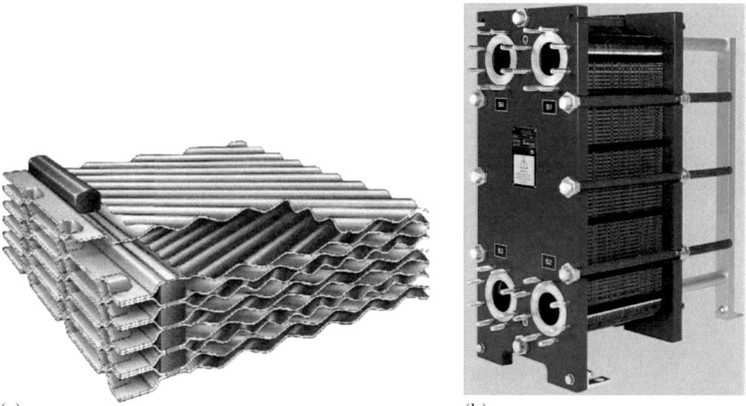

(a) (b)

Figure 3.39 Semiwelded PHE manufactured by Alfa Laval: (a) plate pack from welded plate pairs and (b) semiwelded PHE (courtesy of OAO Alfa Laval Potok).

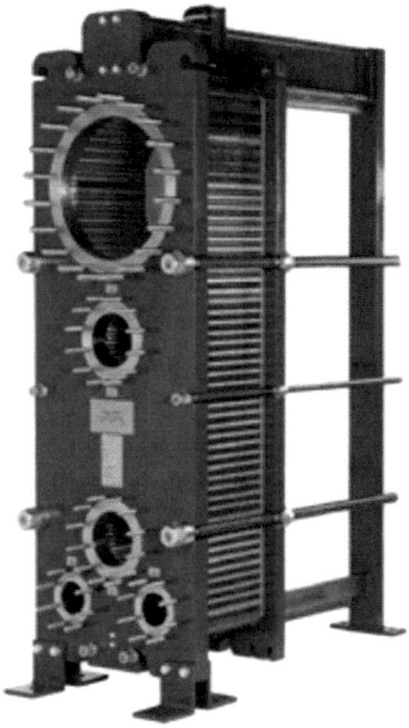

Figure 3.40 Semiwelded PHE AlfaCond manufactured by Alfa Laval for condensing duties (courtesy of OAO Alfa Laval Potok).

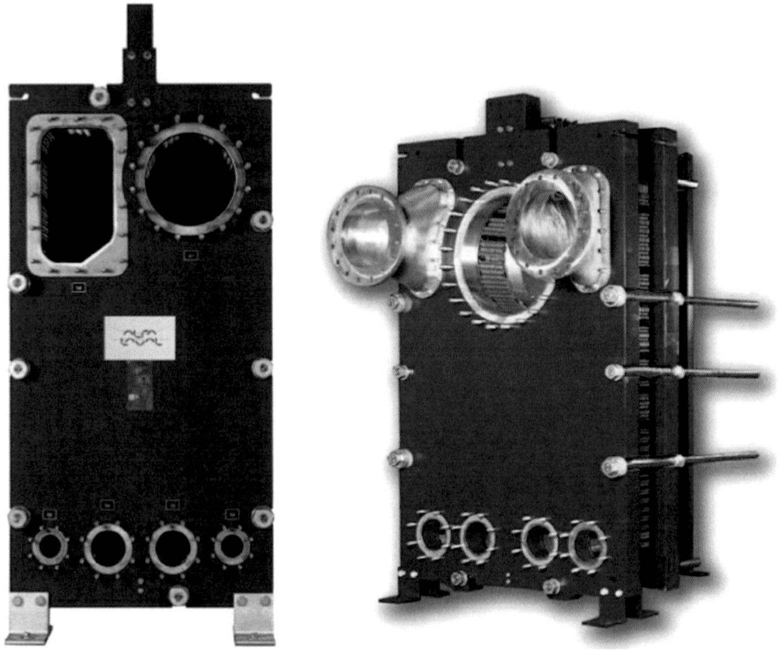

(a) (b)

Figure 3.41 Evaporator semiwelded PHE AlfaVap: (a) AlfaVap-500 and (b) AlfaVap-700 (courtesy of OAO Alfa Laval Potok).

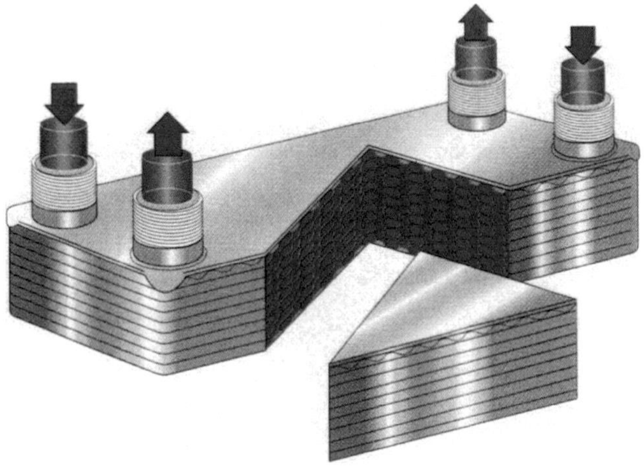

Figure 3.42 Drawing of BPHE showing the fluid flows in brazed plate pack (courtesy of OAO Alfa Laval Potok).

(a) (b)

Figure 3.43 Some of BPHEs produced by Alfa Laval: (a) general-purpose BPHEs and (b) special design for oil cooling (courtesy of OAO Alfa Laval Potok).

Figure 3.45 Graphite PHE (courtesy of OAO Alfa Laval Potok).

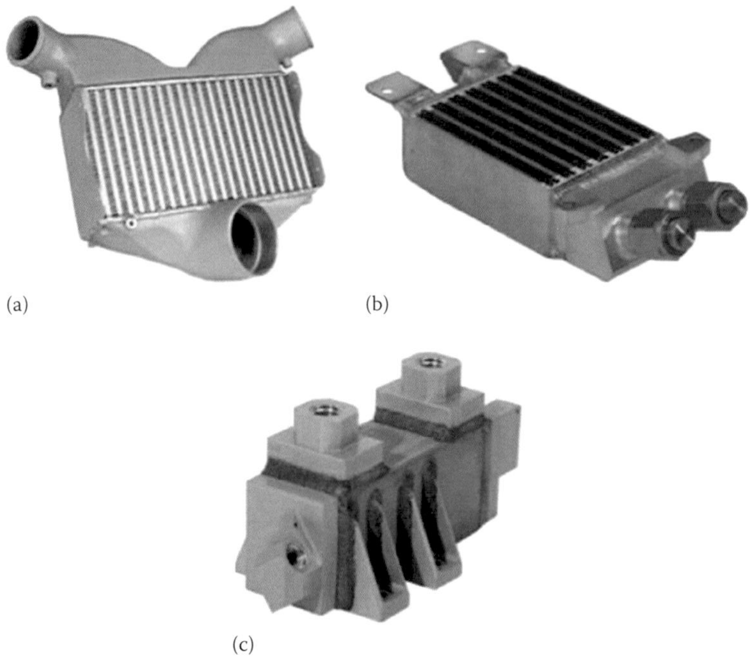

Figure 3.48 Custom-designed PFHE produced by Lytron, USA: (a) PFHE for air-to-air cooling, (b) PFHE for liquid cooling by air in an airplane and (c) liquid-to-liquid oil cooler for commercial aircraft (after Lytron, 2012).

Figure 3.49 Chart's aluminum-brazed PFHE (after Chart, 2014).

(a) (b)

Figure 3.50 Example of (a) finned tube and (b) TFHE made of such tubes (manufactured by Defon Heat Exchanger Co., Ltd, in China, see Defon, 2013).

Figure 3.51 SHEs (courtesy of OAO Alfa Laval Potok).

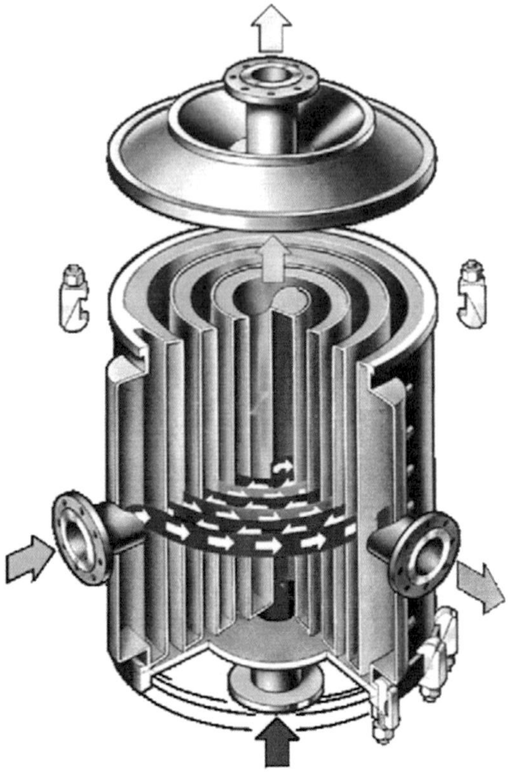

Figure 3.52 Fluid flow in a liquid–liquid SHE (courtesy of OAO Alfa Laval Potok).

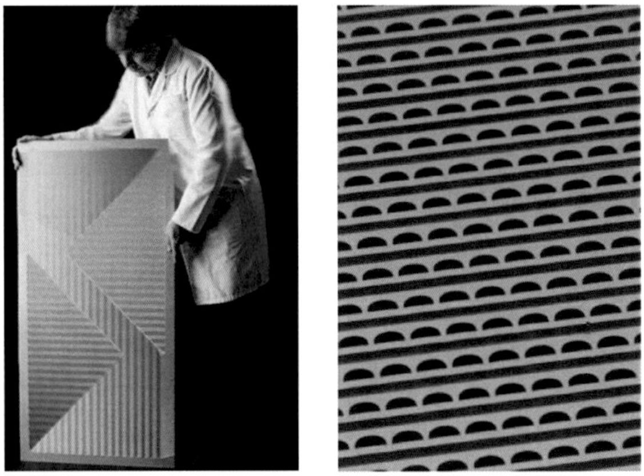

(a) (b)

Figure 3.54 Example of (a) PCHE Heatric™ plate and (b) block cross section (from Heatric, 2014).

Figure 3.55 Example of PCHE Heatric™ for oil and gas processing industry (from Heatric, 2014).

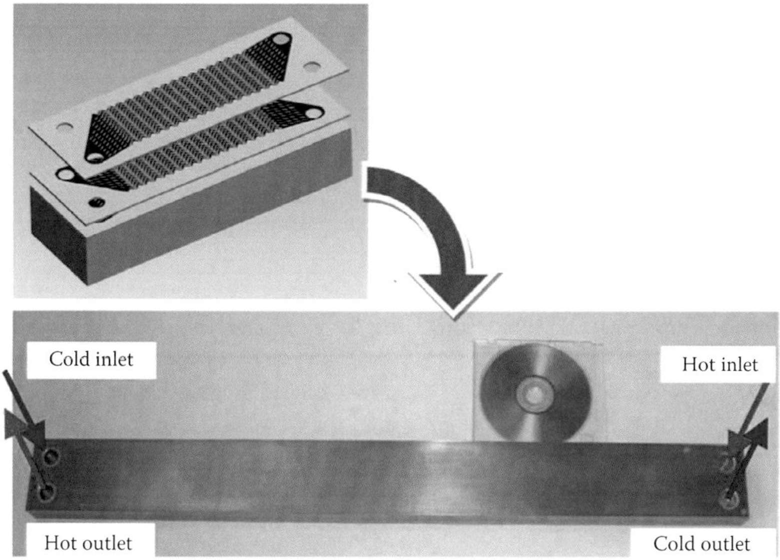

Figure 3.56 PCHE for water heating by CO_2 as heat pump working fluid (after Tsuzuki, 2009).

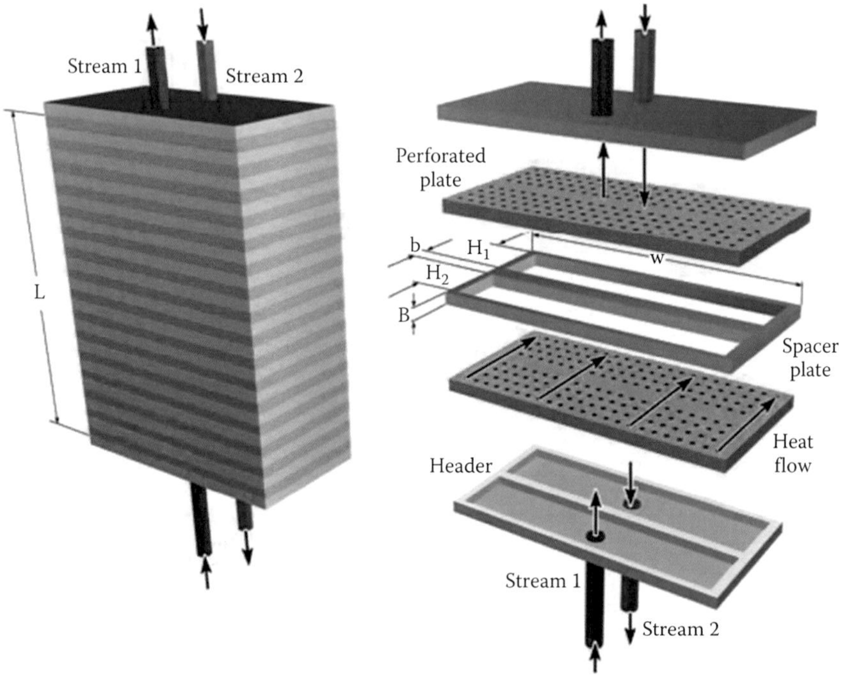

Figure 3.57 The construction of matrix MCHE (after Moheisen, 2009).

Figure 3.60 Monoblock™ heat exchanger produced by AB Segerfrojd (www.segerfrojd.com) (after Zaheed and Jachuck, 2004).

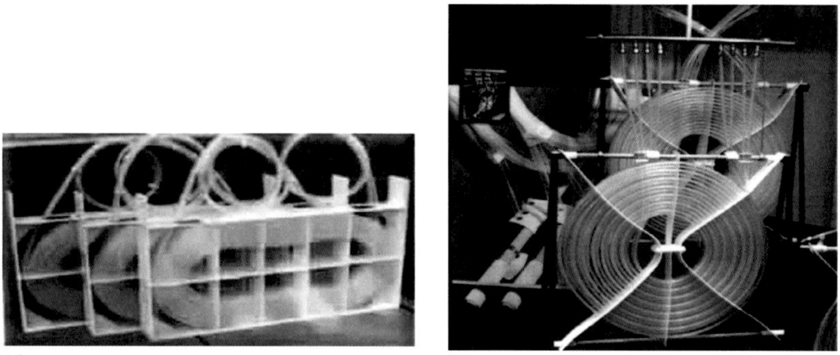

(a) (b)

Figure 3.61 Fluorotherm immersion heating coils. (a) Polypropylene frame and (b) PTFE frame (after Zaheed and Jachuck, 2004).

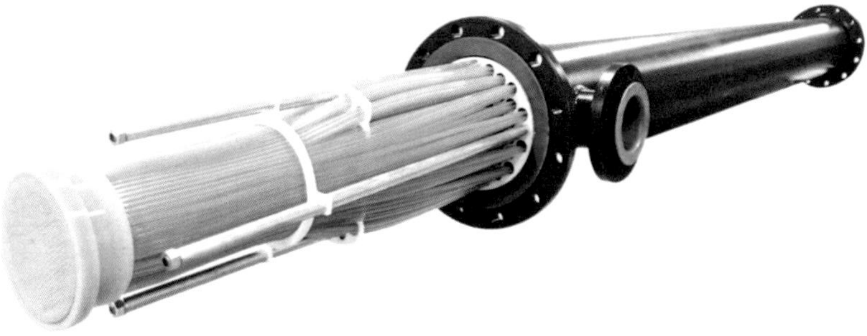

Figure 3.62 Ametek shell and tube teflon heat exchanger (after Ametek, 2010).

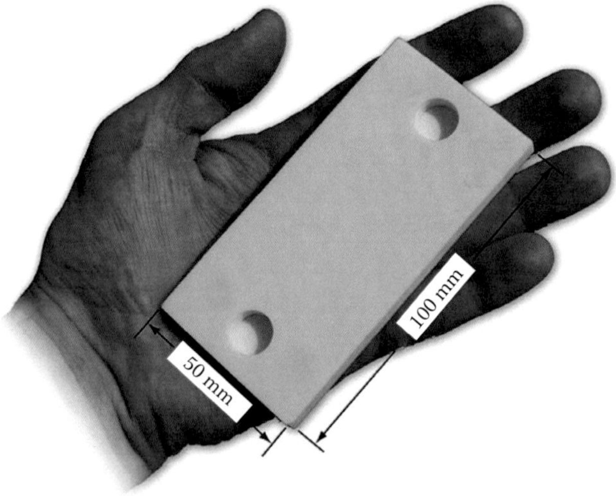

Figure 3.63 Photograph of a compact ceramic microchannel counterflow heat exchanger. This view shows the hot and cold inlets, with the exhaust flows connected from the bottom side (after Kee et al., 2004, 2011).

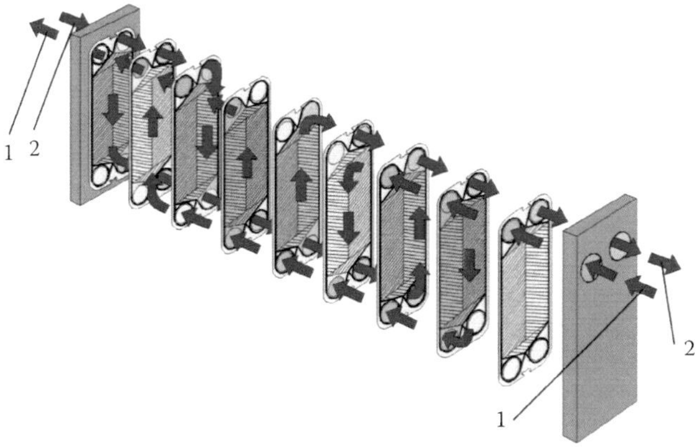

Figure 3.22 Schematic drawing of two-pass frame-and-plate PHE: 1, hot stream and 2, cold stream (courtesy of OAO Alfa Laval Potok).

Figure 3.25 District heating substation module with plate-and-frame PHE produced with Alfa Laval plates by AO Spivdruzhnist-T LLC in Kharkiv, Ukraine.

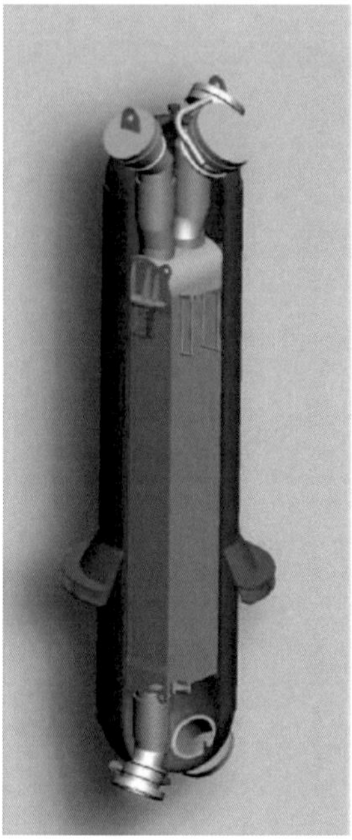

Figure 3.28 Drawing of Alfa Laval Packinox welded PHE (courtesy of OAO Alfa Laval Potok).

Figure 3.29 Alfa Laval AlfaRex welded PHE with cut of welded plate pack (courtesy of OAO Alfa Laval Potok).

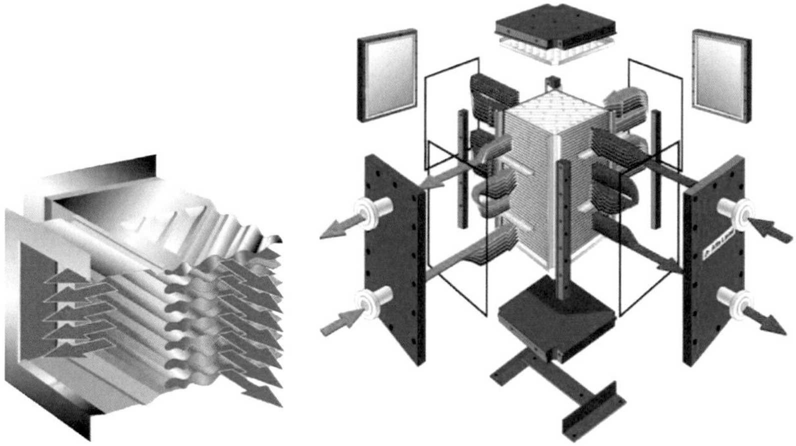

(a) (b)

Figure 3.31 Compabloc™ PHE of Alfa Laval production: (a) schematic drawing of a plate stack and (b) exploded view showing cross-countercurrent arrangement (courtesy of OAO Alfa Laval Potok).

Figure 3.32 Compabloc™ PHE of Alfa Laval production installed by AO Spivdruzhnist-T LLC at CHP unit in Sochi, Russian Federation.

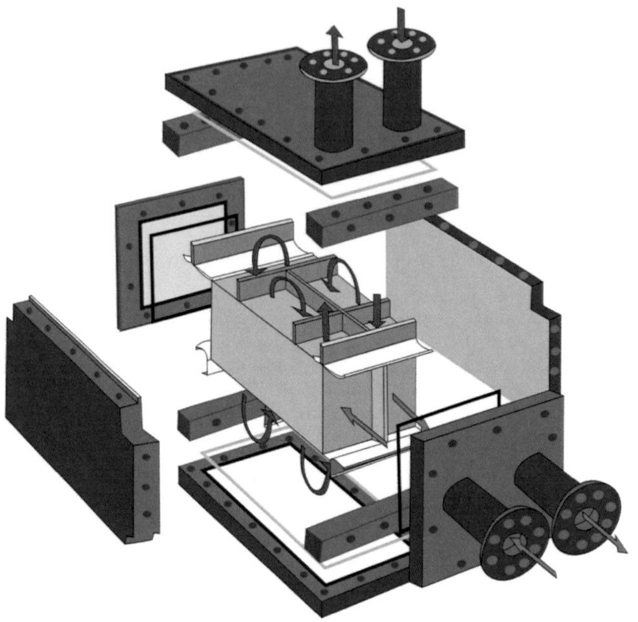

Figure 3.33 APV Hybrid PHE: exploded view showing cross-countercurrent arrangement (after SPX, 2012).

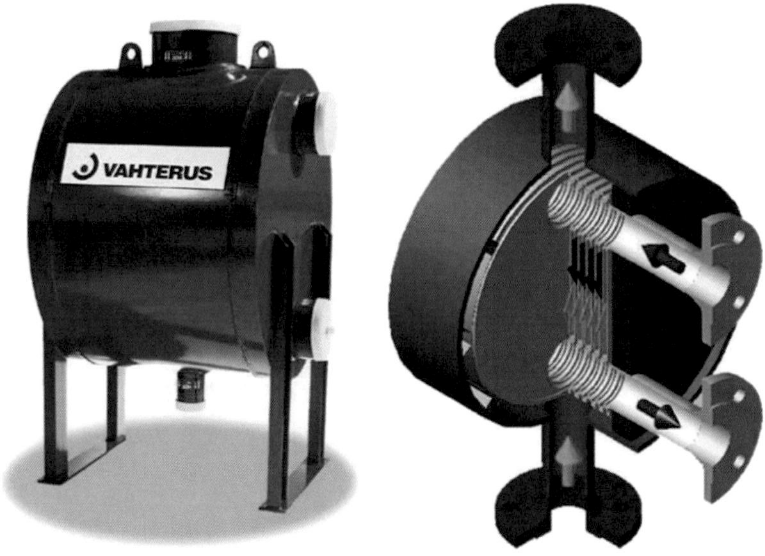

(a) (b)

Figure 3.34 Plate & Shell® HE: (a) assembled PSHE and (b) flow arrangement in one pass unit (after Vahterus Oy, 2013).

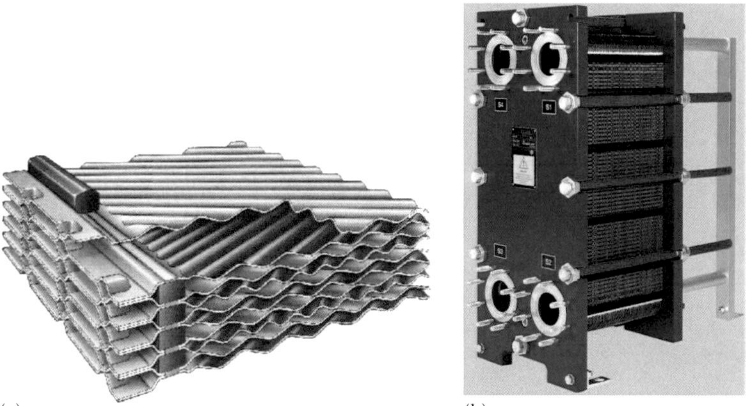

(a) (b)

Figure 3.39 Semiwelded PHE manufactured by Alfa Laval: (a) plate pack from welded plate pairs and (b) semiwelded PHE (courtesy of OAO Alfa Laval Potok).

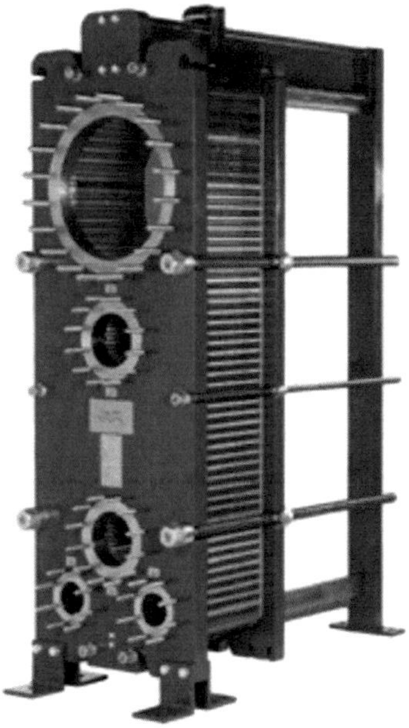

Figure 3.40 Semiwelded PHE AlfaCond manufactured by Alfa Laval for condensing duties (courtesy of OAO Alfa Laval Potok).

(a) (b)

Figure 3.41 Evaporator semiwelded PHE AlfaVap: (a) AlfaVap-500 and (b) AlfaVap-700 (courtesy of OAO Alfa Laval Potok).

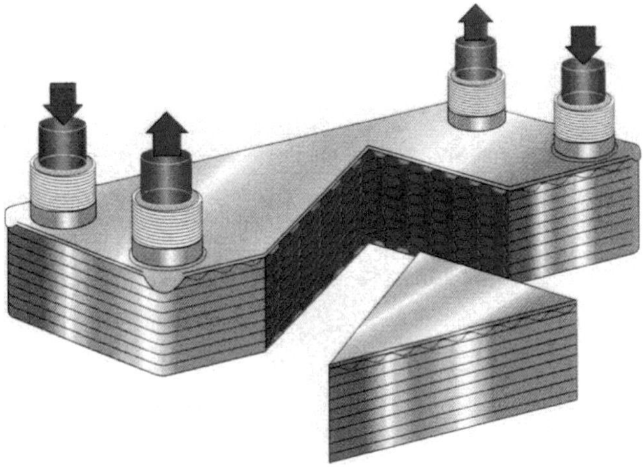

Figure 3.42 Drawing of BPHE showing the fluid flows in brazed plate pack (courtesy of OAO Alfa Laval Potok).

(a) (b)

Figure 3.43 Some of BPHEs produced by Alfa Laval: (a) general-purpose BPHEs and (b) special design for oil cooling (courtesy of OAO Alfa Laval Potok).

Figure 3.45 Graphite PHE (courtesy of OAO Alfa Laval Potok).

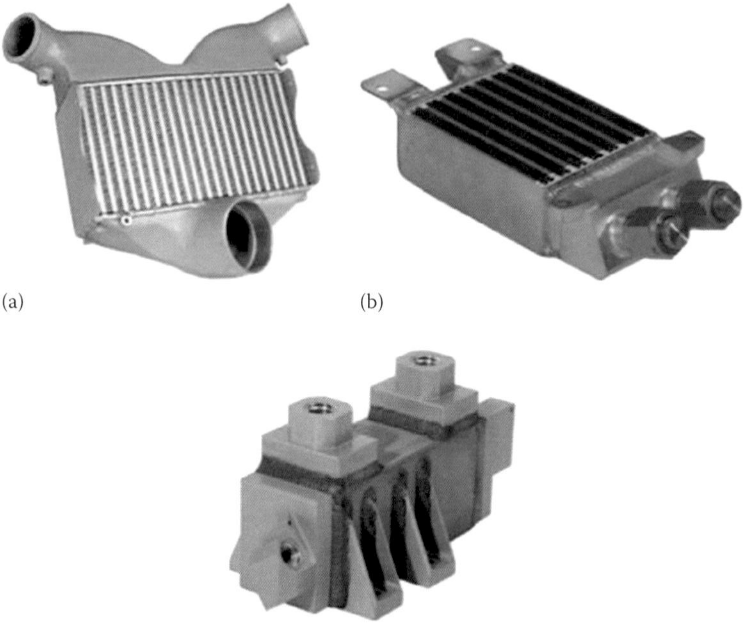
(a) (b) (c)

Figure 3.48 Custom-designed PFHE produced by Lytron, USA: (a) PFHE for air-to-air cooling, (b) PFHE for liquid cooling by air in an airplane and (c) liquid-to-liquid oil cooler for commercial aircraft (after Lytron, 2012).

Figure 3.49 Chart's aluminum-brazed PFHE (after Chart, 2014).

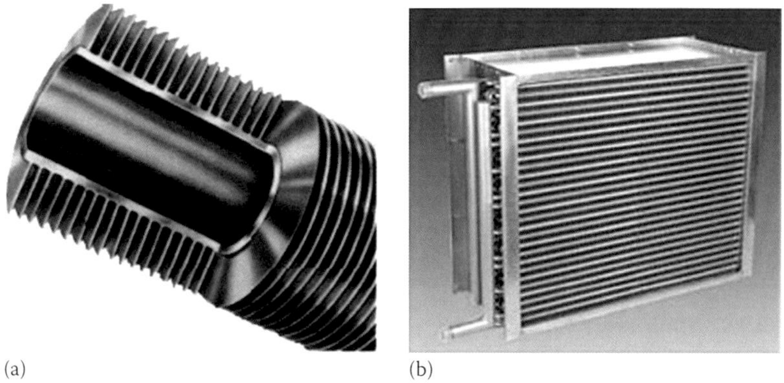

Figure 3.50 Example of (a) finned tube and (b) TFHE made of such tubes (manufactured by Defon Heat Exchanger Co., Ltd, in China, see Defon, 2013).

Figure 3.51 SHEs (courtesy of OAO Alfa Laval Potok).

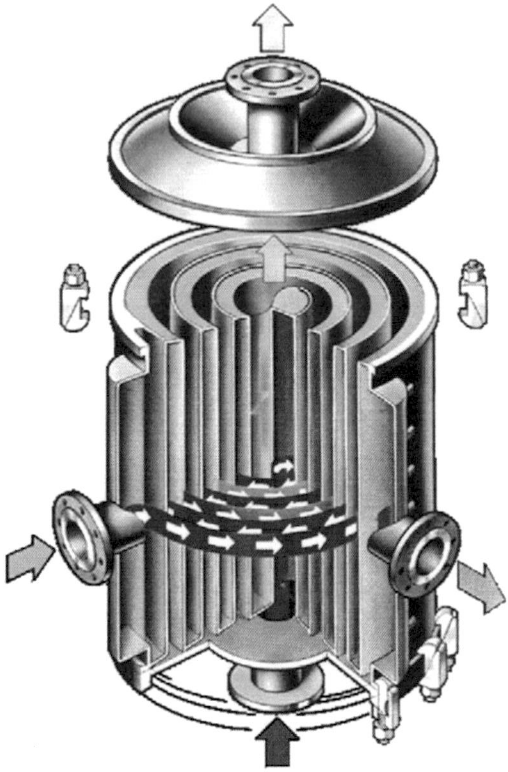

Figure 3.52 Fluid flow in a liquid–liquid SHE (courtesy of OAO Alfa Laval Potok).

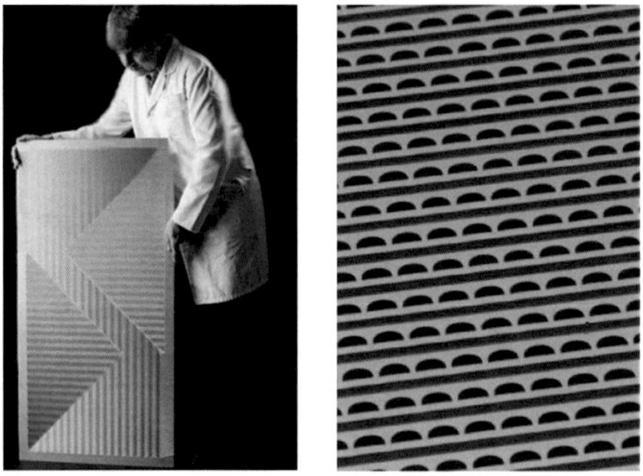

(a) (b)

Figure 3.54 Example of (a) PCHE Heatric™ plate and (b) block cross section (from Heatric, 2014).

Figure 3.55 Example of PCHE Heatric™ for oil and gas processing industry (from Heatric, 2014).

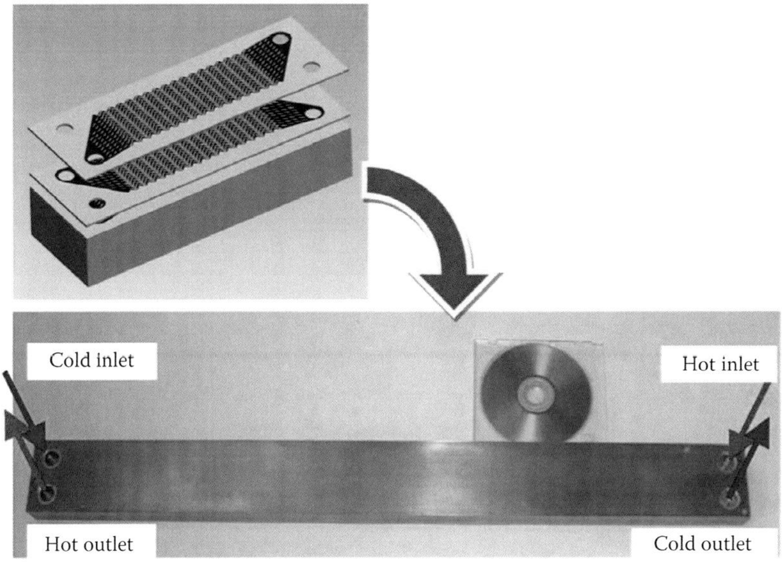

Figure 3.56 PCHE for water heating by CO_2 as heat pump working fluid (after Tsuzuki, 2009).

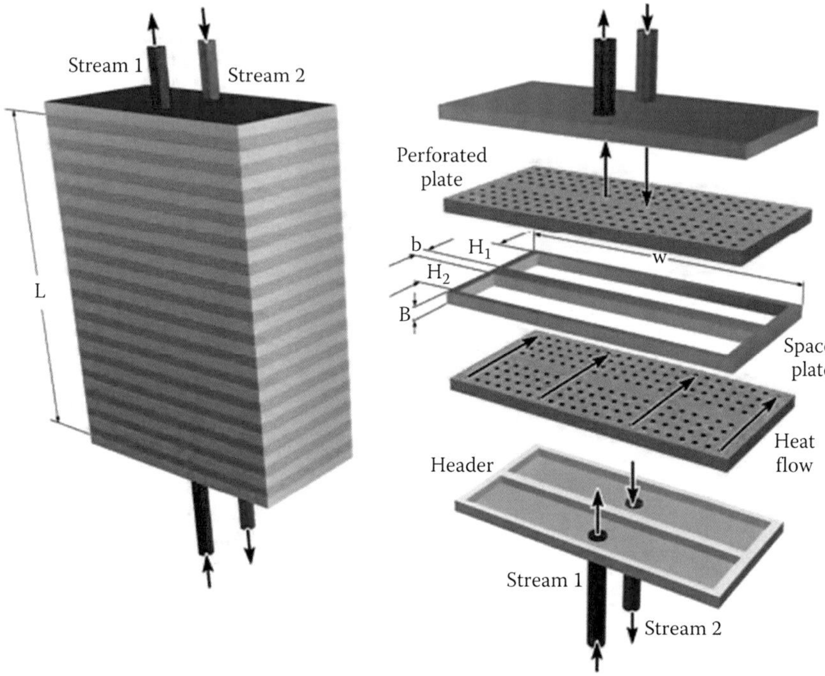

Figure 3.57 The construction of matrix MCHE (after Moheisen, 2009).

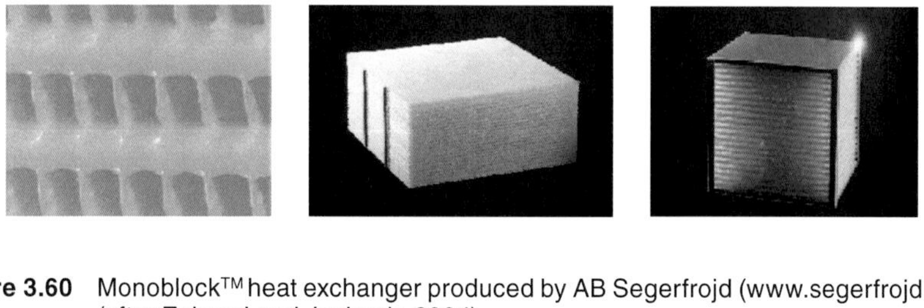

Figure 3.60 Monoblock™ heat exchanger produced by AB Segerfrojd (www.segerfrojd.com) (after Zaheed and Jachuck, 2004).

(a) (b)

Figure 3.61 Fluorotherm immersion heating coils. (a) Polypropylene frame and (b) PTFE frame (after Zaheed and Jachuck, 2004).

Figure 3.62 Ametek shell and tube teflon heat exchanger (after Ametek, 2010).

Figure 3.63 Photograph of a compact ceramic microchannel counterflow heat exchanger. This view shows the hot and cold inlets, with the exhaust flows connected from the bottom side (after Kee et al., 2004, 2011).

Condensation takes place in many fields where heat transfer processes involving gaseous phase undergoing change to liquid phase occur. It involves the utilisation of the latent heat of phase change that would be lost without the condensation. These include refrigeration plants, sugar-refinery evaporators, desalination and evaporation stations at many brunches of chemical industry, top condensers of distillation columns, condensing steam after power generation turbines and vapour in organic power generation cycles, air conditioning systems and a number of other applications. It is very important in utilisation of low-grade heat discharged with exhaust gases which contain steam that can be condensed with cooling to sufficiently low temperature. Enhancement of condensation has significant effect on overall heat exchanger performance when thermal resistance on the condensation side is comparable to or greater than the thermal resistance for discharging the heat through the heat transfer wall to cooling media. In a number of applications, it can be the case with intensified heat transfer on the cooling side. In processes with condensation of steam mixture with non-condensable gases, the condensation heat transfer coefficient can be several times less than that on the cooling side.

When saturated at given pressure vapour comes to a contact with surface which temperature is lower than saturation temperature of this vapour, the condensation of vapour with its transformation to liquid phase happens and latent heat of phase change is released. There exist two different forms of condensation: *film condensation* and *dropwise condensation*. The later one happens on hydrophobic surfaces which are not wetted by condensed liquid, and it occurs when droplets with free spaces between them are formed. The vapour on these spaces is in direct contact with the surface without any considerable additional thermal resistance to flow of released latent heat. On vertical or inclined surfaces, droplets are quickly going down and large parts of the surface remain free for contact with vapour ensuring very high heat transfer coefficients on the condensation side. Dropwise condensation is regarded as enhanced condensation itself and methods of special surface treatments, vapour additives or creating ultra-thin coatings promoting dropwise condensation are generally regarded as methods of condensation enhancement. The review of methods promoting dropwise condensation was presented by Ma et al. (2000). Some new developments with dropwise–filmwise hybrid surfaces are reported by Peng et al. (2014) and heat transfer enhancement in mini-channel by Derby et al. (2014). But still it is hard to be implemented in industry and its application in low-grade heat utilisation is very problematic and is outside of the scope of this book.

In most practical applications, condensate wets the surface and forms a liquid film on it which creates thermal resistance to heat flow from film outer boundary on which vapour is condensing to the surface of the wall, which in heat exchanger is cooled from the other side by cooling media. In case of pure saturated vapour, this thermal resistance is the main factor limiting heat transfer from vapour to the wall surface, except the rather rare cases of very low pressure and slip flows when effects in Knudsen boundary layer should be accounted for. All methods of pure saturated vapour condensation enhancement are directed on decreasing the thermal resistance of condensate film by making it thinner using surface tension forces, disturbing flow in it or acting on it with gas-phase shear forces. When gas phase is a mixture of different vapours or non-condensable gas is present, the process is more complicated with diffusion phenomena from the main gas flow to condensate film surface.

4.2.1.1 Film Condensation of Slow-Moving Saturated Vapour on a Smooth Surface

The simple case of laminar condensate film flow on vertical surface was analysed by Nusselt in 1916 using differential equation for condensate movement in laminar film (see Figure 4.37):

$$g\cdot\left(\rho_L-\rho_G\right)+\mu\cdot\frac{\partial^2 w_x}{\partial y^2}=0 \tag{4.97}$$

Here, the gravity force of condensate elementary $g(\rho_L - \rho_G)$ volume is compensated by viscosity force acting from neighbouring liquid layers. The inertia forces are neglected; g is the specific gravity, m/s^2; ρ_L and ρ_G are densities of liquid and

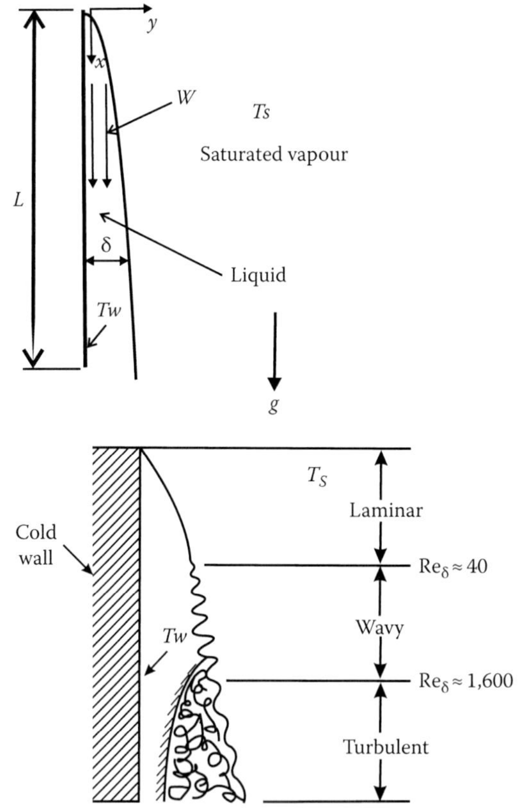

Figure 4.37 Film condensation of saturated slow vapour on a vertical wall surface.

gas phases, respectively, kg/m^3; w_x is the velocity component in "x" direction; μ_L is dynamic viscosity of liquid phase, Pa·s. Integrating Equation 4.97 gives:

$$w_x = \frac{g \cdot (\rho_L - \rho_G)}{2 \cdot \mu_L} \cdot y^2 + C_1 \cdot y + C_2 \tag{4.98}$$

The boundary conditions are as follows: $w_x = 0$ at $y = 0$ and $\partial w_x / \partial y = 0$ at $y = \delta$. From these conditions follow $C_2 = 0$ and $C_1 = g(\rho_L - \rho_G) \cdot \delta / \mu_L$. Substituting these to Equation 4.98 gives

$$w_x = \frac{g \cdot (\rho_L - \rho_G)}{\mu_L} \cdot \left(y \cdot \delta - 0.5 \cdot y^2 \right) \tag{4.99}$$

With this velocity distribution, the average velocity of liquid condensate in cross section with "x" coordinate is

$$\overline{w}_x = \frac{1}{\delta} \int_0^\delta w_x dy = \frac{g \cdot (\rho_L - \rho_G)}{3 \cdot \mu_L} \cdot \delta^2 \tag{4.100}$$

Then the Reynolds number for condensate film is

$$\text{Re}_{Lx} = \frac{\delta \cdot \rho_L \cdot \overline{w}_x}{\mu_L} = \frac{g \cdot \rho_L \cdot (\rho_L - \rho_G)}{3 \cdot \mu_L^2} \cdot \delta^3 \tag{4.101}$$

The local film heat transfer coefficient is

$$h_{Lx} = \frac{\lambda_L}{\delta} = \frac{1}{\sqrt[3]{3}} \cdot \sqrt[3]{\frac{g \cdot \rho_L \cdot (\rho_L - \rho_G) \cdot \lambda_L^3}{\mu_L^2}} \cdot \text{Re}_{Lx}^{\frac{-1}{3}} \tag{4.102}$$

Here, λ_L is the thermal conductivity of liquid phase, W/(m·K).

The flow regime in condensate film remains strictly laminar for the Reynolds numbers Re_L lower than about 10. At higher Reynolds numbers, the waves forming at the liquid–vapour interface is observed although the flow in liquid film remains laminar. It is *wavy laminar* flow at which waves tend to increase heat transfer. The waves also complicate the analytical solutions. As discussed by Kutateladze (1979), the increase in heat transfer due to wave formation is about up to 20% for the Reynolds numbers below 100 and up to 50% and more at the Reynolds numbers 400 and higher when flow becomes turbulent.

The Reynolds number can be expressed using the flow rate G_{Lx} of liquid phase through the film cross section with width "b" at coordinate "x":

$$\text{Re}_{Lx} = \frac{G_{Lx}}{b \cdot \mu_L} \tag{4.103}$$

The flow rate G_{Lx} is gradually increasing along the surface as more vapour is condensed. It can be calculated through distribution along the surface of heat flux $q(x)$ using latent heat of vaporisation r in J/kg and neglecting sensitive heat discharged in the process:

$$G_{Lx} = \int_0^x \frac{q(x) \cdot b \cdot dx}{r} \tag{4.104}$$

With increase of liquid flow rate in the condensate film, the flow regime is changing from laminar to wavy laminar and turbulent. It is much complicating the estimation of average film heat transfer coefficients. These features can be accounted by numerical calculation on increments of heat transfer surface. The analytical solutions are possible for laminar flow with calculation by Equation 4.102 with assumption on heat flux or temperature distributions.

For example, assuming that heat flux q_c through the unit of the surface width is not changing along the surface, from Equation 4.104 follows:

$$G_{Lx} = \frac{q_c \cdot b \cdot x}{r} \tag{4.105}$$

Substituting it into Equation 4.103 and then into Equation 4.102 gives as follows:

$$h_{Lx} = \lambda_L \cdot \sqrt[3]{\frac{r \cdot g \cdot \rho_L \cdot (\rho_L - \rho_G)}{3 \cdot \mu_L \cdot q_c}} \cdot x^{\frac{-1}{3}} \tag{4.106}$$

The average value of film heat transfer coefficient on the surface with height H is:

$$\bar{h}_{LH} = \frac{1}{H} \cdot \int_0^H h_{Lx} \cdot dx = \frac{3}{2} \cdot \lambda_L \cdot \sqrt[3]{\frac{r \cdot g \cdot \rho_L \cdot (\rho_L - \rho_G)}{3 \cdot \mu_L \cdot q_c \cdot H}} \tag{4.107}$$

Expressing this relation through the Reynolds number gives as follows:

$$\bar{h}_{LH} = 1.04 \cdot \lambda_L \cdot \sqrt[3]{\frac{g \cdot \rho_L \cdot (\rho_L - \rho_G)}{\mu_L^2}} \cdot \mathrm{Re}_{LH}^{-1/3} \tag{4.108}$$

In his analysis, Nusselt made an assumption about constant wall temperature T_w, from which follows that the difference between vapour saturation temperature T_s and wall temperature ΔT is also constant. At the position x, the increase of liquid phase flow rate on the element of surface with length dx and width b is equal to the amount of condensed vapour:

$$dG_{Lx} = \frac{\Delta T}{r} \cdot \frac{\lambda}{\delta} \cdot b \cdot dx \tag{4.109}$$

On the other hand, accounting for Equation (4.100):

$$dG_{Lx} = b \cdot d\left(\bar{w}_x \cdot \delta \cdot \rho_L\right) = b \cdot d\left(\frac{g \cdot \rho_L \cdot \left(\rho_L - \rho_G\right)}{3 \cdot \mu_L} \cdot \delta^3\right)$$
$$= \frac{g \cdot \rho_L \cdot \left(\rho_L - r_G\right) \cdot 3}{3 \cdot \mu_L} \cdot b \cdot \delta^2 \cdot d\delta \tag{4.110}$$

From Equations 4.109 and 4.110 follows:

$$\frac{g \cdot \rho_L \cdot \left(\rho_L - \rho_G\right)}{\mu_L} \cdot \delta^3 \cdot d\delta = \frac{\Delta T}{r} \cdot \lambda \cdot dx \tag{4.111}$$

After integration, it follows that

$$\frac{g \cdot \rho_L \cdot \left(\rho_L - \rho_G\right)}{4 \cdot \mu_L} \cdot \delta^4 = \frac{\Delta T}{r} \cdot \lambda \cdot x + C \tag{4.112}$$

The integration constant $C = 0$ as $\delta = 0$ at $x = 0$. Therefore, the local film heat transfer coefficient at the position "x" can be calculated as follows:

$$h_{Lx} = \frac{\lambda_L}{\delta} = \sqrt[4]{\frac{r \cdot g \cdot \rho_L \cdot \left(\rho_L - \rho_G\right) \cdot \lambda_L^3}{4 \cdot \mu_L \cdot \Delta T \cdot x}} \tag{4.113}$$

The average value of film heat transfer coefficient on the surface with height H is

$$\bar{h}_{LH} = \frac{1}{H} \cdot \int_0^H h_{LH} \cdot dx = \frac{4}{3} \cdot \sqrt[4]{\frac{r \cdot g \cdot \rho_L \cdot \left(\rho_L - \rho_G\right) \cdot \lambda_L^3}{4 \cdot \mu_L \cdot \Delta T \cdot H}}$$
$$= 0.943 \cdot \sqrt[4]{\frac{r \cdot g \cdot \rho_L \cdot \left(\rho_L - \rho_G\right) \cdot \lambda_L^3}{\mu_L \cdot \Delta T \cdot H}} \tag{4.114}$$

The average heat transfer coefficient is determined also by the following equation:

$$\bar{h}_{LH} = \frac{Q}{\Delta T \cdot H \cdot b} = \frac{G_{LH} \cdot r}{\Delta T \cdot H \cdot b} \tag{4.115}$$

Expressing from this equation ($\Delta T \cdot H$) and substituting it to Equation 4.114 accounting for Equation 4.103 after some rearranging and taking in power 3/4 it follows that

$$\bar{h}_{LH} = 0.925 \cdot \lambda_L \cdot \sqrt[3]{\frac{g \cdot \rho_L \cdot \left(\rho_L - \rho_G\right)}{\mu_L^2}} \cdot \mathrm{Re}_{LH}^{-1/3} \tag{4.116}$$

The comparison with Equation 4.108 obtained for assumption of constant heat flux shows that the resulting average film heat transfer coefficient is smaller on 12.5%. It is shown that even in this simplest case of condensation process, the attempt to average heat transfer coefficient can lead to substantial error. For accurate prediction of heat exchanger performance with processes of phase change, it is required to consider the actual distribution of process local parameters along the heat transfer surface.

The derived equations are also applicable for condensation on inner and outer surfaces of vertical smooth tubes with slow-moving vapour and diameter high enough to neglect surface tension forces. These equations are also frequently used to estimate the condensation heat transfer enhancement by different methods. There are also correlations developed for film condensation inside horizontal and inclined tubes, different tube bundles of shell-and-tube heat exchangers which can be found in literature and used for design of conventional heat exchangers.

4.2.1.2 Enhancement of the Film Condensation with Vapour Action

The condensation in confined channels of relatively big length like tubes or channels of compact heat exchangers is not feasible with small vapour velocity, especially at high condensation rate. All vapour that will be condensed after some cross section should go through with considerable velocity that is maximal near the channel entrance, decreasing along the channel as gaseous phase transforms to liquid phase of much bigger density. The dynamical interaction of high-velocity vapour with condensate film is leading to considerable increase of heat transfer intensity through that film by making it thinner and catching some liquid to the vapour stream, which becomes a mixture of gaseous and liquid phases. It creates dispersed-annular pattern of two-phase flow on most parts of the channel length with liquid film at the channel walls and gas–liquid mixture at the central part of the channel as shown in Figure 4.38.

The flow structure in turbulent film can be considered similar to the turbulent flow in regions near the wall of liquid flowing in the same channel. As it was discussed

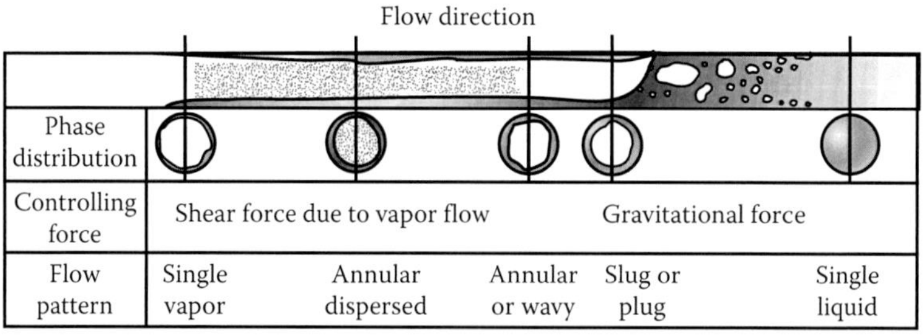

Figure 4.38 Condensation flow structures along the channel.

earlier for the single-phase turbulent flow for Pr ≥ 1 in viscous and buffer sublayers at near-wall region, the main part of the thermal resistance to heat transfer is concentrated. Boyko and Kruzhilin (1967) have developed and experimentally confirmed approximate approach of estimating film heat transfer coefficients for condensation of steam inside tubes based on the Reynolds analogy of heat and momentum transfer which determines the link between heat transfer and wall shear stress.

When all gas phase is liquefied, the velocity of liquid in channel is determined as follows:

$$W_L^* = \frac{G}{\rho_L \cdot A_{ch}} \tag{4.117}$$

where

G is the total mass flow rate of two-phase flow, kg/s

A_{ch} is the channel cross-sectional area, m²

The shear stress at the wall is

$$\tau_w^* = \frac{1}{8} \cdot \zeta^* \cdot \rho_L \cdot W_L^{*2} = \frac{1}{8} \cdot \zeta^* \cdot \frac{G^2}{\rho_L \cdot A_{ch}^2} \tag{4.118}$$

As the rapidly moving vapour affects the condensate film in a channel, under conditions of high friction shear stresses on the interface, there is an intensive carry-over of liquid drops away from the condensate film to the flow core. It is suggestive of a dispersed-annular two-phase flow regime at a high concentration of dispersed liquid phase in the flow core in form of mixture with density ρ_m. The shear stress at the wall in this case is

$$\tau_w = \frac{1}{8} \cdot \zeta_m \cdot \frac{G^2}{\rho_m \cdot A_{ch}^2} \tag{4.119}$$

The expression of the Reynolds analogy by Equation 4.71 accounting for Equation 4.118 and assuming that $\mathrm{Pr}_L = 1$ for totally condensed liquid flow can be rearranged to the following:

$$h^* = \frac{\lambda_L \cdot W_L^* \cdot \zeta^*}{\nu_L \cdot 8} = \sqrt{\frac{\rho_L \cdot W_L^{*2} \cdot \zeta^*}{8}} \cdot \sqrt{\frac{c_{pL}^2 \cdot \zeta^*}{8}} = \sqrt{\tau_w^*}\sqrt{\frac{c_{pL}^2 \cdot \zeta^*}{8}} \tag{4.120}$$

In assumption that turbulent condensate film holds also the Reynolds analogy, it follows that

$$h = \sqrt{\tau_w}\sqrt{\frac{c_{pL}^2 \cdot \zeta_m}{8}} \tag{4.121}$$

The ratio of film heat transfer coefficients h and h^* accounting for expressions of wall shear stress by Equations 4.118 and 4.119 can be written as follows:

$$\frac{h}{h^*} == \sqrt{\frac{\rho_L}{\rho_m} \cdot \frac{\zeta_m}{\zeta^*}} \tag{4.122}$$

The friction factor ζ^* can be determined using single-phase correlation for the channel. There is no information about friction factor ζ_m for gas–liquid mixture, but as a first approximation, it can be assumed $\zeta_m = \zeta^*$. The ratio of liquid density to the density of homogeneous gas–liquid mixture can be expressed using mass vapour quality X (the ratio of gas-phase mass flow rate to total flow rate of two-phase mixture):

$$\frac{\rho_L}{\rho_m} = 1 + X\left(\frac{\rho_L}{\rho_g} - 1\right) \tag{4.123}$$

Therefore, the local film heat transfer coefficient for steam condensation expressed in dimensionless Nusselt number and properties of liquid phase can be determined as follows:

$$\text{Nu} = \text{Nu}^* \cdot \sqrt{1 + X\left(\frac{\rho_L}{\rho_g} - 1\right)} \tag{4.124}$$

The average film heat transfer coefficient in channel or part of its length can be calculated from its local values at inlet ($_{in}$) and exit ($_{ex}$) as follows:

$$\overline{\text{Nu}} = \overline{\text{Nu}^*} \cdot \frac{1}{2} \cdot \left(\sqrt{1 + X_{in}\left(\frac{\rho_L}{\rho_g} - 1\right)_{in}} + \sqrt{1 + X_{ex}\left(\frac{\rho_L}{\rho \lg} - 1\right)_{ex}} \right) \tag{4.125}$$

Here, Nu* is the Nusselt number calculated with correlation for heat transfer in single-phase flow of liquid with flow rate equal to total flow rate of two phases.

As it is shown by Kutateladze (1979), these equations can be derived also based on the Prandtl analogy of heat and momentum transfer with division of turbulent flow on viscous sublayer and turbulent flow core. The accuracy of Equations (4.124) and (4.125) was confirmed experimentally for condensation of high-velocity steam inside horizontal tubes (see, e.g., Boyko and Kruzhilin, 1967) and vertical smooth tubes (see, e.g., Isachenko, 1977). However, for smooth tubes made from different metals, the calculated values of the Nusselt number can be from 12% (for steel) up to 50% (for copper) higher than predicted by the equations using same correlation for single-phase flow. It is accounted by introduction of corresponding correction factor to single-phase correlation for Nu^*. Boyko and Kruzhilin (1967) explained these phenomena that increasing the friction on the wall in two-phase flow compared to two single-phase flows according to Equation 4.122 can cause such effect. It is possible also the influence of different roughness on the walls from different metals.

The validity of Equation 4.124 for channels of compact PHEs with stainless steel plates was experimentally confirmed by Tovazhnyansky and Kapustenko (1978). Contrary to condensation inside smooth tubes, about 7% decrease of heat transfer coefficients compared to predicted with Nu* taken by correlation for single-phase flow was observed. The same observation about somewhat lower values of heat transfer coefficients than predicted by Equation 4.124 was pointed out by Wang et al. (2000) who reported about experiments on steam condensation with seven different PHEs including plate-and-frame and brazed BHEs with different corrugation inclination angle 30°, 45° and 60°. They have proposed to introduce correction factor into power at density ratio:

$$\frac{h}{h^*} = \left(\frac{\rho_L}{\rho_m}\right)^{a+b\cdot \mathrm{Re}_L^c} \tag{4.126}$$

where the constants a, b and c are in the ranges 0.3–0.37, 5.0–6.0 and −0.6 to −0.64, respectively. Such complication can be good for correlation of experimental data, but for preliminary calculations with optimisation of the channel geometry with data only for single-phase flow, it is not much suitable, as it requires some general representation of these constants. To account for comparative reduction of heat transfer coefficients, Arsenyeva et al. (2011) proposed just small change to Equation 4.124 in a form

$$\mathrm{Nu} = \mathrm{Nu}^* \cdot \left[1 + X\left(\frac{\rho_L}{\rho_g} - 1\right)\right]^{0.48} \tag{4.127}$$

It preserves the convenience of using Equation 4.124 for condensation in channels of different geometries which requires only the information about heat transfer in single-phase flow at channel under consideration. As was surveyed by Wang et al. (2007), Equation 4.124 was used by the number of researchers for stepwise calculation of PHE condensers with local heat transfer characteristics on small portions of channel length. With its theoretical background, it enables the calculations for different vapours, like ammonia and other refrigerants, and develops computer models of complex systems using PHE condensers (see, e.g., Ventas et al., 2010). Knowing correlations for enhanced heat transfer in some channel, it enables to estimate the enhancement of condensation process. It should be considered that all discussed correlations earlier are accurately predicting process behaviour for small local zones of the channel. As process parameters are significantly changing with vapour condensation along the heat transfer surface, the calculations should be stepwise by small increments of the surface area along its length.

There are a number of approaches and correlations developed by different authors for estimation of heat transfer during the condensation in tubes and channels. The comprehensive survey of these works is published by Garimella in a book with Kandlikar et al. (2005). His example on comparison of calculations for condensation of refrigerant R134a in a smooth tube of 1 mm diameter using 10 different equations have shown that calculation with Equation 4.124 gives results within ±9% from other six compared with underestimation not more than 25% compared to three remaining.

Therefore, it can be recommended at least for preliminary estimation of heat transfer performance during condensation in channels giving results mostly on a safer side.

4.2.1.3 Enhanced Condensation Heat Transfer Surfaces

To enhance film condensation, it is necessary to reduce thermal resistance of the condensate film. With passive enhancement methods, it is possible with two mechanisms. One of these consist in creating high-flow turbulence with surface artificial roughness or boundary layer disruptions similar to heat transfer enhancement in single-phase flows. Another one is to reduce condensate film thickness by making use of surface tension forces and removing condensate from the surface.

In many applications of heat exchangers for vapour condensation, it is important to enhance heat transfer on both sides of heat transfer surface. Kalinin and Dreitser (1998) have reported about experimental study of condensation on horizontal rolled tubes with transverse ribs shown in Figure 4.8. The experiments were conducted with water, acetone and extraction gasoline vapours. It was concluded that irrespective of the thermophysical properties of the liquid, the presence of annular grooves causes a substantial increase of condensation film heat transfer coefficient from 1.8 to 3 times. The ribs on the horizontal tubes promote drainage of the condensate, a decrease in the film condensate thickness on the projecting parts of tubes and, accordingly, an increase in the intensity of vapour condensation heat transfer. Supported by enhanced heat transfer to the cooling media, it leads to the increase of the overall heat transfer coefficient more than two times. The experiments with mixture of vapours have shown similar results. The significant increase in overall heat transfer coefficients was also reported for condensation on vertical tubes with helical grooves outside the tubes and ribs inside. There are also presented empirical correlations for calculation of condensation film heat transfer coefficients in a range of tube geometrical parameters. Rifert and Smirnov (2004) provided results of the research on enhanced condensation on profiled tubes from the former USSR and also describe different theoretical models of the condensation process.

Nowadays, in-tube condensation of refrigerants is widely used in air-cooled and water-cooled condensers for water chillers, air conditioners and heat pumps. It has become common to enhance heat transfer during condensation inside horizontal tubes in condensers for air conditioners, in which the special microfinned tubes with fins of very small dimensions are employed. Such tubes of outside diameter from 4 to 15 mm are typically manufactured from copper. The helical fins are arranged usually in a single set of 50–70 fins with helix angle from 6° to 30° (see, e.g., Figure 4.39). The fin height can be from 0.1 to 0.25 mm; they have triangular or trapezoidal shapes with an apex angle from 25° to 90°. There are also other enhanced surfaces that have been developed. Among them are cross-grooved tubes with an additional set of grooves with opposite helical orientation and herringbone tubes with V-shaped grooves. Microfin tubes have shown a heat transfer enhancement of 50%–180% compared to smooth tubes at the same operating conditions. It is achieved with much more moderate increase of pressure drop. The enhancement of heat transfer is caused by a number of reasons. First, it is the effective heat transfer area enlargement due to fins.

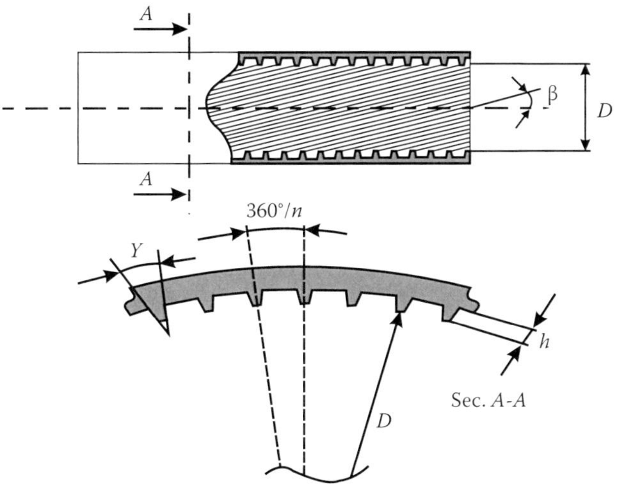

Figure 4.39 Microfin tubes schematic picture.

Another two are turbulence induced by microfins in the liquid film and effect of surface tension forces increasing intensity of liquid drainage. The review of papers with research results for tubes with microfins on internal surface and also externally microfinned tubes for shell and tube condensers, with analysis of correlations for prediction of heat transfer, was published by Cavallini et al. (2003). The flow patterns of vapour and condensate two-phase flows in microfin tubes were investigated and analysed by Doretti et al. (2013).

Tubes with pin fins are another type of enhanced surfaces for condensation, schematically shown in Figure 4.40. Recent publications with research results for such types of microfinned tubes are discussed by Ali and Briggs (2013). The results for condensation of ethylene glycol and R-113 at near atmospheric pressure and low velocity on three identical pairs of pin–fin tubes made of copper, brass and bronze are reported. The correlation for calculation of heat transfer is proposed.

The tendency to increase compactness brought attention for condensation processes in channels of small hydraulic diameters. The number of researchers has proposed the different criteria to estimate what size of channel can be regarded as very small (microchannel) from the point of view that surface tension forces are influencing the two-phase flow patterns causing significant deviations in pressure gradients and heat transfer coefficients from models developed for channels of bigger sizes. One of the first attempts to define the hydraulic diameter below which microchannel phenomena should be considered was made by Suo and Griffith (1963). To express it, Serizawa et al. (2002) have used Laplace constant:

$$Lp = \sqrt{\frac{\sigma}{g \cdot \left(\rho_L - \rho_g\right)}} \tag{4.128}$$

where σ is surface tension, N/m^2.

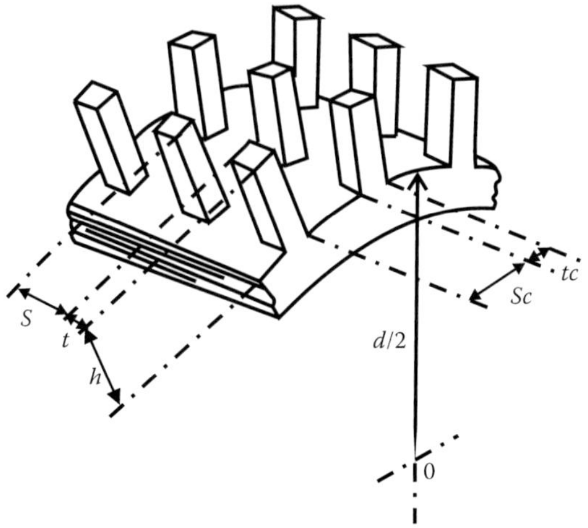

Figure 4.40 The outer surface of pin–fin tube.

Using Laplace constant, the criterion looks like as follows:

$$Lp > C_o \cdot D_h \tag{4.129}$$

The constant $C_0 = 0.94$, as estimated from the paper of Suo and Griffith (1963). Garimella (2014) has taken $C_0 = 1$ as microchannel transition criteria and estimated that for a high-surface-tension fluid pair like air–water at standard temperature and pressure, a channel with $D_h < 2.7$ mm would be a microchannel. For channels of bigger hydraulic diameters, all relations for heat transfer and pressure losses obtained for conventional channels can be used. For vapours of the liquids used as refrigerants at saturation temperature 50 °C, the microchannel transition diameters are as follows: 0.69 mm for R134a, 0.46 mm for R404a, 1.01 mm for propane and 1.63 mm for ammonia. Another approach to quantify the size of microchannels based on research of boiling in single, small-diameter tubes was proposed by Kew and Cornwell (1997). In terms of the confinement of bubbles in channels with the analysis of typical bubble sizes that is more relevant for two-phase flow mechanism during the boiling inside channels. According to this approach, microchannel two-phase flows are featured by separate elongated bubbles occupying the channel cross section, while in cross section of conventional channel, these are featured by multiple bubbles at a time. Following this interpretation of two-phase flow picture, Kew and Cornwell (1997) introduced "confinement number" which is actually calculated as C_0 in Equation 4.129 with recommendation to use it as transition criterion (for boiling applications) that for microchannels it should be smaller than 0.5. For its lower values ($C_0 < 0.5$), microchannel effects are present. According to this recommendation, the microchannel transition diameters are two times bigger than calculated with $C_0 = 1$.

Regarding the use of compact heat exchangers for low-grade heat utilisation applications, the use of very low diameters especially in condensation processes is rather limited. Mostly, it can be heat pumps condensers, where transition hydraulic diameter can be as low as 0.46 mm for R404a or 1.64 mm for ammonia. For condensing of low-pressure steam, it is usually recommended to use channels of the diameter high enough not to create excessive loss of pressure. Therefore, for most applications of compact heat exchangers in low-grade heat utilisation, the correlations were obtained for channels of conventional sizes with hydraulic diameter more than 3 mm.

Compact heat exchangers with condensation in microchannels and minichannels are widely used in small devices in the refrigeration and air-conditioning, in automotive industry, in heat pipes, in thermosyphons and in other applications for system thermal control. Currently, it is a fast-growing field of research. The two-phase flow structure and correlations which can be used for calculation of heat transfer and pressure losses in minichannels and microchannels are presented and discussed by Garimella (2014). Another comprehensive review of state of art in the field of condensation in microchannels with analysis of correlations available in literature and future trends is published by Awad et al. (2014).

4.2.1.4 Condensation of Vapour from Mixture with Non-Condensable Gas

During condensation of clean saturated vapour, it is flowing freely to the surface of condensate film, practically without any resistance. The presence of any gas that is not condensed at this temperature creates resistance to diffusion of vapour from the main flow to the boundary of liquid phase or to the wall in case of dropwise condensation. It leads to significant redistribution of temperature and vapour concentration profiles, as shown in Figure 4.41.

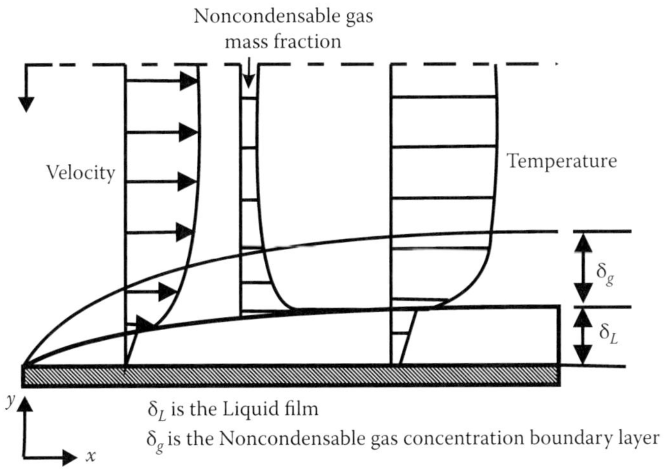

Figure 4.41 Schematic drawing of vapour condensation from gas–vapour mixture.

For the small concentrations of vapour in its mixture with non-condensable gas, the mass transfer intensity from the main flow stream to liquid film boundary can be determined by analogy between heat and mass transfer written in a form

$$\psi_D = \frac{\mathrm{Nu}_D \cdot m_{gf}}{\mathrm{Nu}_{D0}} = 1 \tag{4.130}$$

Here

m_{gf} is the mass fraction of non-condensable gas

Nu_{D0} is the diffusion Nusselt number calculated by correlation for heat transfer Nusselt number with changing Pr number for diffusion Prandtl number $\mathrm{Pr}_D = \nu_{vg}/D_{vg}$

D_{vg} is diffusivity coefficient for gas–vapour mixture, m²/s

ν_{vg} is cinematic viscosity of gas phase, m²/s

$\mathrm{Nu}_D = \beta_m \cdot D_h/D_{vg}$ is the diffusion Nusselt number

β_m is the mass transfer coefficient, kg/(m² · s)

With increase of vapour concentration, the transverse mass flux to liquid boundary that influences the heat and mass transfer increases. As shown by Kutateladze and Leont'ev (1990), it can be accounted by permeability parameters for heat transfer b_T and for mass transfer b_D:

$$b_T = \frac{j_v \cdot \mathrm{Re} \cdot \mathrm{Pr}}{\rho_{vg} \cdot w_{vg} \cdot \mathrm{Nu}_0} \cdot \frac{c_{pv}}{c_{pvg}}; \qquad b_D = \frac{j_v \cdot \mathrm{Re} \cdot \mathrm{Pr}_D}{\rho_{vg} \cdot w_{vg} \cdot \mathrm{Nu}_{D0}} \tag{4.131}$$

Based on experimental research of steam–air mixture condensation in PHE channels, Tovazhnyansky and Kapustenko (1984b) have proposed the following empirical correlation to calculate relative increase of heat ($\psi_T = \mathrm{Nu}/\mathrm{Nu}_0$) and mass transfer ($\psi_D$) coefficients with transverse mass flux:

$$\psi_{T(D)} = 1 + 0.85 \cdot b_{T(D)} \tag{4.132}$$

The Equation 4.132 has discrepancy not more than 10% with theoretical solution by Kutateladze and Leont'ev (1990) for turbulent boundary layer, which was confirmed experimentally for two-component mixtures of steam with helium, air and R12, as also vapour mixtures of ethanol steam, acetone steam, ethanol–propanol and ethanol–butanol. The applicability of this equation for different geometries of channels that allow its use also for channels with heat transfer enhancement knowing correlation for heat transfer in single-phase flow was shown.

The transverse mass flux of vapour to the condensate surface, kg/(m² · s), is

$$j_v = \beta_m \cdot \left(m_{vm} - m_{vs}\right) \tag{4.133}$$

The heat flux from the main flow of gas–vapour mixture to liquid film outer surface, W/m^2, is

$$q = h_v \cdot (T_m - T_s) + r \cdot \beta_m \cdot (m_{vm} - m_{vs}) \tag{4.134}$$

Here

T_m is temperature at the main stream, K

T_s is temperature at the liquid film surface, K

m_{vm} is mass concentration of vapour at the main stream

m_{vs} mass concentration of saturated vapour at temperature of the liquid film surface T_s.

In heat exchanger, this heat flux is discharged from liquid film outer surface to cooling media moving at another side of the wall. There, the thermal resistances to this heat flux are as follows: at the condensate film ($1/h_f$); at the wall (δ_w/λ_w); at fouling deposits (R_{foul}) and at cooling media ($1/h_c$). It is as follows:

$$q = \frac{T_s - T_c}{\left(\frac{1}{h_f} + \frac{\delta_w}{\lambda_w} + R_{foul} + \frac{1}{h_c}\right)} \tag{4.135}$$

Mass concentration of vapour at liquid film outer boundary is linked to film surface temperature by liquid–vapour equilibrium condition $m_{ss} = m_{ss}(T_s)$. The heat transfer coefficient h_v and mass transfer coefficient β_m are dependent from transverse mass flux j_v through Equations 4.132 and through it by Equation 4.134 from heat flux q. In some cross section of the heat exchanger with known flow rates of gas and liquid phases and cooling media, their vapour concentration, temperatures and physical properties, Equations 4.134 and 4.135 can be regarded as a system of two nonlinear algebraic equations with two variables q and T_s. Because of rather complex nonlinearity, this system in general case cannot be solved analytically. Expressing T_s from Equation 4.135 and substituting it to Equation 4.134, the last one becomes nonlinear algebraic equation that can be solved numerically on a computer. It should be noted that method of iterations is not reliable, as not giving convergent solution in some cases and more rigorous methods must be used.

The condensation of the vapour is leading to significant change of gas and liquid phases, flow rates, vapour concentration in gas–vapour mixture, temperatures, pressures and all other process parameters along the heat transfer surface. The reliable and accurate calculations are not possible by simple averaging these parameters. Stepwise calculation by small parts of the channel length or heat transfer surface area that should be implemented as computer software is required.

4.2.1.5 Condensation Pressure Drop

The pressure drop during the condensation in channels is important not only from the point of energy losses but also because of its influence on process intensity as saturation temperature of the vapour can decrease significantly with the change of pressure along the heat transfer surface. In channels of small hydraulic diameters with enhanced heat transfer, this effect can play much bigger role than in smooth tubes and channels as high intensity of condensation can lead to significant increase of vapour velocities at the areas near the channel entrance and also higher flow rates of two-phase mixture along the channel. It is also applying additional restrictions on the use of channels installed in parallel at one heat exchanger, as also temperature and flow conditions in them. The resulting differences in pressure drop at parallel channels (because of passes on cooling media side or use of channels with different geometries) can lead to the effect that significant parts of some parallel channels will be flooded by condensate and not working properly.

The total pressure drop of condensing stream in the heat exchanger channel consists of the several elements including friction losses ΔP_{tp} (for enhanced heat transfer channels of complex geometry, it also includes form drug losses), static pressure change due to gravity and to stream acceleration:

$$\Delta P = \Delta P_{tp} + \Delta P_a + \Delta P_g \tag{4.136}$$

The gravitational static pressure is changing along the channel and depends on two-phase void fraction α. For the small increment of channel length with the difference of height between inlet and outlet Δz (with sign "–" for down flow and "+" for up flow), it can be estimated as

$$\Delta P_g = \left[\alpha \cdot \rho_v + (1-\alpha) \cdot \rho_L\right] \cdot \Delta z \tag{4.137}$$

In condensation process, the velocity of stream decreases (actually, acceleration is negative or deceleration) that causes increase of pressure, the effect alternate to loss of pressure. The increase of static pressure due to deceleration can be calculated as difference between dynamic pressures at outlet and inlet of the channel or part of its length:

$$\Delta P_a = \frac{G^2}{2} \cdot \left\{ \left[\frac{X_0^2}{\alpha_0 \cdot \rho_v} + \frac{(1-X_0)^2}{\rho_L \cdot (1-\alpha_0)} \right] - \left[\frac{X_i^2}{\alpha_i \cdot \rho_v} + \frac{(1-X_i)^2}{\rho_L \cdot (1-\alpha_i)} \right] \right\} \tag{4.138}$$

Here, G is mass velocity (or flux) of two-phase flow in the channel, kg/(m$^2 \cdot$ s).

On analysis of experimental data, Garimella (2014) concluded that the acceleration term is extremely small compared to the overall pressure drop. Lockhart and Martinelly (1949) came to the same conclusion with note that its significance increases at low-pressure vapour condensation. According to Tovazhnyansky et al. (2004), for the small increment of the channel length, it can be approximately estimated using vapour velocity calculated for all cross-sectional area:

$$\Delta P_a = \frac{G^2}{2} \cdot \left(\frac{X_0^2}{\rho_v} - \frac{X_i^2}{\rho_v} \right) \tag{4.139}$$

Pressure losses in heat exchanger also include additional loss of pressure in connections and distribution collectors, which depend on heat exchanger construction. According to Shah and Sekulic (2003), for PHE, it can be approximately calculated as 1.5 times the inlet velocity head. More detailed alternative approach is presented by Wang et al. (2007).

The frictional pressure losses ΔP_{tp} are usually the major contributor in Equation 4.136. The most often used methods of their calculation are based on parameters proposed by Lockhart and Martinelly (1949) for pressure drop per unit length at two-phase flow in pipes:

$$X_{LM} = \sqrt{\frac{\Delta P_L}{\Delta P_g}} \tag{4.140}$$

$$\Phi_L = \sqrt{\frac{\Delta P_{tp}}{\Delta P_L}} \tag{4.141}$$

$$\Phi_g = \sqrt{\frac{\Delta P_{tp}}{\Delta P_g}} \tag{4.142}$$

where

ΔP_L is the pressure drop of liquid phase moving alone in the channel, Pa
ΔP_g is the pressure drop of gas phase moving alone in the channel, Pa

These pressure drops are calculated by correlations for single-phase flow.

Chisholm (1967) has proposed the equation for expressing the correlation between Φ_L and X_{LM} as follows:

$$\Phi_L^2 = 1 + \frac{C}{X_{LM}} + \frac{1}{X_{LM}^2} \tag{4.143}$$

For adiabatic flows in pipes, the parameter C in this equation is determined depending on flow regimes for gas and liquid phases:

- Turbulent–turbulent $C = 20$
- Viscous liquid–turbulent gas $C = 12$
- Turbulent liquid–viscous gas $C = 10$
- Viscous–viscous $C = 5$

For channels of complex geometry, the value of parameter C can be determined using experimental data on condensation in channel of similar form. For example, Wang et al. (2007) have recommended for PHE channels the value $C = 16$ using their experimental data for average pressure drop during steam condensation. They have noticed a relatively big deviation existed between predicted and experimental values. They also pointed out that this single value could not perform well in the stepwise calculation procedure.

During the vapour condensation, the highest pressure gradients are observed near the channel entrance diminishing with the reduction of vapour velocity along the channel and parameter X_{LM} is below 0.5 for most part of the channel length with substantial pressure losses. It is even more emphasised when it is condensing only part of vapour or non-condensable gas is present. Therefore, the pressure drop of condensing two-phase flow is mostly determined by gas phase and it is more convenient to use correlation for Φ_g, which can be written as

$$\Phi_g^2 = X_{LM}^2 + C \cdot X_{LM} + 1 \tag{4.144}$$

Tovazhnyansky and Kapustenko (1984) have presented experimental data for pressure drop on seven parts of PHE channel model along its length in the range of $0 < X_{LM} < 0.5$. They have correlated these data for local pressure drop of two-phase flow during condensation by the following equation (Tovazhnyansky and Kapustenko, 1989):

$$\Phi_g^2 = 1 + 2.9 \cdot X_{LM}^{0.46} \tag{4.145}$$

A number of approaches aimed to improve accuracy of pressure drop in two-phase flow in tubes was developed. Majority of them are using Lockhart–Martinelli parameters to correlate data for adiabatic flows. Garimella (2014) has presented the comparison of results obtained by 12 methods proposed by different researchers in the same conditions of refrigerant R134a condensation inside tube of 1 mm inner diameter. The calculations were made for average process parameters. The interesting fact is that 6 out of 12 methods gave results close to predicted for gas phase moving alone in the channel with total flow rate of gas–vapour mixture. The maximal discrepancies were from −14% to +22%. Results by Lockhart–Martinelli approach are 65% higher. The maximal deviation between two different methods is reaching up to 8.5 times. Garimella attributed this large variation to the big "difference in two-phase multipliers developed by the various investigators. The only recommendation that can be made is to choose a model that is based on the geometry, fluid and operating conditions similar to those of interest for a given application. Another point is that the simple averaging of condensation process parameters, which are considerably changing along the channel, basically cannot give accurate results, that is possible only with stepwise calculations accounting for distribution of local parameters.

The experimental data available in literature for adiabatic and condensing two-phase flows are obtained in channels of different geometrical forms for a variety of vapour–liquid pairs, majority for water–air, water–steam and different liquid–vapour pairs of refrigerants. Kim and Mudawar (2014b) have analysed available experimental data and predictive methods for pressure drop in adiabatic and condensing two-phase flows of different fluids in straight mini- and microchannels of different cross-sectional geometries. The range of hydraulic diameters was from 0.07 to 6.22 mm. They have proposed universal correlation in form of Equation 4.143 with specification of formulas for estimation of parameter C and friction factors for different flow regimes and fluids properties. Estimated mean absolute error of prediction for different data sets with this correlation was from 8.1% to 33.9%. In turbulent–laminar flows, the errors more than 50% were for 10.2% of data and in turbulent–turbulent flows for 3.6% of data. It confirms the difficulties of accurate estimation of pressure drop in two-phase condensing flows for different process conditions and cited earlier recommendation.

4.2.2 Boiling in Compact Heat Exchangers

Boiling is a liquid-to-vapour phase change process which occurs at the solid surface which has temperature higher than the saturation temperature of the liquid. Liquid phase evaporates with consuming latent heat of phase change coming from the heat transfer surface. In recuperative heat exchangers called evaporators or boilers, this heat is supplied through the wall from heating media. The process in confined channels of heat exchangers under influence of flow velocity is referred to as flow boiling, opposite to pool boiling taking place in large volumes of evaporating liquid.

In systems of low-grade heat utilisation, evaporators and boilers are used for producing vapour in steam or organic cycles of power generation, for extracting heat by evaporation of refrigerants in heat pump systems. Generated vapour can also be used for low-grade heat distribution and transport to consumers. In processes of heat extraction, flow boiling has advantage over single-phase cooling because of higher heat transfer coefficients and smaller mass flow rates at the same heat removal capabilities. During the transport of heat with vapour, there is much less losses in temperature potential than with single-phase media.

In compact heat exchangers, the flow boiling occurs inside the channels of small hydraulic diameters. It is a complicated process with flow mechanism and all parameters are considerably changing along the channel length as illustrated in Figure 4.42. The necessary condition for the boiling is that the temperature of the channel wall surface T_w must be higher than saturation temperature T_s of the liquid flowing inside the channel at its actual pressure.

When subcooled below T_s, the liquid is coming at the channel entrance where it is heated until the temperature at the wall surface and nearby liquid becomes higher than T_s. If there are small cavities on the channel surface, they are serving as nucleation sites trapping vapour or gases. As the surface temperature become higher than the saturation temperature on certain ΔT_s, a bubble may grow. It is analysed in more detail in the literature, as, for example, by Kandlikar (2005). After the bubble reaches a certain size, it is washed out to the main stream of liquid. In subcooled liquid, it will shrink, while vapour in it is condensing giving latent heat to liquid and substantially increasing heat transfer intensity (*bubbly flow*). As the bulk of the fluid becomes heated to saturation temperature (maximal heat transfer coefficients), the bubble is not shrinking but merge with other bubbles (the heat transfer coefficient is dropping gradually). Merged bubbles grow till some sections of the stream will be occupied by slugs of vapour (*slug flow*). After some way along the channel, all vapour is moving in the central part and liquid as a film at the wall surface (*annular flow*). As liquid in the film evaporates, it becomes thinner and some patches of the

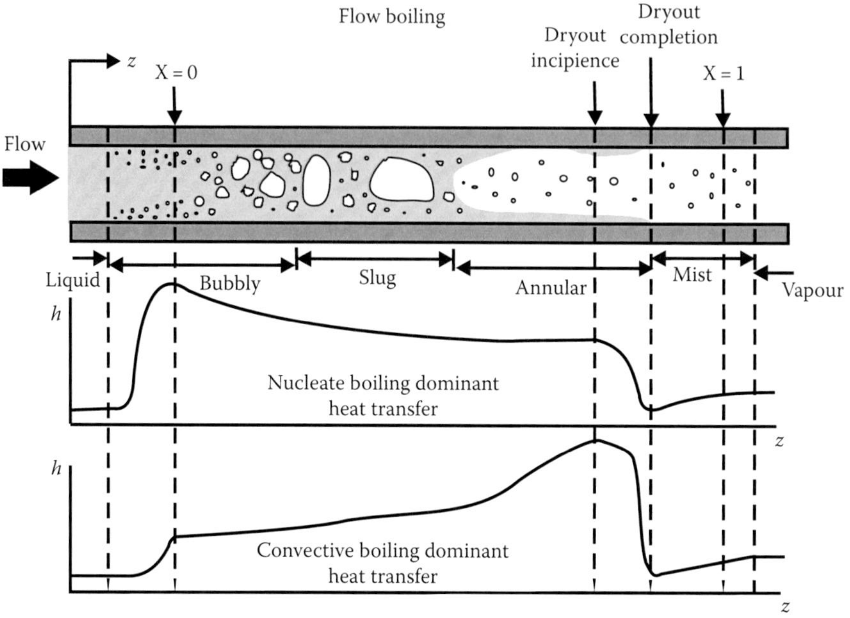

Figure 4.42 Schematic picture of flow boiling in a channel of small hydraulic diameter.

clear surface dry out and start to grow (the moment called *dryout incipience*). The heat transfer coefficient substantially drop and after all liquid film evaporates and surface *dryout*, the vapour overheating starts with heat transfer coefficient as calculated for single-phase vapour flow.

The described boiling mechanism is called *nucleate boiling dominant*. If there are few nucleation sites on the heating surface, the liquid near it becomes overheated and bubbles can appear inside the liquid out of the surface. In this case, the heat transfer intensity is determined by convection near the wall. It is *convective boiling dominant* mechanism of flow boiling. The heat transfer coefficient at first is much smaller than with nucleate boiling but grows with the increase of vapour share in the flow volume and corresponding increase of velocity gradients in liquid. The existence of nucleate or convective boiling mechanisms beside other process parameters much depends on surface material and fluid characteristics. While nucleate boiling was observed by most researchers, the coexistence of both mechanisms with different share in heat transfer process intensity is widely recognised. It is accounted in many correlations for calculation of film heat transfer coefficients in flow boiling (see, e.g., Kandlikar, 2014).

The increase of heat flux over the certain limit can cause significant drop of heat transfer during the bubbly boiling when bubbles merge near the heat transfer surface to vapour film hampering heat transfer. This limiting value of heat flux is called critical heat flux (CHF). As it is discussed by Kandlikar (2014) in channels of small hydraulic diameter, CHF is leading to flow instabilities and significant decrease of heat transfer coefficient. The CHF is especially important when cooling of different devices requires estimating maximum heat flux that can be achieved. The comprehensive survey of different models and recommendations for estimation of CHF in channels of small diameters can be found in review paper published by Konishi and Mudawar (2015).

4.2.2.1 Heat Transfer at Flow Boiling in Compact Heat Exchangers

The extensive researches of flow boiling have resulted in development of a big number of empirical correlations for heat transfer inside tubes and channels. Initially, it was stipulated by interest in development of steam-generating equipment, especially at nuclear power stations. Nowadays, the development of compact refrigeration equipment and special cooling devices has stipulated researches on flow boiling in channels of small hydraulic diameters, in mini- and microchannels. Kim and Mudawar (2014a) have reviewed a number of existing correlations for calculation of heat transfer coefficients in strait tubes and channels with emphasis for flow boiling in mini- and microchannels. Using the form of correlation proposed by Schrock and Grossman (1962), they attempted to develop universal correlation for pre-dryout saturated flow boiling heat transfer in channels of small hydraulic diameter. The film heat transfer coefficient in two-phase flow boiling h_{tp} is determined as average of nucleate boiling h_{nb} and convective boiling h_{cb} heat transfer coefficients as follows:

$$h_{tp} = \left(h_{nb}^2 + h_{cb}^2\right)^{0.5} \tag{4.146}$$

$$h_{nb} = \left[2,345 \cdot \left(\text{Bo}\frac{P_H}{P_F}\right)^{0.7} \cdot P_R^{0.38} \cdot (1-X)^{0.51}\right] \cdot \text{Nu}_{L0} \cdot \frac{\lambda_L}{D_h} \tag{4.147}$$

$$h_{cb} = \left[5.2 \cdot \left(\text{Bo}\frac{P_H}{P_F}\right)^{0.08} \cdot \text{We}_{L0}^{-0.54} + 3.5 \cdot \left(\frac{1}{X_{LM}}\right)^{0.94} \cdot \left(\frac{\rho_g}{\rho_L}\right)^{0.25}\right] \cdot \text{Nu}_{L0} \cdot \frac{\lambda_L}{D_h} \tag{4.148}$$

The dimensionless numbers in these equations are calculated as follows:

$$\text{Bo} = \frac{q''_H}{G \cdot r_{Lg}}; \quad P_R = \frac{P}{P_{crit}}; \quad \text{We}_{L0} = \frac{G^2 \cdot D_h}{\rho_L \cdot \sigma}; \quad X_{LM} = \left(\frac{\mu_L}{\mu_g}\right)^{0.1}\left(\frac{1-X}{X}\right)^{0.9}\left(\frac{\rho_g}{\rho_L}\right)^{0.5}$$

Here

Bo is the boiling number
P_R is reduced pressure
We_{L0} is Weber number
X_{LM} is Lockhart–Martinelli parameter
X is vapour quality
q_H'' is heat flux, W/m^2
G is mass velocity of two-phase flow, kg/(m$^2 \cdot$ s)
P is pressure, Pa

P_{crit} is critical pressure, Pa
r_{Lg} is latent heat of vaporisation, J/kg
σ is surface tension, N/m
P_H is the heated perimeter of channel, m
P_F is the wetted perimeter of channel, m

Nu_{L0} is the Nusselt number determined for single-phase liquid flow rate in tube by Dittus–Boelter correlation

$$Nu_{L0} = 0.023 \cdot \left(\frac{G \cdot (1 - X) \cdot D_h}{\mu_L} \right)^{0.8} \cdot Pr_L^{0.4} \tag{4.149}$$

The correlation is based on consolidated database of 10,805 saturated boiling heat transfer coefficient data points from 37 sources with working fluids: FC72, R11, R113, R123, R1234yf, R1234ze, R134a, R152a, R22, R236fa, R245fa, R32, R404A, R407C, R410A, R417A, CO_2 and water. The range of hydraulic diameters is from 0.19 to 6.5 mm, mass velocity G from 19 to 1,608 kg/(m$^2 \cdot$s), flow quality X from 0 to 1 and reduced pressure P_R from 0.005 to 0.69.

The extensive review of different correlations for heat transfer prediction at flow boiling with emphasis of CO_2 used as refrigerant is presented by Fang et al. (2013). While different experimental correlations obtained by different researchers for specific process conditions can predict their experimental data with better accuracy, the universal correlation (4.146) through (4.148) gives a general picture of the influence of physical properties of working fluids and other parameters on heat transfer, which can be estimated using single-phase data. When there are heat transfer correlations available for the channels of different geometries with heat transfer enhancement, it can be expected also to predict influence of process parameters on flow boiling heat transfer with certain accuracy.

The flow boiling in tubes and channels with enhanced heat transfer has been studied by a number of researchers. Among the tubular geometries, the highest attention was paid to flow boiling in microfin tubes like shown in Figure 4.39. For example, Colombo et al. (2012) reported about their experiments on refrigerant R134a boiling inside micro-finned tubes of four different geometries and inner diameter 8.62 and 8.92 mm; the height of helix microfins was from 0.16 to 0.23 mm. It was shown the increase of heat transfer coefficient from 1.5 to 2.5 times with even smaller increase in pressure gradient. The comparison of experimental results with prediction by some correlations available in literature has shown good agreement with correlation of Yun et al. (2002). The data on pressure drop in a good agreement with pressure drop correlation were proposed by Kuo and Wang (1996). For experimental data on evaporation of refrigerant R417a in microfin tubes of 8.8 mm inner diameter and about 0.16 mm fins, Zhang and Yuan (2008) have found big discrepancies with available in literature correlations. They have proposed new modified correlation to fit their experimental results.

The widening application of compact PHEs as evaporators, especially in refrigeration, has stipulated experimental research of flow boiling in PHE channels of criss-cross flow geometries. Wang et al. (2007) presented a survey of earlier experimental researches on flow boiling in plate evaporators. They came to a conclusion that there are mostly published qualitative results or some empirical correlations for individual heat exchangers for specific fluids. There is no universal correlation for the prediction of the evaporation heat transfer coefficient and even no certain conclusion about the domination of nucleate boiling or convective boiling. The influences on two-phase heat transfer of heat flux, vapour quality, mass velocity and pressure are sometimes contradictory for different experiments depending on geometry, fluid type and accuracy of individual experiments. The important influence on plate evaporator performance has flow distribution in the inlet manifold that can explain contradictory conclusions by different researchers.

As it is discussed for single-phase flows, the main feature of heat transfer intensification in criss-cross flow channels of PHEs is the permanent 3D redistribution of velocity profiles in the main stream with disruption of the boundary layers and active turbulence generation. It leads to significant increase in wall shear stresses and turbulent convective heat transfer enhancement. This flow structure helps to intensify heat transfer during film condensation by thinning liquid film at the channel wall and disrupting its flow with turbulent pulsations. However, to do the same can cause adverse effect on heat transfer at flow boiling. In bubbly and slug two-phase flow regimes, it can suppress nucleate-dominated boiling by premature washing out bubbles. In annular flow regime, the disruptions of liquid film can cause its premature dryout on some parts of the wall surface that lead to decrease of the locally averaged heat transfer coefficients. The local redistribution of static pressure around channel wall can also influence flow boiling. All these processes are much dependent on the plate corrugations and channel geometry. Therefore, to estimate the influence of the channel geometry on heat transfer in flow boiling is much more difficult than for heat transfer in single-phase flows or in film condensation.

Some of the researchers attempted to correlate their experimental data on flow boiling heat transfer coefficients in PHEs using correlations for single-phase heat transfer. Correlating experimental results on evaporation of refrigerant R134a in a channel of PHE, Margat et al. (1997) concluded that the effect of nucleate boiling was not significant and proposed the following equation for heat transfer coefficient:

$$h_{tp} = h_{L0} \cdot F_{tp} \tag{4.150}$$

where h_{L0} is the single-phase liquid heat transfer coefficient calculated for total flow rate of two-phase flow in considered PHE channel. F_{tp} is the enhancement factor calculated through Lockhart–Martinelli parameter with Chisholm (1967) representation (see, e.g., Equation 4.143):

$$F_{tp} = \sqrt{1 + \frac{3}{X_{LM}} + \frac{1}{X_{LM}^2}} \tag{4.151}$$

Tovazhnyansky et al. (2002) reported about experiments on water boiling in the model of PHE channel with spacing between plates 5 mm and corrugation inclination angle 45°. The comparisons with heat transfer coefficients for flow boiling in tubes have shown the increase in PHE channels from 2 up to 3 times. The influence on heat transfer of both heat flux and flow velocity was observed and data were correlated by the following empirical equation:

$$\frac{h_{tp}}{h_{L0}} = 6{,}500 \cdot \left[\left(\frac{q''_H}{r_{Lg} \cdot w_L} \right) \cdot \left(\frac{\rho_g}{\rho_L} \right)^{1.45} \cdot \left(\frac{r_{Lg}}{c_p \cdot T_s} \right)^{0.33} \right]^{0.7} \tag{4.152}$$

Saturated flow boiling of the ozone-friendly refrigerant R-410A (mixture half on half in weight of R32 and R-125) in vertical brazed PHE was investigated experimentally by Hsieh and Lin (2002). The spacing between plates was 2.9 mm and corrugation inclination angle 60°. It was reported the influence on heat transfer of heat flux, which was accounted by introduction in correlation of boiling number Bo as a correction to the all-liquid single-phase heat transfer coefficient h_{L0}:

$$h_{tp} = h_{L0} \cdot \left(88 \cdot \mathrm{Bo}^{0.55} \right) \tag{4.153}$$

The evaporation of ammonia–water mixture in brazed PHE of Alfa Laval with the spacing between plates 2.0 mm and corrugation angle 60° was investigated by Taboas et al. (2012). The research was aimed to develop desorbers for ammonia–water absorption refrigeration machines driven by waste heat or solar energy. The range of parameters was mean vapour quality from 0.0 to 0.22, pressure between 7 and 15 bar and ammonia mass concentration between 0.35 and 0.62. After analysis of available in literature different forms of correlations for heat transfer coefficients Taboas et al. have selected approaches proposed by Margat et al. (1997) and Hsieh and Lin (2002). As a result of correlating experimental data, it was recommended to select the highest of two heat transfer coefficients calculated by the following equations:

$$h_{tp} = \text{the\_greater\_of}: \left\{ \begin{array}{c} h_{np} = \left(5 \cdot \mathrm{Bo}^{0.15} \right) \cdot h_{L0} \\ \hline h_{np} = \left(1 + \dfrac{3}{X_{LM}} + \dfrac{1}{X_{LM}^2} \right)^{0.2} \cdot h_{L0} \end{array} \right. \tag{4.154}$$

This approach is somewhat different from those proposed by other researchers that imply different forms of the averaging of heat transfer coefficients at nucleate-dominated (h_{nb}) and convective-dominated (h_{cb}) flow boiling (see, e.g., Equation 4.146) but proved good for correlating specific set of experimental data.

The evaporation of ammonia in PHE channels have been the subject of a number of experimental researches. Djordjevic and Kabelac (2008) have studied the evaporation of ammonia and also refrigerant R134a in industrial PHEs produced by GEA Ecoflex GmbH, Germany. Two PHEs consisting each of 10 plates of identical dimensions but different chevron angle (27° in one PHE and 63° in another) were studied in experiments with ammonia. The spacing between the plates was 3.2 mm. The saturation temperature of ammonia varied between –5 °C and +10 °C (corresponding pressure from 3.55 bar to 5.73 bar). The feature of experimental setup is the measurement of temperature profiles along a plate that enabled to estimate local heat transfer coefficients on small parts of the heat transfer surface. The results were compared with empirical correlation for average heat transfer coefficients at nucleate-dominated flow boiling of refrigerants R12 and R22 in flat parallel walls channels with spacing about 2 mm presented by Azarskov et al. (1981) and written as follows:

$$\mathrm{Nu}_{tp} = 4.2 \cdot \mathrm{Re}_g^{0.3} \cdot \mathrm{Bn}^{0.33} \cdot \mathrm{Re}_*^{0.2} \tag{4.155}$$

where

$$\mathrm{Nu}_{tp} = \frac{h_{tp} \cdot D_h}{\lambda_L}; \mathrm{Re}_g = \frac{w_g \cdot D_h}{\nu_g}; w_g = \frac{X \cdot G}{\rho_g}$$
$$\mathrm{Re}_* = \frac{q''_H \cdot D_h}{r_{Lg} \cdot \nu_L \cdot \rho_L}; \mathrm{Bn} = \frac{g \cdot \rho_L \cdot D_h^2}{\sigma} \tag{4.156}$$

Here

Bn is the bond number accounting for surface tension forces

Re_* is accounting for the influence of heat flux in nucleate-dominated flow boiling

For convection-dominated boiling that they supposed for slug and annular two-phase flow regimes, Re_* was excluded and correlation has taken another form:

$$\mathrm{Nu}_{tp} = 3.0 \cdot \mathrm{Re}_g^{0.3} \cdot \mathrm{Bn}^{0.33} \tag{4.157}$$

Djordjevic and Kabelac (2008) have found a good agreement of their experimental data on local heat transfer coefficient with calculation by Equation 4.156 scaled up by some correction factor, 1.15 for plate with corrugation inclination angle 63° and 1.36 with angle 27°. It is fairly good and agrees with data of Azarskov et al. (1981) who reported that heat transfer coefficients for channels formed by corrugated plates with spacing 4 mm are up to 1.45 times higher than calculated by Equation 4.156 as it is shown on graph in that paper. This relative increase of heat transfer coefficient is much smaller than such increase for single-phase flow in PHE channels and just opposite influence on heat transfer coefficient of corrugation inclination angle.

The results of another experimental study dealing with flow boiling of ammonia in PHEs with plates of different chevron angle of corrugations is reported in series of three papers by Khan et al. (2012a,b, 2014). The experiments were made with ammonia in saturation temperature range from –25 °C to –2 °C. The plate-and-frame heat exchangers of the type TL-90 manufactured by Thermowave GmbH from Germany has been investigated. Papers describe results of experiments with three PHEs having plates with different geometries of corrugations, but the same other sizes so that plates with different geometry of corrugation can be assembled in one heat exchanger as shown in Figures 3.23 and 5.4 from Chapters 3 and 5, respectively. It were studied three PHEs with channels of the same geometry in each of them. The first one had plates with chevron corrugations oriented at 60° to plate longitudinal axis, channel spacing 2.2 mm and corrugation pitch 6.25 mm. The second PHE was assembled with plates having chevron angle 30° and the same channel spacing and corrugation pitch. In the third paper, Khan et al. (2014) presented results of experiments on the ammonia evaporation in PHE assembled of a mixture of plates with chevron angle 60° and 30°. But the channel spacing and corrugation pitch for the plate with chevron angle 30° were different 3.6 and 13.5 mm. Correlation for all experimental data looks as follows:

$$\mathrm{Nu}_{tp} = \left(257.12 - 173.52 \cdot \beta^*\right) \cdot \left(\mathrm{Re}_{eq} \cdot \mathrm{Bo}_{eq}\right)^{0.0005 - 0.09 \cdot \beta^*} \cdot \left(P_R\right)^{0.624 \cdot \beta^* - 0.822} \quad (4.158)$$

The equivalent Reynolds number Re_{eq}, equivalent Boiling number Bo_{eq} and the reduced pressure P_R are given as follows:

$$\mathrm{Re}_{eq} = \frac{G_{eq} \cdot D_h}{\mu}; \quad \mathrm{Bo}_{eq} = \frac{q''_H}{G_{eq} \cdot r_{Lg}}; \quad P_R = \frac{P}{P_{crit}}$$

$$G_{eq} = G \cdot \left[1 - X_{ex} + X_{ex} \cdot \left(\frac{\rho_L}{\rho_g}\right)\right]^{\frac{1}{2}}; \quad \beta^* = \frac{\beta}{60} \quad (4.159)$$

Here

G_{eq} is the equivalent mass flow rate of two-phase flow, kg/(m$^2 \cdot$ s)
μ is dynamic viscosity, Pa · s
X_{ex} is exit vapour quality
P_{crit} is critical pressure of refrigerant, Pa.

It is rather difficult to judge about the influence of plate corrugation geometry on heat transfer just by the form of correlation (4.158). Unexpectedly the most influence of the chevron angle β appears in power at reduced pressure P_R which is changing from 0.198 at β = 60° to 0.51 at β = 30°. However, Khan et al. (2014) reported that the increase of chevron angle β leads to enhancement in heat transfer coefficient in flow boiling, similar as in single-phase flow. For instance, in comparison to 30°/30° plate configuration at −25 °C, on average, the heat transfer coefficient increased by about 2 times for 30°/60° plate configuration and by about 4 times for 60°/60° plate configuration. In comparison, the pressure drop for the same saturation temperature for 30°/60° plate configuration is almost the same as that of 30°/30° plate configuration but pressure drop increases about 3 times for 30°/60° plate configuration compared to 30°/30° plate configuration. However, comparison with flow boiling in smooth tubes or channels is not presented.

As it was discussed previously, Djordjevic and Kabelac (2008) came to opposite conclusion in their experiments. Their experiments were made on different plate types, much wider (486 mm against 185 mm) and bigger spacing (3.2 mm against 2.2 mm). Besides, the ammonia temperatures were also higher going up to 10 °C. It is just confirming the complexity of flow boiling phenomena in channels of PHEs. It is still a long way to obtaining some universal correlation. The empirical correlations available in the literature can only be used in the range of the process parameters and flow geometries at which these correlations were obtained.

4.2.2.2 Pressure Drop at Flow Boiling in Compact Heat Exchangers

The total pressure drop during the flow boiling in the heat exchanger channel consists of the several elements including friction losses ΔP_{tp} (for enhanced heat transfer channels of complex geometry it also includes form drug losses), static pressure change due to gravity ΔP_g and to stream acceleration ΔP_a. It can be expressed by the same Equation 4.136 as for condensation. The gravity and stream acceleration terms also can be calculated by Equations 4.137 and 4.138, but in case of flow boiling, acceleration term decreases the static pressure and increases overall pressure drop.

The comprehensive survey of published in literature correlations for prediction of pressure drops in two-phase boiling flows in macrochannels and mini- and macrochannels is presented by Kim and Mudawar (2014b). They have analysed available in literature data and flow structures of two-phase flows. It was concluded that fundamental differences in annular flow structure during boiling flow is linked with droplet entrainment. Droplets are formed upstream by shattering of the liquid ridges between slug flow bubbles. These droplets exist in the gas phase even downstream of the wall dryout location. The universal correlation for prediction of flow boiling pressure gradient in a form of Equation 4.143 with Lockhart–Martinelli parameters is proposed for the straight mini- and macrochannels in the range of hydraulic diameters from 0.349 to 5.45 mm. There are proposed correlations for calculation of single-phase friction factors at different flow regimes and of parameter C in Equation 4.143.

For microfin tubes, the increase of pressure drop compared to smooth tubes is usually from 1.3 to 2 times, not very big. Kuo and Wang (1996) have compared their experimental data on pressure drop during evaporation of refrigerant HCFC-22 in 9.52 mm microfin tube with different correlations proposed for smooth tubes. Two of the correlations based on Equation 4.143 and data for single-phase pressure drop with parameter C determined as a function of the Reynolds number gave deviations in the range of ±25%. The third correlation was based on an alternative approach using mean two-phase flow viscosity μ_m calculated as

$$\mu_m = \rho_m \cdot \left[\frac{X \cdot \mu_g}{\rho_g} + (1 - X) \cdot \frac{\mu_L}{\rho_L} \right] \tag{4.160}$$

where

$$\rho_m = \frac{X}{\rho_g} + \frac{1 - X}{\rho_L}$$

The equivalent Reynolds number

$$\mathrm{Re}_{eq} = \frac{G \cdot D_h}{\mu_m} \tag{4.161}$$

The pressure drop in two-phase flow then is calculated by friction factor ζ_{tp} and formula for pressure drop in single-phase flow:

$$\Delta P_{tp} = \frac{2 \cdot \zeta_{tp} \cdot G^2 \cdot \Delta x}{D_h} \tag{4.162}$$

Very small dependence of friction factor from the Reynolds number and experimental data determined for microfin tube ζ_{tp} = 0.0139 and for smooth tube $\zeta_{tp.s}$ = 0.0101 was found.

The accuracy of the approach proposed by Kuo and Wang (1996) was confirmed by Colombo et al. (2012) in experiments on refrigerant R134a boiling inside micro-finned tubes of four different geometries and inner diameter 8.62 and 8.92 mm, with helix microfins of 0.16–0.23 mm height.

The modification of the approach using equivalent mass flow rate and the Reynolds number determined by Equations 4.159 was used by Khan et al. (2012a,b, 2014) for correlation of pressure drop experimental data for ammonia flow boiling in PHE channels of different geometries. They have obtained the following empirical correlations for investigated channels geometries:

For plates 30°/30° (Khan et al. 2012b),

$$\zeta_{tp} = 673.336 \cdot \mathrm{Re}_{eq}^{-1.3} \cdot \left(P_R\right)^{0.9} \tag{4.163}$$

For plates 60°/30° (Khan et al. 2014b),

$$\zeta_{tp} = 305.59 \cdot \mathrm{Re}_{eq}^{-1.26} \cdot \left(P_R\right)^{0.9} \tag{4.164}$$

For plates 60°/60° (Khan et al. 2012a),

$$\zeta_{tp} = 212 \cdot \mathrm{Re}_{eq}^{-0.51} \cdot \left(P_R\right)^{0.53} \tag{4.165}$$

While Khan et al. (2014) stating that the approximation of separate data sets for each of the geometries gives very good accuracy (error not more than 10%), it is rather difficult to find general trends and generalised correlation was not presented.

As it is observed by Wang et al. (2007), after analysis of published results of researches on flow boiling in PHE channels, the most widely accepted correlation type is the Lockhart–Martinelli model in Chisholm reduction, similar to the condensation case (see, e.g., Equations 4.140 through 4.143). However, the selection of parameter C value is quite contradictory in the literature. Thonon et al. (1995) has suggested $C = 8$, but later, it was changed in paper with Magat et al. (1997) for $C = 3$.

Different modifications of Lockhart–Martinelli approach were proposed by other researches. But there still is no general correlation enabling to estimate pressure drop during flow boiling in PHEs for different geometries and boiling media. Further work is thus obviously required to solve this problem.

REFERENCES

Alamgholilou A., Esmaeilzadeh E. 2012. Experimental investigation on hydrodynamics and heat transfer of fluid flow into channel for cooling of rectangular ribs by passive and EHD active enhancement methods. *Experimental Thermal and Fluid Science* 38: 61–73.

Ali H. M., Briggs A. 2013. Condensation heat transfer on pin-fin tubes: Effect of thermal conductivity and pin height. *Applied Thermal Engineering* 60(1): 465–471.

Amano R.S. 1985. Turbulent heat transfer in a channel with two right-angle bends. *International Journal of Heat and Mass Transfer* 28 (11): 2177–2179.

Arsenyeva O., Tovazhnyansky L., Kapustenko P., Demirskiy O. 2012a. Heat transfer and friction factor in criss-cross flow channels of plate-and-frame heat exchangers. *Theoretical Foundations of Chemical Engineering* 46(6): 634–641.

Arsenyeva O., Tovazhnyansky L., Kapustenko P., Khavin G. 2011. The generalized correlation for friction factor in criss-cross flow channels of plate heat exchangers. *Chemical Engineering Transactions* 25: 399–404.

Arsenyeva O., Tovazhnyansky L., Kapustenko P., Khavin G. 2011b. Optimal design of plate-and-frame heat exchangers for efficient heat recovery in process industries. *Energy* 36: 4588–4598.

Arsenyeva O., Tovazhnyansky L., Kapustenko P., Perevertaylenko O., Khavin G. 2011c. Investigation of the new corrugation pattern for low pressure plate condensers. *Applied Thermal Engineering* 31(13): 2146–2152.

Arsenyeva O.P., Tovazhnyanskyy L.L., Kapustenko P.O., Demirskiy O.V. 2012b. Accounting for thermal resistance of cooling water fouling in plate heat exchangers. *Chemical Engineering Transactions* 29: 1327–1332.

Arsenyeva O.P., Tovazhnyanskyy L.L., Petro O. Kapustenko P.O., Demirskiy O.V. 2014. Generalised semi-empirical correlation for heat transfer in channels of plate heat exchanger. *Applied Thermal Engineering* 70: 1208–1215.

Ashmantas, L. A., Dzyubenko, B. V., Dreitser, G. A., Segal, M. D. 1985. Unsteady-state heat transfer and mixing of a heat carrier in a heat exchanger with flow twisting. *International Journal of Heat and Mass Transfer* 28(4): 867–877.

Awad M. M., Dalkiliç A. S., Wongwises S. 2014. A critical review on condensation heat transfer in microchannels and minichannels. *Journal of Nanotechnology in Engineering and Medicine* 5(1): 010904.

Aydin O. 2005. Effects of viscous dissipation on the heat transfer in forced pipe flow. Part 1: Both hydrodynamically and thermally fully developed flow. *Energy Conversion and Management* 46: 757–769.

Ayub Z. H. 2003. Plate heat exchanger literature survey and new heat transfer and pressure drop correlations for refrigerant evaporators. *Heat Transfer Engineering* 24(5): 3–16.

Azarskov V. M., Danilowa G. N., Zemskov B. B. 1981. Heat transfer in plate evaporators of different geometry. *Kholodilnaya Tekhnika* (4): 25–31. (In Russian)

Bilen K., Akyol U., Yapici S. 2001. Heat transfer and friction correlations and thermal performance analysis for a finned surface. *Energy Conversion and Management* 42: 1071–1083.

Bogaert R., Bölcs A. 1995. Global performance of a prototype brazed plate heat exchanger in a large Reynolds number range. *Experimental Heat Transfer* 8: 293–311.

Boyko L. D., Kruzhilin G. N. 1967. Heat transfer and hydraulic resistance during condensation of steam in a horizontal tube and in a bundle of tubes. *International Journal of Heat and Mass Transfer* 10(3): 361–373.

Cavallini A., Censi G., Del Col D., Doretti L., Longo G. A., Rossetto L., Zilio C. 2003. Condensation inside and outside smooth and enhanced tubes—A review of recent research. *International Journal of Refrigeration* 26(4): 373–392.

Chisholm D. 1967. A theoretical basis for the Lockhart-Martinelli correlation for two-phase flow. *International Journal of Heat and Mass Transfer* 10(12): 1767–1778.

Churchill S.W. 1977. Friction-factor equation spans all fluid-flow regimes. *Chemical Engineering* 84(24): 91–92.

Churchill S. W. 1983. The development of theoretically based correlations for heat transfer. *Latin American Journal of Heat and Mass Transfer* 7(2): 207–229.

Cui J., Patel V. C., Lin C. L. 2003. Large-eddy simulation of turbulent flow in a channel with rib roughness. *International Journal of Heat and Fluid Flow* 24(3): 372–388.

Danilov Y. I., Dzyubenko B. V., Dreitser G. A., Ashmantas L. A. 1986. *Heat Transfer and Hydrodynamics in Channels of Complex Form. (Teploobmen i gidrodinamika v kanalah slozhnoy formy).* Mashinostroenie, Moscow, USSR.

Derby M. M., Chatterjee A., Peles Y., Jensen M. K. 2014. Flow condensation heat transfer enhancement in a mini-channel with hydrophobic and hydrophilic patterns. *International Journal of Heat and Mass Transfer* 68: 151–160.

Djordjevic E., Kabelac S. 2008. Flow boiling of R134a and ammonia in a plate heat exchanger. *International Journal of Heat and Mass Transfer* 51(25): 6235–6242.

Doretti L., Zilio C., Mancin S., Cavallini A. 2013. Condensation flow patterns inside plain and microfin tubes: A review. *International Journal of Refrigeration* 36(2): 567–587.

Dović D., Palm B., Švaić S. 2009. Generalized correlations for predicting heat transfer and pressure drop in plate heat exchanger channels of arbitrary geometry. *International Journal of Heat and Mass Transfer* 52: 4553–4563.

Dzyubenko B.V., Dreitser G.A. 1979. Study of heat transfer and hydraulic resistance in a heat exchanger with flow twisting. *Izvestiya Akademii Nauk SSSR, Energetika i Transport* (5): 163–171.

Eiamsa-ard S., Somkleang P., Nuntadusit C., Thianpong C. 2013. Heat transfer enhancement in tube by inserting uniform/non-uniform twisted-tapes with alternate axes: Effect of rotated-axis length. *Applied Thermal Engineering* 54: 289–309.

Eiamsa-ard S., Thianpong C., Eiamsa-ard P. 2010. Turbulent heat transfer enhancement by counter/co-swirling flow in a tube fitted with twin twisted tapes. *Experimental Thermal and Fluid Science* 34: 53–62.

Eiamsa-ard S., Wongcharee K., Thianpong C., Eiamsa-ard P. 2010. Heat transfer enhancement in a tube using delta-winglet twisted tape inserts. *Applied Thermal Engineering* 30: 310–318.

Engelhorn H. R., Reinhart A. M. 1990. Investigations on heat transfer in a plate evaporator. *Chemical Engineering and Processing: Process Intensification* 28(2): 143–146.

Fang X., Zhou Z., Li D. 2013. Review of correlations of flow boiling heat transfer coefficients for carbon dioxide. *International Journal of Refrigeration* 36(8): 2017–2039.

Focke W.W., Zacharadies J., Olivier I. 1985. The effect of the corrugation inclination angle on the thermohydraulic performance of plate heat exchangers. *International Journal of Heat and Mass Transfer* 28: 1469–1479.

García A., Solano J.P., Vicente P.G., Viedma A. 2012. The influence of artificial roughness shape on heat transfer enhancement: Corrugated tubes, dimpled tubes and wire coils. *Applied Thermal Engineering* 35: 196–201.

Garimella S. 2014. Condensation in minichannels and microchannels—Chapter 6. In: S.G. Kandlikar (Ed.), *Heat Transfer and Fluid Flow in Minichannels and Microchannels*, 2nd edn. Elsevier, Oxford, UK, 295–494.

Gawande V.B., Dhoble A.S., Zodpe D.B. 2014. Effect of roughness geometries on heat transfer enhancement in solar thermal systems—A review. *Renewable and Sustainable Energy Reviews* 32: 347–378.

Gherasim I., Taws M., Galanis N., Nguyen C.T. 2011. Heat transfer and fluid flow in a plate heat exchanger part I. Experimental investigation, *International Journal of Thermal Sciences* 50: 1492–1498.

Gnielinski V. 1975. New equations for heat and mass-transfer in turbulent pipe and channel flow. (Neue Gleichungen für den Wärme- und den Stoffübergang in turbulent durchströmten Rohren und Kanälen). Forschung im Ingenieurwessen 41(1): 8–16. (In German).

Gnielinski V. 2009. Heat transfer coefficients for turbulent flow in concentric annular ducts. *Heat Transfer Engineering* 30(6): 431–436.

Green D.W., Perry R.H. 2008. *Perry's Chemical Engineers Handbook*, 8th edn. McGraw-Hill, New York, USA.

Guhman A.A. 1938. Method to compare convective heating surfaces (Metodika sravneniya convectivnyh poverhnostey nagreva). *Technical Physics (Zhurnal tehnicheskoy phisiki)* 8(17): 1584–1602. (In Russian).

Hamlington P.E., Krasnov D., Boeck T., Schumacher J. 2012. Statistics of the energy dissipation rate and local enstrophy in turbulent channel flow. *Physica D* 241: 169–177.

Hasanpour A., Farhadi M., Sedighi K. 2014. A review study on twisted tape inserts on turbulent flow heat exchangers: The overall enhancement ratio criteria. *International Communications in Heat and Mass Transfer* 55: 53–62.

Heavner R.L., Kumar H., Wanniarachchi A.S. 1993. Performance of an industrial plate heat exchanger: Effect of chevron angle. *AICHE Symposium Series, New York*, USA, 89(295): 262–267.

Hsieh Y. Y., Lin, T. F. 2002. Saturated flow boiling heat transfer and pressure drop of refrigerant R-410A in a vertical plate heat exchanger. *International Journal of Heat and Mass Transfer* 45(5): 1033–1044.

Ievlev V.M., Dzyubenko B.V., Dreitser G.A., Vilemas Y.V. 1982. In-line and cross-flow helical tube heat exchangers. *International Journal of Heat and Mass Transfer* 25(3): 317–323.

Isachenko V. P. 1977. *Heat Transfer in Condensation*. Energiya, Moscow, USSR. (In Russian).

Kalinin E.K., Dreitser G.A. 1998. Heat transfer enhancement in heat exchangers. *Advances in Heat Transfer* 31: 159–332.

Kalinin E.K., Dreitser G.A., Yarkho S.A. 1990. *Intensification of Heat Transfer in Channels* (Intensifikatsiya teploobmena v kanalah). Mashinostroenie, Moscow, USSR. (In Russian).

Kanaris A.G., Mouza A.A., Paras S.V. 2009. Optimal design of a plate heat exchanger with undulated surfaces. *International Journal of Thermal Sciences* 48: 1184–1195.

Kandlikar S. 2014. Flow boiling in minichannels and microchannels—Chapter 5. In: *Heat Transfer and Fluid Flow in Minichannels and Microchannels*, 2nd edn. Elsevier, Oxford, UK, 295–494.

Kandlikar S., Garimella S., Li D., Colin S., King M. R. 2005. *Heat Transfer and Fluid Flow in Minichannels and Microchannels*. Elsevier, Oxford, UK.

Kays, W.M and London, A.L. 1984. *Compact Heat Exchangers*, 3rd edn., McGraw Hill, New York, USA.

Kew P. A., Cornwell K. 1997. Correlations for the prediction of boiling heat transfer in small-diameter channels. *Applied Thermal Engineering* 17(8–10): 705–715.

Khan T.S., Khan M.S., Chyu M-C., Ayub Z.H. 2010. Experimental investigation of single phase convective heat transfer coefficient in a corrugated plate heat exchanger for multiple plate configurations. *Applied Thermal Engineering* 30: 1058–1065.

Khan T. S., Khan M. S., Chyu M. C., Ayub Z. H. 2012a. Experimental investigation of evaporation heat transfer and pressure drop of ammonia in a 60° chevron plate heat exchanger. *International Journal of Refrigeration* 35(2): 336–348.

Khan M. S., Khan T. S., Chyu M. C., Ayub Z. H. 2012b. Experimental investigation of evaporation heat transfer and pressure drop of ammonia in a 30° chevron plate heat exchanger. *International Journal of Refrigeration* 35(6): 1757–1765.

Khan M. S., Khan T. S., Chyu M. C., Ayub Z. H. 2014. Evaporation heat transfer and pressure drop of ammonia in a mixed configuration chevron plate heat exchanger. *International Journal of Refrigeration* 41: 92–102.

Kim S. M., Mudawar I. 2014a. Review of databases and predictive methods for heat transfer in condensing and boiling mini/micro-channel flows. *International Journal of Heat and Mass Transfer* 77: 627–652.

Kim S. M., Mudawar I. 2014b. Review of databases and predictive methods for pressure drop in adiabatic, condensing and boiling mini/micro-channel flows. *International Journal of Heat and Mass Transfer* 77: 74–97.

Kirpichev M.V. 1942. About the best form of heating surface (O naivygodneyshey forme poverhnosti nagreva). *Transactions of ENII (Izvestiya ENII im. Krzhizhanovskogo)* 12: 15–19. (In Russian).

Konishi C., Mudawar I. 2015. Review of flow boiling and critical heat flux in microgravity. *International Journal of Heat and Mass Transfer* 80: 469–493.

Kukulka D.J., Smith R. 2013. Thermal-hydraulic performance of Vipertex 1EHT enhanced heat transfer tubes. *Applied Thermal Engineering* 61: 60–66.

Kuo C. S., Wang C. C. 1996. In-tube evaporation of HCFC-22 in a 9.52 mm micro-fin/smooth tube. *International Journal of Heat and Mass Transfer* 39(12): 2559–2569.

Kutateladze S.S. 1979. *Fundamentals of Heat Exchange Theory*. Atomizdat, Moscow, USSR.

Kutateladze S.S. and Leont'ev A.I. 1990. *Heat Transfer, Mass Transfer, and Friction in Turbulent Boundary Layers*. Hemisphere, New York, USA.

Liou T.M., Hwang J.-J., Chen S.-H. 1993a. Simulation and measurement of enhances turbulent heat transfer in a channel with periodic ribs on one principal wall. *International Journal of Heat and Mass Transfer* 36: 507–517.

Liou T.M., Wu Y.Y., Chang Y. 1993b. LDV measurements of periodic fully developed main and secondary flows in a channel with rib-distributed walls. *Journal of Fluids Engineering* 115: 109–114.

Liu S., Sakr M. 2013. A comprehensive review on passive heat transfer enhancements in pipe exchangers. *Renewable and Sustainable Energy Reviews* 19: 64–81.

Lockhart R.W., Martinelli R. C. 1949. Proposed correlation of data for isothermal two-phase, two-component flow in pipes. *Chemical Engineering Progress* 45(1): 39–48.

Lyon R.N. 1951. Liquid metal heat transfer coefficients. *Chemical Engineering Progress* 47(2): 75–79.

Ma X., Rose J. W., Xu D., Lin J., Wang B. 2000. Advances in dropwise condensation heat transfer: Chinese research. *Chemical Engineering Journal* 78(2): 87–93.

Margat L., Thonon B., Tadrist L. 1997. Heat transfer and two phase flow characteristics during convective boiling in a corrugated channel. In: R.K. Shah, K.J. Bell, S. Mochizuki, V.V. Wadekar (Eds.), *Compact Heat Exchangers for the Process Industries*. Begell House, New York, USA, 323–330.

Martin H. 1996. Theoretical approach to predict the performance of chevron-type plate heat exchangers. *Chemical Engineering and Processing* 35: 301–310.

Meyer J.P., Olivier J.A. 2011a. Transitional flow inside enhanced tubes for fully developed and developing flow with different types of inlet disturbances: Part I–Adiabatic pressure drops. *International Journal of Heat and Mass Transfer* 54: 1587–1597.

Meyer J.P., Olivier J.A. 2011b. Transitional flow inside enhanced tubes for fully developed and developing flow with different types of inlet disturbances: Part II–heat transfer. *International Journal of Heat and Mass Transfer* 54: 1598–1607.

Migay V.K. 1981. *Modelling of Heat Transfer Equipment for Power Industry* (Modelirovanie teploobmennogo energeticheskogo oborudovaniya). Energoatomizdat, Leningrad, USSR. (In Russian).

Muley A., Manglik R.M. 1999. Experimental study of turbulent flow heat transfer and pressure drop in a plate heat exchanger with chevron plates. *ASME Journal of Heat Transfer* 121: 110–117.

Muley A., Manglik R.M., Metwally H.M. 1999. Enhanced heat transfer characteristics of viscous liquid flows in a chevron plate heat exchanger. *Journal of Heat Transfer* 121: 1011–1017.

Nagano Y., Hattori H., Houra T. 2004. DNS of velocity and thermal fields in turbulent channel flow with transverse-rib roughness. *International Journal of Heat and Fluid Flow* 25: 393–403.

Peng B., Ma X., Lan Z., Xu W., Wen R. 2014. Analysis of condensation heat transfer enhancement with dropwise-filmwise hybrid surface: Droplet sizes effect. *International Journal of Heat and Mass Transfer* 77: 785–794.

Pethkool S., Eiamsa-ard S., Kwankaomeng S., Promvonge P. 2011. Turbulent heat transfer enhancement in a heat exchanger using helically corrugated tube. *International Communications in Heat and Mass Transfer* 38: 340–347.

Prasad B.N., Behura A.K., Prasad L. 2014. Fluid flow and heat transfer analysis for heat transfer enhancement in three sided artificially roughened solar air heater. *Solar Energy* 105: 27–35.

Prasad B.N., Saini J.S. 1988. Effect of artificial roughness on heat transfer and friction factor in a solar air heater. *Solar Energy* 41: 555–560.

Promvonge P., Pethkool S., Pimsarn M., Thianpong C. 2012. Heat transfer augmentation in a helical-ribbed tube with double twisted tape inserts. *International Communications in Heat and Mass Transfer* 39: 953–959.

Ravigururajan T.S., Bergles A.E. 1996. Development and verification of general correlations for pressure drop and heat transfer in single-phase turbulent flow in enhanced tubes. *Experimental Thermal and Fluid Science* 13: 55–70.

Rifert V. G., Smirnov H. F. 2004. Condensation heat transfer enhancement. *Developments in Heat Transfer*, Vol. 10. P. 392. WIT Press, Southampton, UK.

Savostin A.F., Tikhonov A.M. 1970. Investigation of the characteristics of plate-type heating surfaces. *Thermal Engineering* 17(9): 113–117.

Schrock V. E., Grossman L. M. 1962. Forced convection boiling in tubes. *Nuclear Science and Engineering* 12(4): 474–481.

Serizawa A., Feng Z., Kawara Z. 2002. Two-phase flow in microchannels. *Experimental Thermal and Fluid Science*. 26(6–7): 703–714.

Sheriff N., Gumley P. 1966. Heat-transfer and friction properties of surfaces with discrete roughnesses. *International Journal of Heat and Mass Transfer* 9(12): 1297–1320.

Shah R.K., Seculic D.P. 2003. *Fundamentals of Heat Exchanger Design*, Wiley, New York, USA.

Steiner D., Taborek J. 1992. Flow boiling heat transfer in vertical tubes correlated by an asymptotic model. *Heat Transfer Engineering* 13(2): 43–69.

Stogiannis I.A., Paras S.V., Arsenyeva O.P., Kapustenko P.O. 2013. CFD modelling of hydrodynamics and heat transfer in channels of a PHE. *Chemical Engineering Transactions* 35: 1285–1290.

Suo M., Griffith P. 1963. Two phase flow in capillary tubes. Technical Report No. 8581-24 of MTI, Cambrige, MA, USA. scholar.google.com.ua/scholar?q = Two+phase+flow+in+capillary+tubes&btnG = &hl = ru&as_sdt = 0%2C5 (accessed 5 December 2014).

Táboas F., Vallès M., Bourouis M., Coronas A. 2012. Assessment of boiling heat transfer and pressure drop correlations of ammonia/water mixture in a plate heat exchanger. *International Journal of Refrigeration* 35(3): 633–644.

Thonon B., Vidil R., Marvillet C. 1995. Recent research and developments in plate heat exchangers. *Journal of Enhanced Heat Transfer* 2(1–2): 149–155.

Tovazhnyansky L.L. 1988. The principles of heat transfer processes intensification, development and optimization of new types of plate heat exchangers for chemical industry. (Principy intensifikacii processov teploobmena. Razrabotka I optimizaciya novyh tipov plastinchatyh teploobmennikov dlya himicheskih proizvodstv.) The DSc Thesis, NTU KhPI, Kharkiv, USSR. (In Russian).

Tovazhnyanskii L.L., Kapustenko P.A. 1978. Heat transfer during vapour condensation in slot-type grid-flow channel. In: Kutateladze, S.S. (Ed.), *Proceedings of XXI Siberian Thermophysical Seminar*, Novosibirsk, USSR, 349–350. (In Russian).

Tovazhnyansky L.L., Kapustenko P.A. 1984. Intensification of heat and mass transfer in channels of plate condensers. *Chemical Engineering Communications* 31(6): 351–366.

Tovazhnyanskii L.L., Kapustenko P.A. 1984b. Heat and mass transfer in the condensation of steam from a steam and gas mixture in plate condenser channels. [Teplo-i massoobmen pri kondensatsii para iz parogazovoi smesi v kanalakh plastinchatykh kondensatorov]. *Teploenergetika*. (2): 52–54. (In Russian).

Tovazhnyansky L.L., Kapustenko P.A. 1989. Theoretical foundations for design of welded plate cool exchangers for ammonia synthesis units. [Teoreticheskie ocnovy rascheta i razrabotka svarnyh plaetinchatyh holodoobmennikov dlya agregatov sinteza ammiaka]. *Khimicheskaya Promyshlennost* (8): 17–22. (In Russian).

Tovazhnyansky L. L., Kapustenko P. O., Perevertaylenko O. 2002. The investigation of flow boiling for flows in channels with cross-corrugated walls. *Heat Transfer Engineering* 23(6): 62–69.

Tovazhnyansky L.L., Kapustenko P.O., Nagorna O.G., Perevertaylenko O.Y. 2004. The simulation of multicomponent mixtures condensation in plate condensers. *Heat transfer engineering* 25(5): 16–22.

Tovazhnyansky L.L., Kapustenko P.A., Tsibulnic V.A. 1980. Heat transfer and hydraulic resistance in channels of plate heat exchangers. *Energetika* (9): 123–125. (In Russian).

Ventas R., Lecuona A., Zacarías A., Venegas M. 2010. Ammonia-lithium nitrate absorption chiller with an integrated low-pressure compression booster cycle for low driving temperatures. *Applied Thermal Engineering* 30(11): 1351–1359.

Vicente P.G., García A., Viedma A. 2004. Mixed convection heat transfer and isothermal pressure drop in corrugated tubes for laminar and transition flow. *International Communications in Heat and Mass Transfer* 31: 651–662.

Wang L., Christensen R., Sunden B. 2000. An experimental investigation of steam condensation in plate heat exchangers. *International Journal of Heat Exchangers* 1(2): 125–150.

Wang L., Sunden B., Manglik R.M. 2007. *PHEs. Design, Applications and Performance*, WIT Press, Southhampton, UK.

Webb R.L. 1981. Performance evaluation criteria for use of enhanced heat transfer surfaces in heat exchanger design. *International Journal of Heat and Mass Transfer* 24: 715–726.

Webb R.L., Eckert E.R.G., Goldstein R.G. (1971), Heat transfer and friction in tubes with repeated-rib roughness. *International Journal of Heat and Mass Transfer* 14: 601–607.

Webb R.L., Kim N.H. 2005. *Principles of Enhanced Heat Transfer*, 2nd edn. Taylor & Francis, New York, USA.

Weihing P., Younis B.A., Weigand B. 2014. Heat transfer enhancement in a ribbed channel: Development of turbulence closures. *International Journal of Heat and Mass Transfer* 76: 509–522.

Wen-Tao J., Ding-Cai Z., Ya-Ling H., Wen-Quan T. 2012. Prediction of fully developed turbulent heat transfer of internal helically ribbed tubes—An extension of Gnielinski equation. *International Journal of Heat and Mass Transfer* 55: 1375–1384.

Xiang-hui T., Dong-sheng Z., Guo-yan Z., Li-ding Z. 2013. Heat transfer and pressure drop performance of twisted oval tube heat exchanger. *Applied Thermal Engineering* 50: 374–383.

Xiao-Wei L., Ji-An M., Zhi-Xin L. 2007. Experimental study of single-phase pressure drop and heat transfer in a micro-fin tube. *Experimental Thermal and Fluid Science* 32: 641–648.

Yang S., Zhang L., Xu H. 2011. Experimental study on convective heat transfer and flow resistance characteristics of water flow in twisted elliptical tubes. *Applied Thermal Engineering* 31: 2981–2991.

Yu W., Ya-Ling H., Yong-Gang L., Jie Z. 2010. Heat transfer and hydrodynamics analysis of a novel dimpled tube. *Experimental Thermal and Fluid Science* 34: 1273–1281.

Yun R., Kim Y., Seo K., Young Kim H. 2002. A generalized correlation for evaporation heat transfer of refrigerants in micro-fin tubes. *International Journal of Heat and Mass Transfer* 45(10): 2003–2010.

Zimmerer C., Gschwind P., Gaiser G., Kottke V. 2002. Comparison of heat and mass transfer in different heat exchanger geometries with corrugated walls. *Experimental Thermal and Fluid Science* 26: 269–273.

CHAPTER 5

Advanced and Compact Heat Exchangers for the Specified Process Conditions

5.1 INFLUENCE OF GEOMETRICAL PARAMETERS ON HEAT EXCHANGER PERFORMANCE

The design of a heat exchanger for specified process conditions consists in the determination of the geometrical parameters corresponding to type of heat exchanger and its heat transfer surface that insure the required temperature program and corresponding heat load within the limits of allowable pressure drops of the streams. Depending on the construction type of exchanger and heat transfer technology in it, there can be a different set of geometrical parameters that must be determined prior to the start of the design. For shell and tube heat exchangers such parameters include type of baffles, their spacing, diameter of the tubes, passes on the tube side, tube pitch, layout of the bundle, etc. After a preliminary selection of geometrical parameters, these parameters are consequently optimised in search of such their values that minimise the cost of heat exchanger, including capital investment and operating maintenance.

In a simplified way based on the averaged characteristics, the heat transfer area of heat exchanger can be expressed using Equations 3.9 through 3.12 of Chapter 3 as follows:

$$F = \frac{Q}{C_{FT} \cdot \Delta T_{\ln}} \cdot \left[\frac{1}{h_1} + \frac{1}{h_2} \cdot \left(\frac{F}{F_2} \right) + \frac{\delta_w}{\lambda_w} + R_{f1} + R_{f2} \right] \tag{5.1}$$

Here F_2 is the effective surface area on the secondary side according to which the film heat transfer coefficient h_2 is determined.

In Equation 5.1, the values of heat load Q and LMTD (logarithmic mean temperature difference) $\Delta T_{\ln}$ are determined by process conditions, which should be satisfied by correctly designed heat exchanger. All other variables in this Equation are strongly influenced by heat exchanger construction and its geometrical parameters.

The correction factor C_{FT} for LMTD depends on the HE construction and relative flow scheme of heat exchanging streams inside heat exchanger. Its maximal value equal to unity can be achieved only with pure counter current scheme of streams, which is preferable for compact heat exchangers. All other schemes of streams flow lead to decreasing of C_{FT} and hence to deterioration of overall heat transfer. The values of LMTD correction factor for different flow configurations are presented in the literature see, for example, book by Shah and Sekulić (2003) or for compact plate-fin heat exchangers book by Kays and London (1984).

The thermal resistance of the wall δ_w/λ_w and the ratio of effective heat transfer surface area on extended side to the surface area on primary side F_2/F are dependent on construction of heat exchanger and material of which heat transfer wall and possible surface extensions are manufactured. These variables are known from construction type features mostly prior to thermal and hydraulic design of heat exchanger.

The thermal resistance of fouling at both sides of heat transfer surface R_{f1} and R_{f2} is dependent on a number of factors. First it is a nature of the streams media, which can be characterised as process conditions. Some process streams are clean and not inclined for forming the deposits on heat transfer surface. In that case, the thermal resistance of fouling at such stream side can be neglected. But in many practical applications, the streams media contains suspended solid particles or substances that can form deposits on heat transfer surface depending on its temperature and other conditions. In such circumstances, the construction of heat exchanger and its geometrical parameters can significantly influence the value of fouling thermal resistance by eliminating stagnation zones promoting fouling, enhancing flow turbulence and shear stress on the surface wall. These phenomena is discussed in more detail in the next chapter. In all discussions of Chapter 5 will be assumed zero fouling to consider more clear just heat transfer effects.

The film heat transfer coefficients in single phase flow at forced convection are a function of fluid physical properties, stream velocity and channel geometry. For given streams media, the film heat transfer coefficients h_1 and h_2 are the main variables in Equation 5.1 that are influenced by construction features of heat exchanger and geometrical parameters of heat transfer surface. For the specific channel geometry, the main influencing film heat transfer coefficient variable is flow velocity in channel w, which is directly related to the cross sectional area A_c, m^2, of the channels at one side of heat exchanger:

$$w = \frac{G}{\beta \cdot A_c}, \text{ m/s} \tag{5.2}$$

where

G is the stream mass flow rate, kg/s

ρ is the fluid density, kg/m^3

The heat transfer area of the heat exchanger is related to cross section area through its length L, m:

$$F = A_c \cdot L, \text{ m}^2 \tag{5.3}$$

To increase flow velocity, the cross-sectionals area should be decreased that is leading to higher values of film heat transfer coefficient and smaller heat transfer area *F*. For most cases of flow, the increase of film heat transfer coefficient is relatively smaller than increase in velocity (or decrease in cross sectional area). Therefore, the decrease of heat transfer area *F*, required to meet heat load and determined by Equation 5.1, is also relatively smaller than the change in cross section area. To compensate this effect, the length of heat transfer area should be increased following Equation 5.3.

The length of exchanger heat transfer area obtained when the unit is designed with only requirement of full satisfaction to the specified heat load can be called *thermal length*, following Picón-Núñez et al. (2013).

On the other hand, it is obvious that cross section area of heat exchanger cannot be made too small, as the pressure drop is limited. In correct design, the thermal length must be equal or smaller than *hydraulic length* which corresponds to the length of heat transfer surface of the same cross section area with pressure loss equal to specified allowable pressure drop. The design of heat exchanger requires the selection of heat exchanger geometrical parameters which in the best way reconcile thermal and hydraulic lengths on sides for both streams.

The mathematical approaches for optimal design of heat exchangers have been proposed by the number of authors. For example, Caputto et al. (2011) proposed for heat exchanger optimisation method based on life cycle cost with fouling analysis. This method allows for the determination of both the heat exchanger geometrical parameters and its cleaning schedules. Azad and Amidpour (2011) used constructal theory, a new method for optimal design in engineering applications, together with genetic algorithms to find the heat exchanger geometrical parameters that results in more than 50% reduction of the costs of the shell and tube heat exchanger.

The approach with multi-objective optimisation of the heat transfer area and pumping power of a shell and tube heat exchanger was presented by Fettaka et al. (2013). It provides the designer with multiple Pareto-optimal solutions to analyse the trade-off between the two objectives. Nine decision variables were considered: tube layout pattern, number of tube passes, baffle spacing, baffle cut, tube-to-baffle diametrical clearance, shell-to-baffle diametrical clearance, tube length, tube outer diameter and tube wall thickness. The optimisation was performed using the fast and elitist non-dominated sorting genetic algorithm available in the multi-objective genetic algorithm module of MATLAB. The multi-objective optimisation approach for design of shell and tube heat exchangers was proposed also by Sanaye and Hajabodollahi (2010). They have applied similar sorting genetic algorithm with seven continuous and discrete variables to obtain the maximum heat recuperation and the minimum total cost as two objective functions.

Another multi-objective optimisation approach for design of shell and tube heat exchangers was proposed by Rao and Patel (2013) based on a new heuristic algorithm using the natural phenomenon of the teaching–learning process.

For the optimal design of compact heat exchangers, Yousefi et al. (2012) developed a robust parameter-setting-free evolutionary approach using a learning particle swarm optimisation. The optimisation variables included seven design parameters, discreet and continuous ones. The obtained numerical results indicated that the presented approach generates the optimum configuration with higher accuracy and a higher success rate when compared to some other genetic algorithms particle swarm optimisation. The stochastic-based search algorithm for optimisation of heat exchangers was proposed by Hadidi and Nazari (2013) who introduced an optimisation design approach using the concept of biogeography-based optimisation.

The application of genetic algorithm for optimisation of a plate-fin type Compact Heat Exchanger (CHE) is discussed by Xie et al. (2008). The structure sizes of CHE are combined and optimised using the minimum total volume or/and total annual cost of the CHE as objective functions. With fixed geometries of the fins, three shape parameters are varied with or without pressure drop constraints. The resulting optimised CHE provides up to 49% lower volume or about 16% lower annual cost than compared CHEs presented in the literature.

The simplified approach for optimisation of compact plate heat exchanger (PHE) was proposed by Wang and Sunden (2003) with full utilisation of pressure drop as a design objective. Beside considerable simplifications in assumptions about correlations for heat transfer and friction factor as well as in problem definition, discussed by Polley (2003), the approach is giving clear view about the influence of geometrical parameters on the performance of PHE, as specific type of compact heat exchangers. More rigorous procedures for design of commercial PHEs were presented by Shah and Focke (1988) and Shah and Wanniarachchi (1991). These methods were described by Shah and Sekulić (2003) as (1) quite involved, (2) missing reliable data for thermal and hydraulic performance of commercial plates and (3) less rigorous methods can be used as it is easy to change the number of plates if the designed PHE does not confirm to the specification.

For design of PHE, Gut and Pinto (2004) presented an optimisation approach with heat transfer surface area as objective function accompanied by constraints on the number of channels, pressure drops, flow velocities and thermal effectiveness. They produced a screening procedure that identifies the optimal design from within the feasible region. For the case of multi-pass compact PHEs, Arsenyeva et al. (2011) present an approach for the optimisation of the number of passes and the number of plates with different form of corrugations per pass. The recommendations on obtaining correlations for commercial plates are given with examples of such correlations for some specific plates.

In recent times, the powerful tool for the estimation of the effect of geometrical parameters on heat exchanger performance became Computational Fluid Dynamics (CFD). It has been employed for different areas of study in various types of heat exchangers, such as fluid flow maldistribution, fouling, analysis of flow structure, pressure drop and thermal analysis in the design and optimisation phase. The comprehensive review of CFD applications in these areas for various types of heat exchangers is published by Bhutta et al. (2012).

5.2 PARAMETER PLOTS FOR THE PRELIMINARY DESIGN OF COMPACT HEAT EXCHANGERS

The complex, detailed methods for optimal design of heat exchangers implemented as software packages for computer are the powerful tool for heat exchanger selection on a final stage of ordering and manufacturing this equipment. For compact heat exchangers most of such software packages are the property of the manufacturing companies which are bearing responsibility for the operation of manufactured heat exchanger in accordance to requirements specified by customer.

At the preliminary design stage when rough estimation of the exchanger geometrical parameters and size of surface area is needed for decision making, the graphical tools can be a quick and effective aid. It is also helpful for better understanding of the exchanger type and main parameters influence on the efficiency of heat transfer equipment and the ways of its better satisfaction for specific process requirements.

The concept of parameter plots was first introduced by Poddar and Polley (1996) and was developed for the design of shell and tube heat exchangers. It illustrates graphically the relationship between geometrical dimensions of heat exchanger like shell diameter and tube length for specified heat duty and pressure drops. This type of analysis determines the possible range of exchanger geometries that meet the specified process conditions such as heat load and the available pressure drops. The area at the plot at which various combinations of geometrical parameters correspond to heat exchanger that meets the process specifications is called design space. With a parameter plot, a designer can easily estimate the influence of allowable pressure drop on the heat exchanger dimensions and also identify the degree of overdesign for selected values of geometrical parameters. It can be used for preliminary decisions and analysis of the link between different geometrical parameters such as tube count versus tube length or shell diameter versus tube length for shell and tube heat exchanger. A typical parameter plot for shell and tube heat exchanger looks like shown in Figure 5.1. The simplicity of the method and its illustrative character make it helpful in exchanger optimisation.

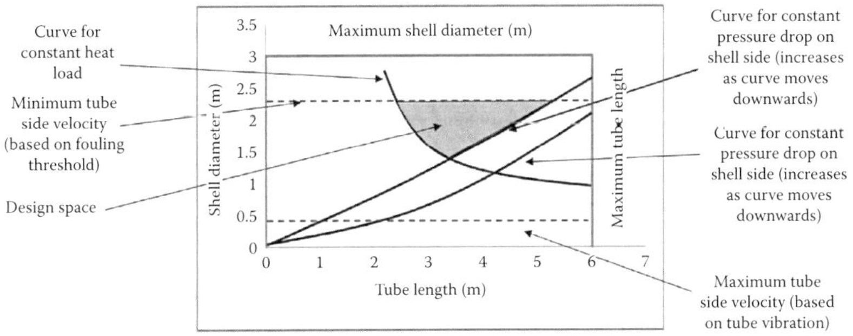

Figure 5.1 Example of parameter plot for the design of shell and tube heat exchanger (after Poddar and Polley, 1996).

The procedure of developing parametric plot requires the computer software for rating calculations of selected type of heat exchanger. It enables to use the constraints placed upon the design to reduce the number of possible design options to some design space corresponding to feasible heat exchangers satisfying required process conditions. To illustrate the method Poddar and Polley (1996) have presented the example of parametric plot of shell and tube heat exchanger for specific process requirements as follows. Naphtha with flow rate of 32.22 kg/s is to be heated from 90 °C to 110 °C by light oil flowing with the rate 19.44 kg/s. Inlet temperature of light oil is 180 °C. By the heat balance of Naphta stream, the heat load is 1.511 MW and light oil should be cooled down to 148 °C. Light oil as higher viscosity liquid is directed into shell side of the heat exchanger. The fouling resistance equal to 0.007 m^2 K/W is used in the design. The straight tube and bundle allowable pressure drops are taken both 4,500 Pa.

The tubes are taken of outside diameter 25.4 mm with wall thickness 2 mm. From the constraint of maximum tube velocity 2 m/s is estimated a tube count of 62 and corresponding shell diameter 0.34 m. The minimum Reynolds Number is set at 10,000 which occur at a velocity of 0.35 m/s. This happens at a tube count of 355 corresponding to a shell diameter of 0.76 m. The baffle cut was set at 25% and the baffle spacing equal to the shell diameter. Using commercial computer program is then used to rate exchangers with the shells of diameters: 0.34 m (minimum), 0.44 m, 0.54 m, 0.64 m and 0.76 m (maximum). As a result the parameter plot depicted in Figure 5.1 was obtained. The shadow area is corresponding to possible space of geometrical parameters that corresponds to process requirements and limitations on shell and tube heat exchanger construction.

With smaller tubes diameters and rating software utilising correlations for enhanced tubes, the method can be used for compact heat exchangers of shell and tube type. Picón-Núñez et al. (2010) extended the concept of parameter plots to plate and frame heat exchangers. In this case, the design space is shown in the form of a bar chart where the number of plates (instead of exchanger length) necessary to meet the required heat load, the specified pressure drop on the hot side and the required pressure drop on the cold side are displayed for different sizes of standard plates. More recently, Picón-Núñez et al. (2013) have further developed this approach for spiral heat exchangers and all welded multi-pass 'compablock' heat exchangers with cross flow in one pass and overall multi-pass counter current flow. The described graphical tools can become helpful in the development of optimal solutions and even the development of software for a number of different types of compact heat exchanger technologies.

5.3 THE INFLUENCE OF PLATE CORRUGATIONS GEOMETRY ON PLATE HEAT EXCHANGER PERFORMANCE IN SPECIFIED PROCESS CONDITIONS

Plate heat exchangers (PHEs) are one of the most efficient types of heat transfer equipment. These compact heat exchangers are relatively much smaller and require much less material for heat transfer surface production than conventional shell and tubes units. PHEs have a number of advantages over shell and tube heat exchangers, such as compactness, low total cost, less fouling, flexibility in changing the heat transfer surface area, accessibility and what is very important for low-grade heat utilisation, a close temperature approach – down to 1 °C. However, due to the differences in construction principles from conventional shell and tube heat exchangers, PHEs require substantially different methods of thermal and hydraulic design.

One of the inherent features of PHEs is their flexibility. The heat transfer surface area can be changed discretely with a step equal to heat transfer area of one plate. All major producers of PHEs manufacture a range of plates with different sizes, heat transfer surface areas and geometrical forms of corrugations. This enables the PHE to closely satisfy required heat loads and pressure losses of the hot and cold streams.

The thermal and hydraulic performance of a PHE with plates of certain size and type of corrugation can be varied in two ways: (1) by adjusting the number of passes for each of exchanging heat streams and (2) by proper selection of plate corrugation pattern. For the most common chevron plates, it is the angle of corrugations inclination to the plate longitudinal axis. One of early attempts to find the patterns that minimise the surface area required for heat transfer was made by Focke (1986). The optimal design of a PHE by adjusting corrugation pattern on plate surface was reported by Wang and Sunden (2003). Picón-Núñez et al. (2010) presented an alternative design approach based on graphical representation, which facilitates the choice from the options calculated for the range of available plates with different geometries. They have estimated correlations for heat transfer and hydraulic resistance from available literature data. Similar estimations were also made by Mehrabian (2009), who proposed a manual method for the thermal design of plate heat exchangers. Wright and Heggs (2002) have shown how the operation of a two stream PHE can be approximated after the plate rearrangement has been made, using the existing PHE performance data. Their method can help when adjusting PHE, which is already in operation, for better satisfaction to required process conditions. Kanaris et al. (2009) have estimated parameters in correlations for Nusselt and friction factor using CFD modelling of the flow in a PHE channel of special geometry. However, their results are still a long way from practical application.

Currently for most PHEs, the effect of varying plate corrugation pattern is achieved by combining chevron plates with different corrugation inclination angle in one PHE. The design approach and advantages of such a method were shown by Marriot (1977) for a one pass counter-current arrangement of PHE channels. A one pass channel arrangement in a PHE has many advantages compared to a multi-pass one, especially in view of piping and maintenance (all connections can be made on not movable frame plate from one side of PHE). But in certain conditions, the required heat transfer load and pressure drops can be satisfied more efficiently by application of multi-pass arrangement of PHE's channels.

Until the 1980s, the proper adjustment of the number of passes was the only way to satisfy the required heat load and pressure drops in PHE consisting of plates of a certain type. The multi-pass arrangement enables increased flow velocities in channels and thus to achieve higher film heat transfer coefficients if allowable pressure drop permits. But, for unsymmetrical passes, the problem of diminishing effective temperature differences has arisen. Most of the authors who have proposed design methods for PHEs have used LMTD correction factors (see, e.g., Cocks 1969 and later Kumar 1994; in Russian literature see Zinger et al. 1988). Initially such correction factors could be taken from handbooks on heat exchanger design. After development of methods for analysis of complicated flow arrangements (see Pignotti and Shah 1992), it became possible to develop closed-form formulas for two-fluid recuperators. Using a matrix algorithm and the chain rule, Pignotti and Tamborenea (1988) developed a computer program to solve the system of linear differential equations for the numerical calculation of the thermal effectiveness of arbitrary flow arrangements in a PHE. Kandlikar and Shah (1989) analysed different flow arrangements and proposed formulas for up to four passes. These and other similar formulas can be found in books on heat exchangers thermal design, for example, Wang et al. (2007), for PHEs and Shah and Seculić (2003) for different heat exchanger types. The main assumptions made on deriving such formulas are (1) constant fluids properties and overall heat transfer coefficients, (2) uniformity of fluid flow distribution between the channels in same pass and (3) sufficiently large number of plates.

Because of the increase in computational power of modern computers, the difference of heat transfer coefficients between passes can be accounted for in the design by solving the system of algebraic equations, as was proposed by Tovazshnyansky et al. (1992) and further developed by Arsenyeva et al. (2011). The flow maldistribution between channels was investigated among others by Rao et al. (2002), who noticed the significant effect of heat transfer coefficient variation, which was not accounted for in previous works. However, we can adhere to the conclusion made earlier by Bassiouny and Martin (1984), based on analytical study of velocity and pressure distribution in both the intake and exhaust conduits of PHE, that plate heat exchanger can be designed with equal flow distribution regardless of the number of plates. The correct design of manifolds and flow distribution zones is also very important for tackling fouling problems in PHEs, as shown by Kukulka and Devgun (2007). However, the design should take account of the limitations imposed by the percentage of port and manifolds pressure drops in the total pressure losses, as well as for flow velocities. Along with flow maldistribution, Heggs and Scheidat (1992) have studied end-plate effect. They concluded that critical number of plates is dependent on the required accuracy of performance, for example, 19 can be recommended for an inaccuracy of only 2.5%.

The significant feature of PHE design is the fact that the required conditions of certain heat transfer process can be satisfied by a number of different plates. But it is achieved with different level of success in terms of material and cost for production. The plate, which is the best for certain process conditions, should be selected from the available set of plates according to some optimisation criterion. Therefore, the design of optimal PHE, for given process conditions, should be made by selection of the best option from available alternative options of plates with different geometrical characteristics. To satisfy requirements of different process conditions any PHE manufacturer is producing not just one plate type, but the sets of different types of plates. To make a right selection, we need the mathematical model of PHE to estimate performance of the different alternative options. It should be accurate enough and at the same time to have small number of parameters, which can be identified on a data available for commercial plates.

5.3.1 Mathematical Modelling and Design of Industrial PHEs

A plate-and-frame PHE consists of a set of corrugated heat transfer plates clamped together between fixed and moving frame plates and tightened by tightening bolts (see Figure 3.21). The plates are equipped with the system of sealing gaskets, as shown in Figure 3.22, which also separate the streams and organising their distribution over the inter-plate channels. In multi-pass PHE, the plates are arranged in such a way that they form groups of parallel channels. An example is shown in Figure 5.2. The temperature distribution in passes can vary and in different groups of channels both counter-current and co-current flows may occur.

The mathematical model of a PHE can be derived based on the following assumptions:

- The heat transfer process is stationary.
- No change of phases in streams.
- The number of heat transfer plates is big enough not to consider the differences in heat transfer conditions for plates at the edges of passes and of total PHE.
- Flow misdistribution in collectors can be neglected.
- The streams are completely mixed in joint parts of PHE collectors.

With these assumptions PHE can be regarded as a system of one pass blocks of plates. The conditions for all channels in such block are equal. For example, an arrangement with three passes for the hot stream ($X_1 = 3$) and two for the cold stream ($X_2 = 2$) is shown in Figure 5.3. The heat transfer area of the block is given by $F_b = F/(X_1X_2)$, where F is the total heat transfer area of PHE. The change of hot stream temperature in each block is δt_i, $i = 1, 2, \ldots, 6$.

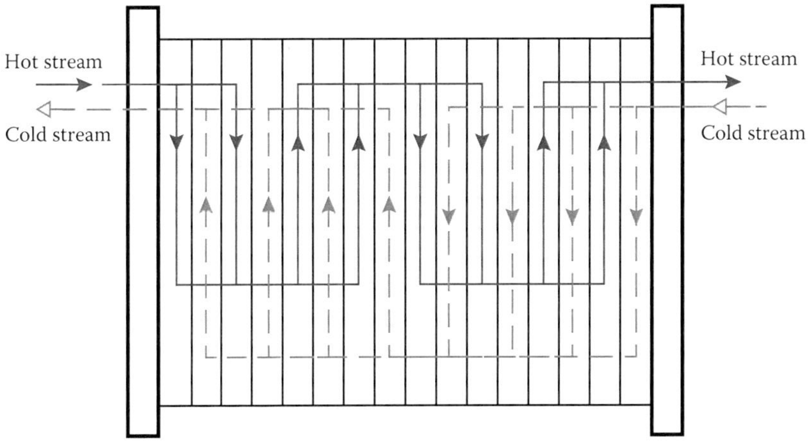

Figure 5.2 An example of streams flows through channels in multi-pass PHE.

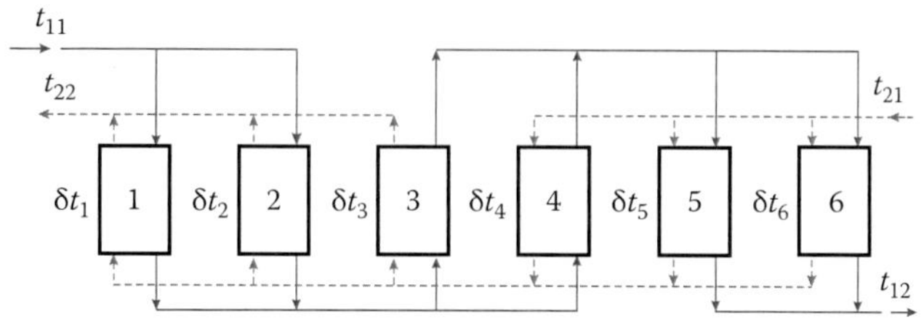

Figure 5.3 The presentation of PHE as a system of plate blocks (X_1 = 3, X_2 = 2).

The total number of blocks is $n_b = X_1X_2$ and the number of heat transfer units in one block, counted for hot stream is

$$NTU_b = U_b \cdot F_b \cdot X_2/(G_1 \cdot c_1) \tag{5.4}$$

where

U_b is the overall heat transfer coefficient in block, W/(m$^2 \cdot$K)
G_1 is the mass flow rate of hot stream, kg/s
c_1 is the specific heat of hot stream, J/(kg · K).

If we assume $G_1 \cdot c_1/X_2 < G_2 \cdot c_2/X_1$, then block heat exchange effectiveness ε_b for counter current flow is:

$$\varepsilon_b = \frac{1 - \exp(NTU_b \cdot R_b - NTU_b)}{1 - R_b \cdot \exp(NTU_b \cdot R_b - NTU_b)} \tag{5.5}$$

where

$R_b = G_1 \cdot c_1 \cdot X_1/(G_2 \cdot c_2 \cdot X_2)$ is the ratio of going through block heat capacities of streams
G_2 and c_2 are mass flow rate [kg/s] and specific heat [J/(kg · K)] of cold stream

If $R_b = 1$, then $\varepsilon_b = NTU_b/(1 + NTU_b)$.
In case of co-current flow:

$$\varepsilon_b = \frac{1 - \exp(-NTU_b \cdot R_b - NTU_b)}{1 + R_b} \tag{5.6}$$

On the other hand, the heat exchange effectiveness of block i is:

$$\varepsilon_{bi} = \delta t_{1i} / \Delta t_i \tag{5.7}$$

where

δt_{1i} is the temperature drop in block i
Δt_i is the temperature difference of streams entering block i

The temperature change of the cold stream:

$$\delta t_{2i} = \delta t_{1i} \cdot R_b \tag{5.8}$$

The above relations also hold true at $G_1 \cdot c_1/X_2 > G_2 \cdot c_2/X_1$. In that case the physical meaning of ε_b and NTU_b are different, as shown by Shah and Seculić (2003). Thus, these relations can be regarded as a mathematical model of a block, which describes the dependence of temperature changes from the characterising block values of F_b and U_b.

For each block, can be written the equation, which describes the link of temperature change in this block to the temperature changes in all other blocks of the PHE. For example, let us consider the first block in Figure 5.3. The difference of temperatures for streams entering the block can be calculated by deducting the average temperature rise of cold stream in blocks 4, 5 and 6 from the initial temperature difference Δ of the streams entering the PHE:

$$\Delta t_1 = \Delta - \left(\delta t_4 \cdot R_b + \delta t_5 \cdot R_b + \delta t_6 \cdot R_b\right)/3 \tag{5.9}$$

After substituting this into Equation 5.7 and rearranging we obtain:

$$\delta t_1 + \delta t_4 \frac{\varepsilon_{b1}}{3} R_b + \delta t_5 \frac{\varepsilon_{b1}}{3} R_b + \delta t_6 \frac{\varepsilon_{b1}}{3} R_b = \varepsilon_{b1}\Delta \tag{5.10}$$

In this way equations can be obtained for every block in the PHE. Consequently we can obtain a system of 6 linear algebraic Equations with 6 unknown variables δt_1, $\delta t_2, \ldots, \delta t_6$.

We have built these systems of equations for the number of passes up to $X_1 = 7$ and $X_2 = 6$, with an overall counter current flow arrangement. The analysis of results has shown that, for any number of passes, the system may be presented in matrix form:

$$[Z][\delta t_i] = [\varepsilon_{bi}\Delta] \tag{5.11}$$

where

- $[\delta t_i]$ is the vector-column of temperature drops in blocks
- $[\varepsilon_i \Delta]$ is the vector-column of the right hand parts of the system
- $[Z]$ is the matrix of system coefficients, whose elements are:

$$z_{ij} = \begin{bmatrix} \dfrac{\varepsilon_{bi} R_b}{2X_1}\left\{1 \cdot \operatorname{sign}\left(j - \left(\operatorname{int}\left[\dfrac{i-1}{X_i}\right] + 1\right) X_1 + 0.5\right) + 1\right\}, \quad \text{if } j > i \\ 1, \quad \text{if } i = j \\ \dfrac{\varepsilon_{bi}}{2X_2}\left\{1 \cdot \operatorname{sign}\left(\operatorname{int}\left[\dfrac{i-1}{X_2}\right] X_2 - j + 0.5\right) + 1\right\}, \quad \text{if } j < i \end{bmatrix} \tag{5.12}$$

Here

- i is the row number
- j is the column number

The numerical solution of this type of linear algebraic Equations system 5.12 can be easily performed on a PC, after which the total temperatures change in the PHE can be calculated as

$$\delta t_{\Sigma 1} = \sum_{i=1}^{X_1} \left(\frac{1}{X_1} \sum_{ii=1}^{X_2} \delta t_{(i-1)X_2+ii} \right), \quad \delta t_{\Sigma 2} = \frac{(G_1 c_1)}{(G_2 c_2)} \delta t_{\Sigma_1} \tag{5.13}$$

The total heat load of PHE is

$$Q = \delta t_{\Sigma 1} \cdot G_1 \cdot c_1 \tag{5.14}$$

This system should be accompanied by equations for the calculation of the overall heat transfer coefficient U, W/m$^2 \cdot$K, as follows:

$$U = 1 \Big/ \left(\tfrac{1}{h_1} + \tfrac{1}{h_2} + \left(\tfrac{\delta_w}{\lambda_w} \right) + R_f \right) \tag{5.15}$$

where

h_1, h_2 is the film heat transfer coefficients for hot and cold streams, respectively, W/(m$^2 \cdot$K)

δ_w is the wall thickness, m

λ_w is the heat conductivity of the wall material, W/(m$\cdot$K)

$R_f = R_{f1} + R_{f2}$ is the sum of fouling thermal resistances for streams, m$^2 \cdot$K/W

For plate heat exchangers the film heat transfer coefficients are usually calculated by empirical correlations:

$$\mathrm{Nu} = f(\mathrm{Re}, \mathrm{Pr}) = A \cdot \mathrm{Re}^n \, \mathrm{Pr}^{0.4} (\mu/\mu_w)^{0.14} \tag{5.16}$$

where μ and μ_w are the dynamic viscosity at stream and at wall temperatures, Pa$\cdot$s

The Nusselt number is

$$\mathrm{Nu} = \alpha \cdot d_e / \lambda \tag{5.17}$$

where

λ is the heat conductivity of the respective stream, W/(m$\cdot$K)

d_e is the equivalent diameter of PHE channel, m

$$d_e \cong \frac{4b\delta}{2(b+\delta)} \approx 2\delta \tag{5.18}$$

where

δ is the inter-plate gap, m

b is the channel width, m

The Reynolds number is given by

$$\mathrm{Re} = w \cdot d_e \cdot \rho/\mu \tag{5.19}$$

where w is the stream velocity in channel, m/s

The Prandtl number is given by

$$\mathrm{Pr} = c \cdot \mu/\lambda \tag{5.20}$$

where

ρ is the stream density, kg/m^3

c is the specific heat capacity of the stream, J/(kg · K)

The streams velocities are calculated as

$$w = g/(f \cdot \rho) \tag{5.21}$$

where

g is the flow rate of the stream through one channel, kg/s

f is the cross section area of inter-plate channel, m^2

The pressure drop in one PHE channel is given by

$$\Delta p = \zeta \cdot \frac{L_p}{d_e} \cdot \frac{\rho \cdot w^2}{2} + \Delta p_{p-c} \tag{5.22}$$

where

$\Delta p_{p-c} = 1.3 \cdot \rho \cdot w_{port}^2/2$ is pressure losses in ports and collector part, Pa

L_p is the effective plate length, m

w_{port} is the velocity in PHE ports and collectors, m/s

ζ is the friction factor

which is usually determined by empirical correlations of following form:

$$\zeta = B/\mathrm{Re}^m \tag{5.23}$$

For multi-pass PHE, the pressure drop in one pass is multiplied by the number of passes X.

In modern PHEs, plates of one type are usually made with two different corrugation angles, which can form three different channels, when assembled in PHE, as shown in Figure 5.4.

H type plates have corrugations with larger angles (about 60°) which form the *H* channels with higher intensity of heat transfer and larger hydraulic resistance. *L* type plates have a lower angle (about 30°) and form the *L* channels which have lower heat transfer and smaller hydraulic resistance. Combined, these plates form *M* channels

Figure 5.4 Channels formed by combining plates of different corrugation geometries: (a) Channel *L* formed by *L*-plates, (b) Channel ML-MH formed by *L*- and *H*-plates and (c) Channel *H* formed by *H*-plates.

with intermediate characteristics (see Figure 5.4). This design technique allows the thermal and hydraulic performance of a plates pack to be changed with a level of discreteness equal to one plate in a pack.

In one PHE two groups of channels are usually used. One is of higher hydraulic resistance and heat transfer (x-channel), another of lower characteristics (y-channel). When the stream is flowing through a set of these channels, the temperature changes in the different channels differ. After mixing in the collector part of the PHE block, the temperature is determined by the heat balance. The heat exchange effectiveness of the plates block with different channels is given by

$$\varepsilon_b = \left(g_x \cdot n_x \cdot \varepsilon_x + g_y \cdot n_y \cdot \varepsilon_y\right) / \left(g_x \cdot n_x + g_y \cdot n_y\right) \tag{5.24}$$

where

n_x and n_y are the numbers of x and y channels in a block of plates

$g_{x,y} = w_{x,y} \cdot \rho \cdot f_{ch}$ is the mass flow rates through one channel of type x or y, kg/s.

These flow rates should satisfy the equation $\Delta p_x = \Delta p_y$ and the material balance:

$$g_x \cdot n_x + g_y \cdot n_y = G_b \tag{5.25}$$

where G_b is the flow rate of the stream through the block of plates, kg/s.

The principle of plate mixing in one heat exchanger gives the best results with symmetrical arrangement of passes ($X_1 = X_2$) and G_b equal to the total flow rate of the respective stream. The unsymmetrical arrangement $X_1 \neq X_2$ is usually used when all channels are the same (any of the three available types).

When the numbers of channels are determined, the numbers of plates can be calculated by

$$N_{pl} = \sum_{i=1}^{X_1} \left(n_{x1i} + n_{y1i}\right) + \sum_{j=1}^{X_2} \left(n_{x2i} + n_{y2i}\right) + 1 \tag{5.26}$$

The total heat transfer area of the PHE (with two end plates not included) is given by

$$F_{PHE} = (N_{pl} - 2) \cdot F_{pl} \tag{5.27}$$

where
F_{pl} is the heat transfer area of one plate, m^2

The above algebraic Equations 5.4 through 5.27 describe the relationship between variables that characterise a PHE and the heat transfer process contained within the PHE. These equations can be presented as a mathematical model of a PHE, and the solution allows the calculation of the pressure and temperature change of streams entering the heat exchanger. It is a problem of PHE rating (analysis).

The problem of PHE sizing (synthesis) requires finding characteristics such as plate type, the numbers of passes and the number of plates with different corrugations, which will best satisfy to the required process conditions. Here, the optimal design with pressure drop specification is considered.

The most important and costly, counting the tools for stamping at the manufacture, parts of plate-and-frame PHE are plates with gaskets. The plates can be made of stainless steels, titanium, and other even more expensive alloys and metals. All other component parts of PHE (frame plates, bars, tightening bolts, etc.) usually are made from less expensive construction steels and have a smaller share in a cost of PHE. The plates constitute the heat transfer area of PHE and there is strong dependence between the cost of PHE and its heat transfer area. Therefore, for optimisation of PHE heat transfer area F can be taken as objective function, which is also characterising the PHE cost and the need in sophisticated materials for plates and gaskets.

For specific process conditions, when temperatures and flow rates of both streams are specified, required heat transfer surface area F_{PHE} is determined through solution of mathematical model presented above by Equations 5.4 through 5.27. It is implicit function of plate type T_{pl}, number of passes X_1, X_2 and composition of plates with different corrugations pattern $[N_H/N_L]$. We can formulate the optimisation problem for PHE design as a task to find the minimum of the following objective function:

$$F_{PHE} = f(T_{pl}, X_1, X_2, \left[N_H/N_L\right]) \tag{5.28}$$

It should satisfy to constraints imposed by required process conditions:

Heat load Q has to be not less than required Q^0:

$$Q \geq Q^0 \quad \text{or} \quad \Delta t_1 \geq \Delta t_1^0 \tag{5.29}$$

The pressure drops of both streams must not accede allowable:

$$\Delta p_1 \leq \Delta p_1^0 \tag{5.30}$$

$$\Delta p_2 \leq \Delta p_2^0 \tag{5.31}$$

There also constraints imposed by the features of PHE construction.

On a flow velocity in ports:

$$w_{port} \leq 7 \text{ m/s} \tag{5.32}$$

The share of pressure losses in ports and collector Δp_{p-c} in total pressure drop for both streams:

$$(\Delta p_{p-c}/\Delta p)_{1,2} \leq 0.3 \tag{5.33}$$

The number of plates on one frame must not exceed the maximum allowable for specific type of PHE plates $n_{max}(T_{pl})$:

$$N_H + N_L \leq n_{\max}(T_{pl}) \tag{5.34}$$

In the PHE, the numbers of channels and their form for both sides must be the same, or differ only on 1 channel:

$$abs\left[\sum_{i=1}^{X_1} n_{x1i} - \sum_{j=1}^{X_2} n_{x2i}\right] \leq 1 \tag{5.35}$$

$$abs\left[\sum_{i=1}^{X_1} n_{y1i} - \sum_{j=1}^{X_2} n_{y2i}\right] \leq 1 \tag{5.36}$$

Analysis of the relations described in Equations 5.4 through 5.36 shows that we have a mathematical problem of finding the minimal value for implicit non-linear discrete/continues objective function with inequality constraints. It does not permit analytical solution without considerable simplifications, but can be readily solved on modern computers numerically. The basis of the developed algorithm is the fact that optimal solution has to be situated in the vicinity of the border, by which constraints on the pressure drop in PHE are limiting the space of possible solutions. Usually in one PHE, three possible types of channel can be used. For limiting flow rates of i-th stream in one channel of the j-type from constraints on pressure drop (5.30) and (5.31), using formulas 5.21 through 5.23, we get

$$g_{ij}^0 = \left[\frac{2\cdot(\Delta p_i^0 - 0.65\cdot X_i\cdot\rho_i\cdot w_{port}^2)\cdot d_{eq}\cdot\rho_i\cdot f_j^2}{L_p\cdot B_j\cdot X_i}\left(\frac{d_e}{f_j\cdot\rho_i\cdot\nu_i}\right)^{m_j}\right]^{1/(2-m_j)} \tag{5.37}$$

Due to constraints (5.35) and (5.36), the required pressure drops for both streams cannot be exactly satisfied simultaneously. We should correct the flow rates for one of the streams, using constraints (5.35) and (5.36) with assumptions that they are both strict equalities and all passes have the same numbers of channels:

$$G_1/g_{1j} = G_2/g_{2j} \tag{5.38}$$

The possible difference in amount of one channel can be accounted for at rating design.

At known values of g_{ij}, the film and overall heat transfer coefficients and all coefficients of the system of linear algebraic equations are directly calculated by equations of the above mathematical model. The system is solved by standard utility programs. When the required Δt_1^0 drops between the values of calculated Δt_{1j} for two channels, the required numbers of these channels and after corresponding plates of different corrugations are calculated using formulas 5.24 through 5.26. In case of Δt_1^0 lower than the smallest Δt_{1j}, all L-plates are used and constraint (5.25) is satisfied as inequality. Margin on heat transfer load can be calculated:

$$M = \Delta t_{1L}/\Delta t_1^0 - 1 = Q/Q^0 - 1 \tag{5.39}$$

If Δt_1^0 higher than the biggest Δt_{1j}, all H-plates are taken and their number increased until the constraint (5.29) is satisfied. But in this case appears the big margin on constraints (5.30) and (5.31). Allowable pressure drops are not completely utilised and configurations with the increased passes numbers must be checked.

The calculations are starting from $X_1 = 1$ and $X_2 = 1$. The number of passes is increased until the calculated heat transfer surface area is lowered. If the area increases, the calculations terminated, all derived surface areas compared and the option with smallest area selected. The procedure can be applied to all available for design plate types and after the best option is selected, but nearby options are also available for designer decision.

The algorithm outlined above is inevitably leading to the best solution for any number N_T of available plate types T_{pl}. It is implemented in developed software for IBM compatible PC. However, the mathematical model contains some parameters, namely, coefficients and exponents in empirical correlations, which are not readily available.

As a rule the empirical correlations for design of industrially manufactured PHEs are obtained during tests on such heat exchangers at specially developed test rigs. Such tests are made for every type of new developed plates and inter-plate channels. The results are proprietary of manufacturing company and not usually published. Every PHE manufacturing company has its own design software according to which the final calculations and selection of PHE for production is made on ordering by the customer of the equipment. But for calculations of multiple options needed to be analysed on a stage of preliminary selection of heat transfer equipment, especially when optimising heat exchanger networks, the independent software that can be integrated into more general software packages is required. There are two ways of obtaining correlations for implementation of described above procedure of PHE design.

First one is based on the presented above mathematical model. A numerical experiment technique can be used which enables the identification of the model parameters by comparison with results obtained for the same conditions with the use of PHE calculation software, which is now available on the Internet from most PHE manufacturers. Such procedure is described in Appendix 5.A with the example of the obtained correlations for prediction of film heat transfer and friction factor for the set of commercial plates with different sizes and other geometrical parameters.

The second one is to employ the semi empirical correlations for prediction of friction factor and film heat transfer coefficients for the main corrugated field of the criss-cross flow PHE channels formed by plates with different geometrical parameters of corrugations. This approach is described below.

5.3.2 Prediction of Heat Transfer and Pressure Drop in Channels Formed by Commercial Plates

Let us consider the plate's surface of industrial plate-and-frame PHE, which is schematically shown in Figure 4.10. The working part of that surface, washed by flowing inside the channel heat exchanging stream, consists of the main corrugated field (4) and zones of flow distribution at the inlet (2) and outlet (5). The most part of heat is being transferred on the main corrugated field, which area comes to 80%–85% of the total plate's heat transfer surface area. On the distribution zones considerably less heat is transferred, but their influence on the overall hydraulic resistance of the channel is significant. In these zones, the velocities of flow are increasing from the velocity value at the main operating field up to the velocity on the outlet from the channel to the gathering collector of PHE (the same is for inlet from the distributing collector of PHE).

The design of corrugations at distribution zones can considerably vary for different plates, what affects their hydraulic resistance and evenness of flow distribution across the inter-plate channel width. As it can be judged from the data of Tovazhnyansky et al. (1980), the share of pressure drop in entrance and exit zones of some PHE channels, formed by commercial plates of chevron type, can reach 50% and more, especially at lower angles β of corrugations inclination to plate vertical axis.

For the mathematical modelling of thermal and hydraulic performance of PHE, the inter-plate channel is considered as consisting of its main corrugated field, which causes major effect on heat transfer, and distribution zones, which are influencing mostly the hydraulic performance. The empirical equation for calculation of friction factor ζ in criss-cross flow channels was reported in previous chapter. It enables to predict friction factors at corrugated field of criss-cross flow channels formed by plates with inclined corrugations in wide range of corrugation parameters.

To account for the pressure losses in distribution zones (2 and 5 in Figure 4.10) of the inter-plate channel between commercial plates it has been introduced the coefficient of local hydraulic resistance in those zones ζ_{DZ}, assuming that these coefficients for inlet and outlet zones are equal. The average velocity of stream at the main corrugated field w is characteristic velocity in determining these coefficients. The total pressure loss in a PHE channel can be calculated as follows:

$$\Delta p = \zeta \cdot \frac{L_F}{d_e} \cdot \frac{\rho \cdot w^2}{2} + \zeta_{DZ} \cdot \rho \cdot w^2 \tag{5.40}$$

where

ρ is the fluid density, kg/m^3

L_F is the length of corrugated field, m

$d_e = 2 \cdot b$ is the equivalent diameter of the channel, m

The detailed data about heat transfer and hydraulic performance of commercially manufactured plates are rather limited, as mostly it is proprietary information of the manufacturing companies. Using available information for plates manufactured by Pavlogradchmmash factory in Ukraine, Tovazhnyansky et al. (1980) have estimated the coefficient of local hydraulic resistance ζ_{DZ} in distribution zones for some plates with triangular shape of inclined at angle β herringbone corrugations. The values of ζ_{DZ} were estimated using correlations for friction factor on the models of corrugated field ζ and friction factor ζ_{total} for channels formed by commercial plates by following equation:

$$\zeta_{DZ} = \frac{1}{2} \cdot \left(\zeta_{total} - \zeta \right) \cdot \frac{L_{EF}}{d_h} \tag{5.41}$$

where

L_{EF} is effective length of the plate, m

d_h is hydraulic diameter of the PHE channel, m

The values of local hydraulic resistance coefficients obtained after averaging in the range of Reynolds numbers $1{,}200 < Re < 10{,}000$ are presented in Table 5.1. The discrepancies of calculations for pressure drop by Equation 5.40 and by empirical correlations using friction factor for channels formed by commercial plates included distribution zones were not exciding 15%.

The data in Table 5.1 show that for the presented there plates with different forms of corrugations, but with similar constructed distribution zones, the coefficients ζ_{DZ} are close in values. For different plates there were obtained the values of such coefficients from 25 to 110. For four out of six investigated plates the average value was 27.25. For those plates the ratio of the length of corrugated field L_F to effective plate length L_{EF} used in calculations of ζ_{DZ} was equal to 0.72. In Equation 5.40, L_F is the length of corrugated field not including distribution zones and equivalent diameter is determined as $d_e = 2 \cdot b$. As some approximation for use in Equation 5.40, the values of ζ_{DZ} in Table 5.1 should be divided by 0.72. It gives $\zeta_{DZ} = 38$. Arsenyeva et al. (2013) have used this value for commercial plates modelling, considering the construction of distribution zones in these four plates of Table 5.1 as more preferable for reproduction in process of new plate development. The particular design of distribution zones is a complex problem which is not considered here. It is assumed that performance of these zones is the same as for better distribution zones on commercial plates which data are presented in the literature.

The experimental data for some commercial plates with different angle of corrugation inclination to plates vertical axis angle $\beta = 60°$ and $\beta = 30°$ are reported by Muley and Manglik (1999). They have obtained the following correlation for friction factor at different inclination angles of sinusoidal shape corrugations based

Table 5.1 Coefficients of Local Hydraulic Resistance in Distribution Zones of some Commercial Plates according Tovazhnyansky et al. (1980)

Type of Plate	0.5E	0.3	0.6	0.5 × 2	0.63	0.8
ζ_{DZ}	110	80	30	28	25	26
β, degrees	60	60	60	60	30	30

on experiments with two differently corrugated commercial plates of one size for Reynolds numbers 1,000 ≤ Re ≤ 10,000:

$$\zeta = 4\cdot\left[2.917-0.1277\cdot\beta+2.016\cdot10^{-3}\cdot\beta^2\right]\cdot \mathrm{Re}^{-\left[0.2+0.0577\cdot\sin\left\{(\pi\cdot\beta/45)+2.1\right\}\right]} \tag{5.42}$$

The comparison of calculations by Equation 5.40 with correlation (5.42) is presented on graph in Figure 5.5 (filled squares and circles). All corrugations parameters are taken from the paper by Muley and Manglik (1999) and the length of the main corrugated field from the drawing presented there L_F = 278 mm. The maximal deviation of calculated results from experimental data not exceeded ±15%. More detailed examination of the graph in Figure 5.5 leading to conclusion about the influence of Reynolds number on the value of ζ_{DZ}. The flow at the distribution zone of the channel has a complex character, but the existence of contact points between adjacent plates makes possible the assumption that the influence of Reynolds number on hydraulic resistance in this zone is similar to that one on a main corrugated field, especially at higher approximated by Equation 4.65 values of β, let us take β = 60°. The closest match of calculated and experimental pressure losses in Figure 5.5 is at Re = 2,700. Then the correction factor for ζ_{DZ} may be introduced as the ratio of ζ_{60}(Re), calculated by Equation 5.1 with β = 60° at examined Re and all other parameters same as for the main corrugated field, to that value of ζ_{60}(2700) at Re = 2,700:

$$\zeta_{DZ} = 38\cdot\frac{\zeta_{60}(\mathrm{Re})}{\zeta_{60}(2700)} \tag{5.43}$$

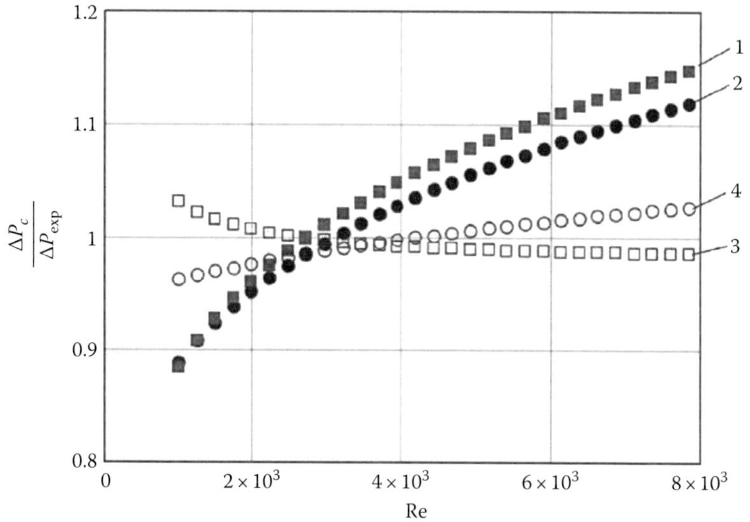

Figure 5.5 The comparison of calculated pressure losses with experimental data presented by Muley and Manglik (1999) for PHEs formed between commercial plates: 1 – β = 30°; 2 – β = 60° (for ζ_{DZ} = 38); 3 – β = 30°; 4 – β = 60° (for ζ_{DZ} calculated by Equation 4.70).

The comparison of calculations by Equation 5.40 with ζ_{DZ} determined from Equation 5.43 presented in Figure 5.5 as hollow squares and circles. The discrepancies not more than 3%, that is well within experimental error reported in cited paper. The maximal error reported in paper by Tovazhnyansky et al. (1980) not exceeded 15% for four other different commercial plates and Equation 5.43 allows improving accuracy even more for commercial plates investigated by Muley and Manglik (1999). Therefore, Equation 5.40 together with Equation 4.65 for friction factor at the main corrugated field and Equation 5.43 for coefficient of local hydraulic resistance can be used for estimation of the hydraulic performance for PHEs assembled from commercial plates of different geometrical characteristics.

5.3.3 Best Geometry of Plate for Specific Process

Performing design of heat exchanger for cooling of process fluid an engineer must strictly satisfy thermal and hydraulic conditions on the process side. The initial temperature of cooling water as well as its quality also cannot be influenced by designer. These parameters are the same for any type of heat exchanger which can be utilised on this process position – tubular, different kinds of PHEs or others. In a general case, the flow rate and hence the outlet temperature of cooling water can be varied, but it is also frequently strictly specified due to constraints on a whole cooling system of the enterprise. In case of recuperative PHE, which exchanging heat between two process streams, the conditions on both sides as a rule cannot be changed. Only internal parameters of PHE construction can be varied, like number, size and corrugation pattern of plates, streams passes numbers. It can lead to different channel geometries, flow velocities and wall temperatures inside the heat exchanger.

Let us assume that the following heat transfer process conditions are specified:

t_{11}, t_{12} – inlet and outlet temperatures of the hot stream, °C
t_{21}, t_{22} – inlet and outlet temperatures of the cold stream, °C
G_1, G_2 – mass flow rates of the hot and cold streams, kg/s
ΔP_1°, ΔP_2° – allowable pressure drops for the hot and cold streams, Pa

The heat load of heat exchanger has to be not less than specified:

$$Q = (t_{11} - t_{12}) \cdot c_{p1} \cdot G_1 = (t_{22} - t_{21}) \cdot c_{p2} \cdot G_2 \tag{5.44}$$

where

c_{p1} and c_{p2} are specific heat capacities of hot and cold fluids, J/(kg · K)
ρ_1 and ρ_2 are fluid densities, kg/m^3

The pressure losses in heat exchanger must be not bigger than allowable for each stream. In PHE the numbers of channels for two heat exchanging streams differ not more than on one. It permits to exactly satisfy pressure drop condition only for one stream, for another one pressure drop will be less than allowable. Let us assume that it is hot stream (if it is opposite, the procedure is analogues).

When designing PHE the task is to obtain heat exchanger with minimal heat transfer area satisfying required process conditions. It is fairly clear that the best result can be obtained with one pass counter current PHE, as for the multiple passes increases the number of distribution zones through which the stream flows. Also, the additional pressure losses for turning flow in collectors between passes are added. The special case when unsymmetrical passes numbers may be necessary, as for example, one example in paper by Arsenyeva et al. (2011), is not considered here. Let us assume that to satisfy process conditions for given geometry of corrugations on a main heat transfer field of the plate we can change the plate length. It is similar to the design approach proposed by Picón-Núñez et al. (2010) in graphical form.

When pressure drop condition for hot stream is exactly satisfied, from Equation 5.40 one can get

$$\frac{L_F}{b} = \frac{4}{\zeta_1(w_1)} \cdot \left(\frac{\Delta P_1^o}{\rho_1 \cdot w_1^2} - \zeta_{DZ} \right) \tag{5.45}$$

where

w_1 is the flow velocity of hot stream in PHE channels, m/s
ζ_{DZ} is the calculated by Equation 5.43
$\zeta_1(w_1)$ by Equation 4.65 at this flow velocity

By the process conditions, it is required that the number of heat transfer units for hot stream in heat exchanger has to be not less than

$$NTU^0 = \frac{(t_{11} - t_{12})}{\Delta t_{\ln}} \tag{5.46}$$

where

$\Delta t_{\ln}$ is the mean logarithmic temperature difference

$$\Delta t_{\ln} = \frac{(t_{11} - t_{22}) - (t_{12} - t_{21})}{\ln\left((t_{11} - t_{22})/(t_{12} - t_{21})\right)} \tag{5.47}$$

On the other hand, the number of heat transfer units, which can be obtained in one PHE channel:

$$NTU = \frac{2 \cdot U \cdot F_{pl}}{c_{p1} \cdot w_1 \cdot \rho_1 \cdot f_{ch}} \tag{5.48}$$

where

U is the overall heat transfer coefficient, W/(m$^2 \cdot$ K)
$f_{ch} = W \cdot b$ is the channel cross section area, m^2
W is the width of the channel, m
F_{pl} is the heat transfer area of one plate, m^2

Estimating approximately, that the heat transfer area of distribution zones is about 15% of the total plate heat transfer area, can obtained

$$F_{pl} = L_F \cdot W \cdot Fx/0.85 \tag{5.49}$$

where Fx is the ratio of developed area to projected area

For PHE exactly satisfying process conditions NTU calculated by Equation 5.48 should be equal to that required by process (Equation 5.46) and we can estimate the plate length to fulfill this:

$$\frac{L_F}{b} = \frac{NTU^0 \cdot 0.85 \cdot c_{p1} \cdot w_1 \cdot \rho_1}{2 \cdot U \cdot Fx} \tag{5.50}$$

When both conditions, for heat load and pressure drop of hot stream, are satisfied Equations 5.45 and 5.50 are forming a system of two algebraic Equations with two unknown variables L_F and w_1. Excluding L_F/b, one equation with one unknown is as follows:

$$w_1 = \sqrt{\frac{\Delta P_1^0}{\rho_1} \cdot \frac{1}{\zeta_{DZ}(w_1) + \zeta_1(w_1) \cdot \dfrac{NTU^0 \cdot 0.85 \cdot c_{p1} \cdot w_1 \cdot \rho_1}{8 \cdot U(w_1) \cdot Fx}}} \tag{5.51}$$

For given geometrical parameters of corrugations (β, Υ, Fx), the friction factor $\zeta_1(w_1)$ is dependent only from velocity w_1 and determined by non-linear Equation 4.65. Same is valid for ζ_{DZ} (w_1), which is determined by Equations 5.43 and 4.65. The overall heat transfer coefficient can be expressed as

$$\frac{1}{U} = \frac{1}{h_1} + \frac{1}{h_2} + \frac{\delta_w}{\lambda_w} + R_{foul} \tag{5.52}$$

where

δ_w is the wall thickness, m

λ_w is the thermal conductivity of the wall, W/(m · K)

R_{foul} is the thermal resistance of fouling (m^2 · K)/W

h_1 and h_2 are film heat transfer coefficients for hot and cold stream, W/(m^2 · K)

Film heat transfer coefficients are determined by Equations 4.96 and 4.75, and for known physical properties of streams and channels geometry they are dependent only from flow velocities in PHE channels. Using the intrinsic feature of PHE, we assume that the numbers of channels for two streams are equal. The geometry of the channels is also the same. So flow velocity in channels of the cold stream can be expressed via flow velocity in channels of the hot stream (and vice versa):

$$w_2 = w_1 \cdot \frac{q_{m2} \cdot \rho_1}{q_{m1} \cdot \rho_2} \tag{5.53}$$

The thermal resistance of fouling in PHEs is much lower than in shell and tube heat exchangers. As it is discussed later in the next chapter, for some fouling mechanisms, it depends from flow velocity. Here, that phenomenon is not considered and for clarity assumed $R_{foul} = 0$.

Based on above discussion, it can be concluded that for given geometrical parameters of corrugations (β, ϒ, *Fx*) the overall heat transfer coefficient is dependent only from flow velocity w_1 and can be calculated using Equations 4.96, 4.75, 5.52 and 5.53).

The non-linear algebraic Equation 5.51, accompanied by Equations 4.65, 5.43, 4.96, 4.75, 5.52 and 5.53 cannot be solved analytically, but its solution is easily obtained numerically on a computer using iterative procedure. The solution gives the flow velocity w_1 in channels. After that the corresponding plate length, at which specified heat transfer load and pressure drop of one stream are exactly satisfied, is calculated from Equation 5.45. For the specified process conditions and fixed corrugations parameters, due to full utilisation of pressure drop, the maximum overall heat transfer coefficients and minimal heat transfer area are achieved. It enables to determine the influence of plate corrugations geometrical parameters on ability of PHE to satisfy specific process conditions with minimal required heat transfer area. The example of such analysis is presented in the following case study.

5.3.4 Illustrative Example of Plate Geometry Selection

As an example here are taken the conditions of PHE application in District Heating (DH) system. The heat exchanger is heating radiator water in the house internal circuit from $t_{21} = 70$ °C to $t_{22} = 95$ °C by DH water, coming at $t_{11} = 120$ °C and discharged with temperature $t_{12} = 75$ °C. The allowable pressure losses $\Delta P_1^\circ = 20$ kPa, $\Delta P_2^\circ = 70$ kPa. Two cases with different heat load are considered: $Q_1 = 600$ kW, typical for medium size apartment block, $Q_2 = 3{,}000$ kW, as for big apartment house or complex of such houses. The flow rates of the streams are readily calculated from heat balance given by Equation 5.44.

To study the influence of plate geometry on ability of PHE to satisfy this specific process conditions, it is used the model presented above in subsection 'The Best Geometry of Plate for Specific Process'. It enables to calculate the plate length at which process parameters will be strictly satisfied. This length is varies significantly with changing of the corrugations geometry, as can be seen from graphs in Figure 5.6. With that also vary the corresponding flow velocity in channels, overall heat transfer coefficient and required heat transfer area. Analysing graphs on Figures 5.6 through 5.8, one can conclude, that the required heat transfer area F_a decreases for the higher corrugations inclination angle β and for smaller plate spacing *b*. The minimal F_a values can be observed at the edge of model application limits. For examined two cases, the minimal heat transfer area is $F_{a1} = 6.06$ m² and $F_{a2} = 30.28$ m² at $\beta = 65°$ and $b = 1.5$ mm (see Figure 5.6). At the same time, the length of the required plates corrugated field L_F also decreases significantly down to 300 mm. To organise equal distribution of flow across the width of inter-plate channels, it is preferable to have bigger ratio of plate length to its width or L_F/b. In majority of

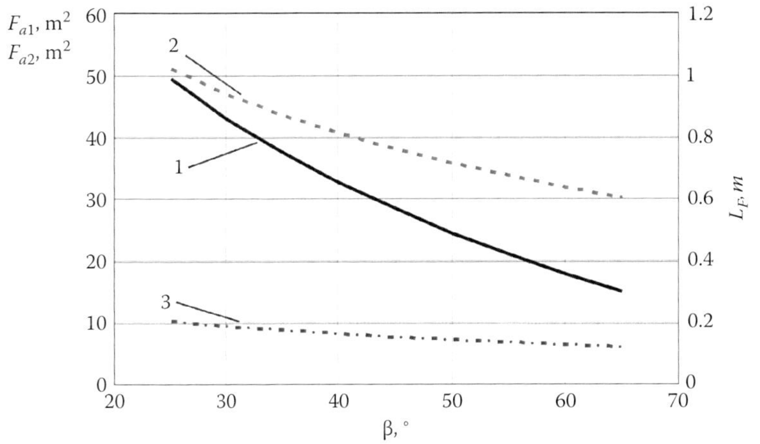

Figure 5.6 The dependence of required plate length L_F and PHE heat transfer area from corrugations angle β with b = 1.5 mm. 1 – L_F; 2 – F_{a2} for Q_2 = 3,000 kW; 3 – F_{a1} for Q_1 = 600 kW.

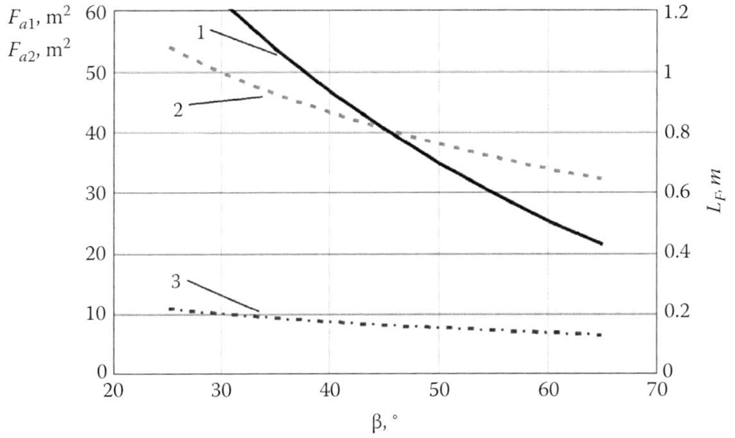

Figure 5.7 The dependence of required plate length L_F and PHE heat transfer area from corrugations angle β with b = 2 mm. 1 – L_F; 2 – F_{a2} for Q_2 = 3,000 kW; 3 – F_{a1} for Q_1 = 600 kW.

commercially produced plates, it varies from 1.5 to 2.5. To estimate the heat transfer area of the plate let's take $L_F/b = 2$. Then the heat transfer area of one plate:

$$F_{pl} = L_F^2 \cdot Fx/(2 \cdot 0.85) \tag{5.54}$$

The heat transfer area of plate with L_F = 300 mm should be F_{pl1} = 0.066 m² and PHE for the first case requires 92 such heat transfer plates. These plates can be used

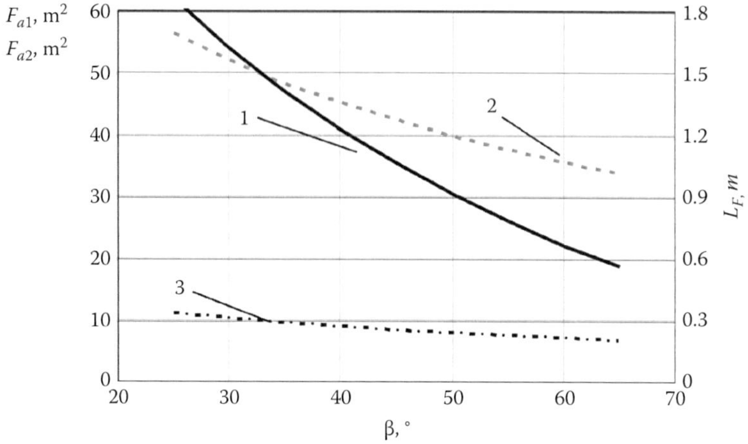

Figure 5.8 The dependence of required plate length L_F and PHE heat transfer area from corrugations angle β with b = 2.5 mm. 1 – L_F; 2 – F_{a2} for Q_2 = 3,000 kW; 3 – F_{a1} for Q_1 = 600 kW.

efficiently for heat loads less than 600 kW, with less number of plates in PHE. The number of heat transfer plates in one PHE is limited mainly because the limited diameter of inlet and outlet ports, which can lead to excessive flow velocities in ports and unnecessary pressure losses there. Let us assume that the maximum number of plates in one PHE that should be designed for regarded process conditions is 120. For the first case with heat load 600 kW, 92 plates are inside that limit, but for 3,000 kW we will require 460 plates. That amount cannot be handled in one frame and therefore we should consider plate of bigger size. For this purpose plate with bigger length L_F has to be taken. As follows from graph in Figure 4, for the same considered plate spacing b = 1.5 mm it can be changed the corrugations angle β. For β = 37° the length of plate is L_F = 710 mm and its heat transfer area F_{pl1} = 0.37 m². The required total PHE heat transfer area F_{a2} = 42.4 m² can be assembled from 115 plates. It is within imposed limit for number of plates, but F_{a2} on 40% larger than previously found minimal 30.28 m².

Another option to increase required plate length is to take wider plate spacing, as one can judge from graph in Figure 5.7. For plate spacing 2 mm the minimal heat transfer areas in two cases under consideration are F_{a1} = 6.43 m² and F_{a2} = 32.15 m² at β = 65°, that means 6% increase compare to these values at spacing 1.5 mm. At the same time for β = 50° the length of plate is L_F = 700 mm and its heat transfer area F_{pl1} = 0.35 m². The required total PHE heat transfer area F_{a2} = 38 m² (25% more than minimal for b = 1.5 mm) can be assembled from 109 plates.

With the further increase the plate spacing to 2.5 mm (see Figure 5.8), the minimal heat transfer areas in two cases under consideration are F_{a1} = 6.74 m² and F_{a2} = 33.72 m² at β = 65°, or 11% more than these values at spacing 1.5 mm. With this spacing can finded the plate length L_F = 640 mm and its heat transfer area F_{pl1} = 0.30 m² at β = 61°. The heat transfer area here 35.1 m² consists from 117 plates and larger than minimal at $b = 1.5$ mm on 16%. Another option is the plate with $L_F = 910$ mm and heat transfer area 0.61 m^2 at $\beta = 50°$, but it requires $F_{a2} = 39.6$ m^2 with 65 plates (31% larger than 30.28 m^2).

The presented case study shows, that the geometry of corrugations, as well as plate length, are considerably influencing the performance of PHE. To satisfy process conditions with minimal heat transfer area, there are required plates of different size with different geometrical parameters. For the smaller heat transfer load (in our case less than 600 kW), the plates with smaller spacing and $\beta = 65°$ give the highest overall heat transfer coefficient and smallest heat transfer area in one PHE. To accommodate bigger heat loads in one PHE, it is necessary to increase the plate length and hence heat transfer area of one plate. To maintain full utilisation of allowable pressure drop with exact satisfaction of heat load, it is possible by decreasing β or by increasing plate spacing b. The second method is preferable, as gives less decrease in overall heat transfer coefficient and smaller heat transfer area. For specified process conditions, the corrugation geometry and size of plate, giving minimal PHE heat transfer area, can be found.

The presented mathematical model enables to predict the influence of plate corrugation parameters on performance of PHE. For the specified process conditions (pressure drop, temperature program, heat load and streams physical properties), the geometrical parameters of plate and its corrugations, which enable to make PHE with minimal heat transfer area, can be found. It requires the development of optimisation algorithm and software, which is the subject of further research.

For different heat transfer process conditions, the best values of plate geometrical parameters significantly varying. It is require designing and manufacturing the plates of different sizes and geometrical form of corrugations, which can better satisfy certain ranges of applications.

In commercial PHEs, the varying of corrugation inclination angle is achieved by combination of two plates with different corrugation angle on one frame. The number of plate sizes and plates with different spacing is limited. All of this leads to situation that optimal corrugation parameters cannot be reached with already available types of plates. But having the optimal plate geometry for given process, we can use this data as a reference point of what can be achieved with PHE on certain position. Besides, the developed mathematical model can be used for designing of new plates with geometry, which is in the best way satisfying process conditions or some range of them.

5.3.5 Illustrative Examples of Plate Heat Exchanger Design with Available Range of Plates

The method of plate geometry selection described in two previous subsections can be used when designing the plates for certain range of process conditions, optimisation of heat exchanger networks with PHE or approximate estimation of heat transfer and friction factor correlations for existing commercial plates with known geometry. For such plates another method of more precise estimation of correlations for heat transfer and hydraulic performance is presented in Appendix 5.A. In practice due to the high cost of tools for plates production, the number of types and sizes of commercial plates is limited and design of commercial PHEs is made based on the range of available plates as described in subsection 'Mathematical Modelling and Design of Industrial PHEs'.

To illustrate the influence of passes, plate type and plates arrangement on PHE performance we can consider two case studies. The first is for a PHE which has been designed to work in a distillery plant. The second is taken from paper by Wang and Sunden (2003). The calculations are made with the developed software.

5.3.5.1 Case Study 1

Example 1a. It is required to heat 5 m³/h of distillery wash fluid from 28 °C to 90.5 °C using hot water which has a temperature of 95 °C and a flow rate 15 m³/h. The pressure of both fluids is 5 bar. The allowable pressure drops for the hot and cold streams are both 1.0 bar. The properties of the wash fluid are considered constant and are as follows: density – 978.4 kg/m³, heat capacity – 3.18 kJ/(kg · K) and conductivity – 0.66 W/(kg · m). Dynamic viscosity at temperatures t = 25 °C, 60 °C, 90 °C is taken as μ = 19.5; 16.6; 9.0 cP.

The optimal solution is obtained for plate type M6M with spacing of plates δ = 3 mm (see Table 5.A.1). The results of calculations for different passes numbers X_1 and X_2 and optimal for those passes plates arrangements are presented in Table 5.2. The total numbers of plates in PHE, required to satisfy process conditions, is shown on diagram in Figure 5.9. The analysis shows that the optimum (38 plates, F_{PHE} = 5.04 m²) is achieved at X_1 = 2 and X_2 = 4 with all medium channels (19 H and 19 L plates in PHE). The next closest option (41 plates, F_{PHE} = 5.46 m²) is at X_1 = X_2 = 2 with mixed channel arrangement in one pass. If we would have only one plate type in the PHE, the minimal number of plates would be 44 (F_{PHE} = 5.88 m²) for both H and L plates, or 15% higher than with mixed channels.

The closest other option for described above conditions is the PHE with plate type M6, having smaller spacing δ = 2 mm (see Table 5.A.1). This PHE has 36 H-plates (F_{PHE} = 5.1 m²) with one pass channels arrangement 1*17H/1*17H. When analysing

Table 5.2 The Influence of Passes Numbers and Plate Arrangement on Heat Transfer Area in M6M PHE for Conditions of Example 1a

	X_1			
X_2	1	2	3	4
1	7.56 m² 1*28H/1*27H	32.34 m² 2*58H/1*116H	21.70 m² 3*26H/1*78H	25.48 m² 4*23H/1*91H
2	9.80 m² 1*35H/2*18H	6.58 m² 2*(6H+6*M)/2*(6H+6*M)	8.26 m² 3*10M/ 2*15M	8.68 m² 4*8L/1*15L+1*16L
3	5.88 m² 1*21H/2*7H+1*8H	6.58 m² 2*12M/3*8M	5.74 m² 3*(4M+3*L)/3*(4M+3*L)	6.72 m² 3*6L+1*7L/3*8L
4	8.54 m² 1*31M/1*7M+3*8M	5.04 m² 2*9M/1*4M+3*5M	6.44 m² 3*8L/1*5L+3*6L	6.72 m² 3*6L+1*7L/4*6L

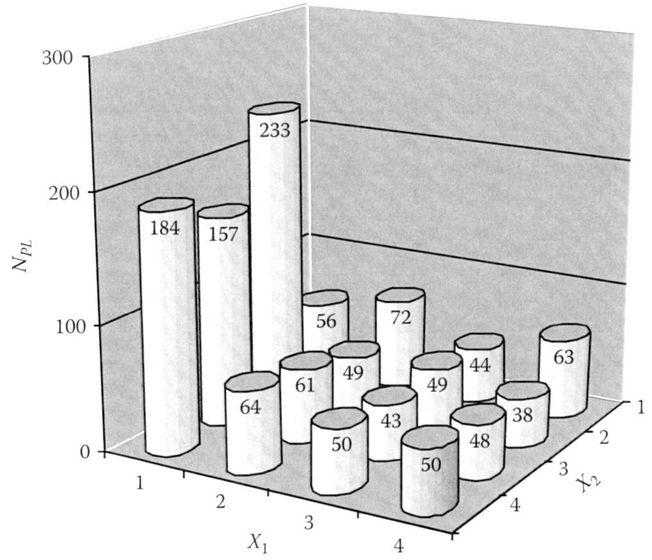

Figure 5.9 The influence of passes arrangement on total number of M6M plates in PHE, which satisfy the process conditions of Example 1a.

other aspects of PHE application, such as maintenance, piping arrangement, the length of plate pack, this option can be probably chosen by the process engineer. In the next two examples, we study how the process conditions can influence an optimal solution.

Example 1b. Consider the case when wash fluid should be heated from 28 °C to 92.5 °C with pressure drops of both streams not more than 0.4 bar. All other conditions are the same as in *Example 1a.* The optimal solution is PHE with 47 M6 plates (F_{PHE} = 6.9 m^2) and one pass channels arrangement 1*23H/1*23H. For PHE with M6M plates the best solution is for 61 plates (F_{PHE} = 8.54 m^2), there two and four passes for streams and channels arrangement 2*15M/(2*7H+2*8M). One can see that the plate M6 much better suitable for examined conditions than M6M. The PHE with M6 plates has heat transfer area smaller on 23% and one pass channels arrangement.

Example 1c. Wash fluid of previous examples has to be heated from 28 °C to 75 °C with allowable pressure drop for both streams 0.1 bar. The best option for this case is one pass PHE with 29 M6M L-plates (F_{PHE} = 3.78 m^2). The smallest for this process conditions PHE from M6 plates has 43 L-plates and heat transfer area bigger on 63% (F_{PHE} = 6.15 m^2) and margin on heat transfer load M = 0.8. It shows that, for different process conditions, the plates with different geometrical characteristics are required to satisfy those conditions in a best way. It is urging the leading PHE manufacturers to produce a wide range of plates that can satisfy any process requirements. However, counting for the cost of tools to manufacture new plate type, this way is rather expensive. On the other hand, the process design engineer can optimise process conditions, like allowable pressure drops and temperature program, accounting for characteristics of available types of plates. The data of one representative set of plates, presented in this paper, can facilitate such approach.

5.3.5.2 *Case Study 2*

This example is that which has been presented by Wang and Sunden (2003). It is required to cool 40 kg/s of hot water with the initial temperature 70 °C down to 40 °C. The flow rate of incoming cooling water is 30 kg/s, with a temperature of 30 °C.

The allowable pressure drop on the hot side, ΔP_1, is 40 kPa, on the hot side 60 kPa. The results of the calculations are presented in Table 5.3. In the first five cases (rows in Table 5.3) the calculations are made for clean plates.

The minimal heat transfer surface area is achieved with the combination of *H* and mixed channels – case #1. Here, the allowable pressure drop on hot side is completely used and the heat load is exactly equal to the specified (the margin $M = (Q - Q°)/Q°$ equal to zero). The option with identical mixed channels has on 5% bigger area and margin of 1%.

For the PHE with identical plates, the surface area is much larger. With *H* plates (case #2) it is 73% larger, but at considerable (68%) margin. With *L* plates (case #4) the surface area is larger by 150% with no margin for heat load. The pressure drop is much lower than allowable in this case. The increase in the number of passes for *L* plates (case #5) also produces a larger surface area (125%), but the pressure drop is utilised and the margin is 62%.

In the example given by Wang and Sunden (2003), the fouling thermal resistance equal to 0.5×10^{-4} m^2 K/W was taken for both sides (compare with data for PHEs on fouling reported by Wang et al. (2007), this value is the biggest one). The total $R_f = 1 \times 10^{-4}$ m^2 K/W was taken for calculations in case #6. This leads to an increase of surface area of 60%, to 59.04 m^2 (Wang and Sunden 2003; obtained 68.8 m^2 for plates considered in their paper). The results also show, compared to case #2, that such allowance for fouling is diminishing advantages of optimal solution to 10%.

Table 5.3 Calculations Results for Case Study 2 (Two M10B PHEs Installed in Parallel)

#	Total Area, m^2	Grouping	ΔP_1, kPa	ΔP_2, kPa	R_f 10^4, m^{2}K/W	M_Q, %	*F/Fmin*
1	36.96	(3H+35M)/(3H+35M)	39	26	0	0	1
2	63.92	64H/64H	40	21	0	68	1.73
3	38.88	40M/40M	38	25	0	1	1.05
4	91.68	95L/95L	14	8	0	0	2.50
5	83.04	2*43L/2*43L	40	24	0	62	2.25
6	59.04	(57H+4M)/(57H+4M)	40	23	1.0	56	1.60
7	41.76	(15H+28M)/(15H+28M)	40	27	0.18	10	1.13
8	46.56	(26H+22M)/(26H+22M)	40	26	0.37	20	1.26

It is also results in a decrease of flow velocities in the channels, as their number in case #6 also increased by 60%. As it is discussed in the next chapter of this book, the excessive allowance for fouling can lead to increase of fouling in real conditions by lowering the flow velocity and wall sheer stress. In cases #7 and #8 (see Table 5.3), the calculations were made with specification of margins 10% and 20% to overall heat transfer coefficient. The increase of surface is 13% and 26%. The corresponding calculated values of R_f much lower and for margin 20% very close to those reported in book by Hesselgreaves (2001) for water of towns.

As can be seen from Table 5.3, in all cases with mixed grouping of channels the specified values of allowable pressure drop for one stream (hot in our example) are satisfied exactly, as is the condition for the heat load (when margin is specified, then with margin). It means that by using the mixed grouping of plates with different corrugation pattern, the thermal and hydraulic characteristics can change of a plate pack in a way close to continues one. The level of discreteness is equal to one plate in a pack. It allows us to satisfy specified conditions very close to equality.

On a final stage of ordering for manufacture, the PHE calculations should be made by PHE manufacturers, which are permanently developing the design procedures and performance of produced equipment, and most probably can offer PHE with even better performance.

5.A APPENDIX: IDENTIFICATION OF MATHEMATICAL MODEL PARAMETERS FOR PHE DESIGN

Based on the PHE mathematical model described in Chapter 5, a numerical experimental technique is developed for the identification of the model parameters by comparison with results obtained the use of PHE calculation software, which is available on the Internet from PHE manufacturers.

The computer programs for thermal and hydraulic design of PHEs in result of calculations give the information about following parameters of designed heat exchanger: Q – heat load, W/m^2; t_{11}, t_{12} – hot stream inlet and outlet temperatures, °C; t_{21}, t_{22} – cold stream inlet and outlet temperatures, °C; G_1, G_2 – flow rates of hot and cold streams, kg/s; ΔP_1, ΔP_2 – pressure losses of respective streams, Pa; n_1, n_2 – numbers of channels for streams and N_{pl} – number of heat transfer plates and heat transfer area F, m^2. One set of such data can be regarded as a result of experiment with calculated PHE. The overall heat transfer coefficient U, W/(m$^2 \cdot$K), even if not presented, can be easily calculated on these data, as also the thermal and physical properties of streams.

The PHE thermal design is based on empirical correlations like (5.16). When all plates in PHE have the same corrugation pattern, the formulas for heat transfer coefficients are the same for both streams and can be written in following form:

$$\frac{h_{1,2}d_e}{\lambda_{1,2}} = A \cdot \left(\frac{G_{1,2}d_e}{f \cdot \mu_{1,2} \cdot N_{1,2}} \right)^n \cdot \left(\frac{c_{1,2}\mu_{1,2}}{\lambda_{1,2}} \right)^{0.4} \cdot \left(\frac{\mu_{1,2}}{\mu_{w1,2}} \right)^{0.14} \tag{5.A.1}$$

In one pass PHE with equal numbers of channels ($N_1 = N_2$) for both streams the ratio of the film heat transfer coefficients is:

$$a = \frac{h_1}{h_2} = \left(\frac{G_1}{G_2}\right)^n \cdot \left(\frac{\lambda_1}{\lambda_2}\right)^{0.6} \cdot \left(\frac{\mu_2}{\mu_1}\right)^{n-0.54} \cdot \left(\frac{c_1}{c_2}\right)^{0.4} \cdot \left(\frac{\mu_{w1}}{\mu_{w2}}\right)^{0.14} \quad (5.A.2)$$

The overall heat transfer coefficient for clean surface conditions is determined by formula 5.15 with $R_f = 0$. When fluids of both streams are same and they have close temperatures, we can take $\mu_{w1} = \mu_{w2}$. Assuming initial value for $n = 0.7$, from the last two equations the film heat transfer coefficient for hot stream:

$$h_1 = \frac{1+a}{\frac{1}{U} - \frac{\delta_w}{\lambda_w}} \quad (5.A.3)$$

For the cold stream

$$h_2 = h_1/a \quad (5.A.4)$$

In case of equal flow rates ($G_1 = G_2$), the plate surface temperature at hot side:

$$t_{w1} = \frac{t_{11} + t_{12}}{2} - \frac{Q}{F \cdot h_1} \quad (5.A.5)$$

At cold side

$$t_{w2} = \frac{t_{21} + t_{22}}{2} + \frac{Q}{F \cdot h_2} \quad (5.A.6)$$

At these temperatures the dynamic viscosity coefficients μ_{w1}, μ_{w2} are determined.

By determined values of film heat transfer coefficients, calculated Nusselt numbers for hot and cold streams (5.16) and the dimensionless parameters as:

$$K_{1,2} = \frac{\mathrm{Nu}_{1,2}}{\mathrm{Pr}_{1,2}{}^{0.4} \cdot (\mu_{1,2}/\mu_{w1,2})^{0.14}} \quad (5.A.7)$$

Making calculations of the same PHE for different flow rates, which ensure the desired range of Reynolds numbers, the relationship is obtained as:

$$K = \Phi(\mathrm{Re}) \quad (5.A.8)$$

Plotted in logarithmic coordinates it enables to estimate parameters A and n in correlation (5.16). To determine these parameters Least Squares method can be used. If the value of n much different from initially assumed 0.7, the film heat transfer coefficients recalculated and new relationship (5.A.8) obtained. The corrected values of A and n can be regarded as final solution.

The pressure drop in PHE is determined by formula 5.22 with friction factor determined by Equation 5.23.

Using these equations, the values of friction factors for hot and cold streams can be obtained from the same data of PHE numerical experiments for calculation of heat transfer coefficients, using data on pressure drops Δp_1 and Δp_2. It gives the relation between friction factor and Reynolds number. From this relation parameters A and m are easily obtained using Least Squares method.

To obtain the representative data the numerical experiments have to satisfy the following conditions:

1. The calculations are made in 'rating' or 'performance' mode for equal flow rates of hot and cold streams. In case of 'rating' the outlet temperatures should be adjusted to have margin equal to zero.
2. The PHE is having one type of inter-plate channels – L (low duty), H (high duty) or M (medium duty). These channels are formed by L-plates, H-plates or M- mixture of L and H plates.
3. To eliminate end-plate effect the number of plates in PHE should be more than 21.
4. All numerical experiments made for water as both streams. The inlet temperature of hot stream 50 °C. The inlet temperature of cold stream in the range of 30 °C–40 °C.

To illustrate the procedure we performed three sets of numerical experiments for plate M-10B (see Alfa-Laval Corporate AB, 2015) with the use of computer program CAS-200 as described by Arsenyeva et al. (2011). The calculations are made for PHE with total 31 plates for three plate arrangements: (1) H-plates only, (2) L-plates only and (3) M-the mixture of 15 L-plates and 16 H-plates. The hot water inlet temperature is 50 °C. Cold water comes with temperature 40 °C. The geometrical parameters of plate are given at Table 5.A.1. These parameters are estimated from information available at CAS-200 and also by measurements on the samples of real plates. The results for heat transfer calculations according to described above procedure are presented in Figure 5.A.1. The obtained sets of parameters in correlation (5.16) permit to calculate heat transfer coefficients with mean square error 1.3% and maximum deviation ±3.5%. The values of these parameters are presented in Table 5.A.2.

Table 5.A.1 Estimations for Geometrical Parameters of some Alfa Laval PHE Plates

Plate Type	δ, mm	d_e, mm	b, mm	F_{pl}, m²	$D_{connection}$, mm	$f_{ch}\cdot 10^3$, m²	L_p, mm
M3	2.4	4.8	100	0.032	36	0.240	320
M6	2.0	4.0	216	0.15	50	0.432	694
M6M	3.0	6.0	210	0.14	50	0.630	666
M10B	2.5	5.0	334	0.24	100	0.835	719
M15B	2.5	5.0	449	0.62	150	1.123	1381

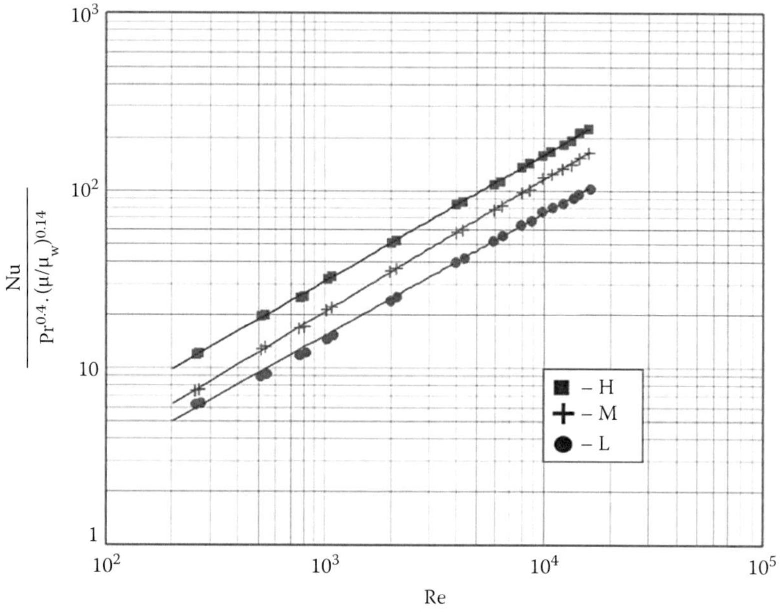

Figure 5.A.1 Heat transfer relations for channels formed by different sets of M10B plates.

The friction factor data are presented in Figure 5.A.2. The change of lines slopes is obvious. The obtained parameters of correlation (5.23) are given in Table 5.A.2. The mean square error of this correlation fitting the data is 1.5% and maximum deviation ±3.8%.

The geometrical characteristics and parameters obtained in the same manner for four other types of plates are presented in Tables 5.A.1 and 5.A.2 (for description of PHEs with these plate types see Alfa-Laval Corporate AB, 2015. They can be used for statistics when generalising the correlations for PHEs thermal and hydraulic performance, for modelling of PHEs when making multiple calculations for heat exchangers network design and also for education of engineers specialising on heat transfer equipment selection.

The error estimation have shown that the obtained sets of parameters, which are presented in Table 5.A.2, permit to calculate heat transfer coefficients and friction factors with mean square error 1.5% and maximum deviation not more than ±4%. The error for calculation of heat transfer area not more than 4%.

The correlations and developed software can be used only for preliminary calculations, when optimising PHEs or heat exchanger network. The final calculations when ordering the PHE should be made by its manufacturer.

Table 5.A.2 Estimations by Proposed Method for Parameters in Correlations for some Alfa Laval PHEs (20,000 > Re > 250)

Plate Type	Channel Type	*A*	*n*	Re	*B*	*m*
M3	*H*	0.265	0.7	<520	33	0.25
				≥520	10.7	0.07
	L	0.12	0.7	<1,000	18.8	0.33
				≥1,000	8.8	0.22
	M	0.18	0.7	<1,000	44	0.4
				≥1,000	5.1	0.1
M6	*H*	0.25	0.7	<1,250	10	0.2
				≥1,250	2.4	0.0
	L	0.12	0.7	<1,500	5.1	0.3
				≥1,500	1.7	0.15
	M	0.165	0.7	<930	9.3	0.3
				≥930	2.72	0.12
M6M	*H*	0.27	0.7	<1,300	11.7	0.13
				≥1,300	4.55	0.0
	L	0.11	0.71	<2,200	4.23	0.23
				≥2,200	1.88	0.12
	M	0.14	0.73	<2,100	5.61	0.16
				≥2,100	1.41	0.0
M10B	*H*	0.224	0.713	<2,100	12.7	0.17
				≥2,000	3.53	0.0
	L	0.126	0.693	<1,600	9.18	0.32
				≥1,500	2.43	0.14
	M	0.117	0.748	<2,150	6.56	0.2
				≥2,150	2.09	0.05
M15B	*H*	0.26	0.7		5.84	0.05
	L	0.085	0.74	<2,900	5.2	0.28
				≥2,900	1.57	0.13
	M	0.13	0.74	<3,500	4.3	0.15
				≥3,500	1.25	0.0

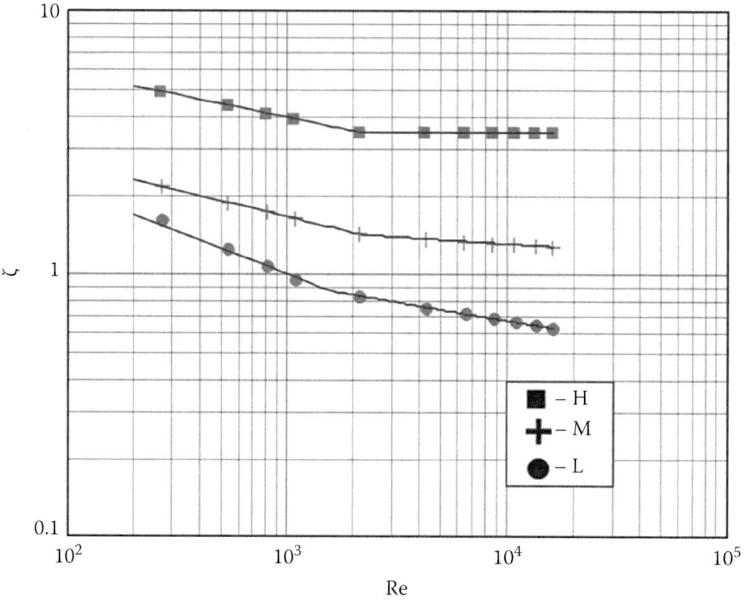

Figure 5.A.2 Friction factor data for channels formed by M10B plates.

REFERENCES

Alfa-Laval Corporate AB. 2015. Gasketed industrial range PHE. Documentation. www.alfalaval.com/solution-finder/products/gasketed-industrial-range-phe (accessed 12 April 2015).

Arsenyeva O., Kapustenko P., Tovazhnyanskyy L., Khavin G. 2013. The influence of plate corrugations geometry on plate heat exchanger performance in specified process conditions. *Energy* 57: 201–207.

Arsenyeva O.P., Tovazhnyansky L.L., Kapustenko P.O., Khavin G.L. 2011. Optimal design of plate-and-frame heat exchangers for efficient heat recovery in process industries. *Energy* 36(8): 4588–4598.

Azad A.V., Amidpour M. 2011. Economic optimization of shell and tube heat exchanger based on constructal theory. *Energy* 36(2): 1087–1096.

Bassiouny M.K., Martin H. 1984. Flow distribution and pressure drop in plate heat exchangers-I. *Chemical Engineering Science* 39(4): 693–700.

Bhutta M.M.A., Hayat N., Bashir M.H., Khan A.R., Ahmad K.N., Khan S. 2012. CFD applications in various heat exchangers design: A review. *Applied Thermal Engineering* 32: 1–12.

Caputo A.C., Pelagagge P.M., Salini P. 2011. Joint economic optimization of heat exchanger design and maintenance policy. *Applied Thermal Engineering* 31(8–9): 1381–1392.

Cocks A.M. 1969. Plate heat exchanger design by computer. *Chemical Engineer (London)* 228: 193–198.

Fettaka S., Thibault J., Gupta Y. 2013. Design of shell-and-tube heat exchangers using multi-objective optimization. *International Journal of Heat and Mass Transfer* 60: 343–354.

Focke W.W. 1986. Selecting optimum plate heat exchanger surface patterns. *Journal of Heat Transfer* 108(1): 153–160.

Gut J.A.W., Pinto J.M. 2004. Optimal configuration design for plate heat exchangers. *International Journal of Heat and Mass Transfer* 47(22): 4833–4848.

Hadidi A., Nazari A. 2013. Design and economic optimization of shell-and-tube heat exchangers using biogeography-based (BBO) algorithm. *Applied Thermal Engineering* 51(1–2): 1263–1272.

Heggs P.J., Scheidat H.J. 1992. Thermal performance of plate heat exchangers with flow maldistribution. In Shah R.K., Rudy T.M., Robertson J.M., Hostetler K.M. (Eds.) *Compact Heat Exchangers for Power and Process Industries*. ASME, New York, USA, vol. 201, 87–93.

Hesselgreaves J.E. 2001. *Compact Heat Exchangers. Selection, Design and Operation*. Elsevier, Oxford, UK.

Kanaris A.G, Mouza A.A, Paras S.V. 2009. Optimal design of a plate heat exchanger with undulated surfaces. *International Journal of Thermal Sciences* 48: 1184–1195.

Kandlikar S.G, Shah R.K. 1989. Asymptotic effectiveness-NTU formulas for multipass plate heat exchangers. *Transactions of the ASME* 111(5): 314–321.

Kays, W.M., London, A.L. 1984. *Compact Heat Exchangers*, 3rd edn. McGraw Hill, New York, USA.

Kukulka D.J., Devgun M. 2007. Fluid temperature and velocity effect on fouling. *Applied Thermal Engineering* 27: 2732–2744.

Kumar H. 1994. Plate heat exchanger: construction and design. *Institution of Chemical Engineers Symposium Series* 86: 1275–1288.

Marriott J. 1977. Performance of an Alfaflex plate heat exchanger. *Chemical Engineering Progress* 73(2): 73–78.

Mehrabian M.A. 2009. Construction, performance, and thermal design of plate heat exchangers. *Proceedings of the Institution of Mechanical Engineers, Part E: Journal of Process Mechanical Engineering* 223(3): 123–131.

Muley A., Manglik R.M. 1999. Experimental study of turbulent flow heat transfer and pressure drop in a plate heat exchanger with chevron plates. *ASME Journal of Heat Transfer* 121: 110–117.

Picón-Núñez M., Polley G.T., Jantes-Jaramillo D. 2010. Alternative design approach for plate and frame heat exchangers using parameter plots. *Heat Transfer Engineering* 31(9): 742–749.

Picón-Núñez M., Polley G.T., Martínez-Rodríguez G. 2013. Graphical tool for the preliminary design of compact heat exchangers. *Applied Thermal Engineering* 61: 36–43.

Pignotti A., Shah R.K. 1992. Effectiveness-number of transfer units relationships for heat exchangers complex flow arrangements. *International Journal of Heat and Mass Transfer* 35(5): 1275–1291.

Pignotti A., Tamborenia P.I. 1988. Thermal effectiveness of multipass plate exchangers. *International Journal of Heat and Mass Transfer* 31(10): 1983–1991.

Poddar T.K., Polley G.T. 1996. Heat exchanger design through parameter plotting. *Chemical Engineering Research and Design* 74(8): 849–852.

Polley G.T. 2003. Letter to the editor. *Applied Thermal Engineering* 23: 2147–2148.

Rao B.P., Kumar P.K., Das S.K. 2002. Effect of flow distribution to the channels on the thermal performance of a plate heat exchanger. *Chemical Engineering and Processing* 41: 49–58.

Rao R.V., Patel V. 2013. Multi-objective optimization of heat exchangers using a modified teaching-learning-based optimization algorithm original. *Applied Mathematical Modelling* 37(3): 1147–1162.

Sanaye S., Hajabdollahi H. 2010. Multi-objective optimization of shell and tube heat exchangers. *Applied Thermal Engineering* 30(14–15): 1937–1945.

Shah R.K., Focke W.W. 1988. Plate heat exchangers and their design theory. In Shah R.K., Subbarao E.C., Mashelkar R.A. (Eds.) *Heat Transfer Equipment Design*. Hemisphere, Washington, DC, USA, 227–254.

Shah R.K., Sekulić D.P. 2003. *Fundamentals of Heat Exchanger Design*. Wiley & Sons, New York.

Shah R.K., Wanniarachchi A.S. 1991. Plate heat exchanger design theory. In: Buchlin J.-M. (Ed.) *Industrial Heat Exchangers. Lecture Series*, vol. 4. von Karman Institute for Fluid Dynamics, Rhode-St-Genèse, Belgium.

Tovazhnyansky L.L., Kapustenko P.A., Pavlenko V.F., Derevyanchenko I.B., Babak T.G., Lupyr V.F. 1992. The optimum design of multi-pass dismountable PHEs. *Chemical and Petroleum Engineering* 28(6): 354–359.

Tovazhnyansky L.L., Kapustenko P.A., Tsibulnic V.A. 1980. Heat transfer and hydraulic resistance in channels of plate heat exchangers. *Energetika* 9: 123–125.

Wang L., Sunden B. 2003. Optimal design of plate heat exchangers with and without pressure drop specifications. *Applied Thermal Engineering* 23(3): 295–311.

Wang L., Sunden B., Manglik R.M. 2007. *PHEs. Design, Applications and Performance*. WIT Press, Southampton, U.K.

Wright A.D, Heggs P.J. 2002. Rating calculation for plate heat exchanger effectiveness and pressure drop using existing performance data. *Transactions of IChemE 2002* 80(Part A, April): 309–312.

Xie G.N., Sunden B., Wang Q.W. 2008. Optimization of compact heat exchangers by a genetic algorithm. *Applied Thermal Engineering* 28(8–9): 895–906.

Yousefi M., Enayatifar R., Darus A.N., Abdullah A.H. 2012. A robust learning based evolutionary approach for thermal-economic optimization of compact heat exchangers. *International Communications in Heat and Mass Transfer* 39(10): 1605–1615.

Zinger N.M, Barmina L.S, Taradai A.M. 1988. Design of plate heat exchangers for heat supply systems. *Thermal Engineering (English Translation of Teploenergetika)* 35(3): 141–145.

CHAPTER 6

Fouling and Heat Transfer Intensity

6.1 EFFECT OF FOULING ON HEAT EXCHANGER PERFORMANCE

In many industrial applications of heat transfer equipment, the surfaces washed by the streams involved in heat transfer process are susceptible to fouling, which refers to any change at the heat transfer surface, whether by deposits of fouling substances or other means, which results in a deterioration of heat transfer intensity across that surface. As it was mentioned in previous chapter with discussion of Equation 5.1, this phenomenon can significantly deteriorate the intensity of heat transfer process and heat exchanger performance by creating additional thermal resistance of fouling layer. Even more over, the decrease of channels cross section area partly blocked by the fouling deposits can lead to significant increase of pressure drop in heat exchanger and finally to clogging of the channels. According to the analysis of different publications presented by Crittenden and Yang (2011), conservative estimation of heat exchanger fouling leads to conclusion that additional costs for fouling in industrialised countries is in the order of 0.25% of Gross Domestic Product (GDP). Fouling is also the cause of around 2.5% of the total equivalent anthropogenic emissions of carbon dioxide. In most of processing industries, fouling creates a severe operational problem that compromises energy recovery and creates additional negative impact on environment. There is loss or reduction in production, increased consumption of energy and pressure losses, flow maldistributions, cost of anti-fouling chemicals, cleaning cost, etc. Inadequate detailed knowledge of the fouling mechanisms is frequently limiting the problem solution, even though the basic principles are understood for some time. The economic and environmental importance of fouling can be classified by main four categories (1) capital expenditure, (2) energy cost and environmental impact, (3) production loss during shutdowns due to fouling and (4) maintenance cost.

Extra capital cost involves surplus heat transfer surface area to compensate the fouling effect, heavier foundations cost, extra space for larger heat exchangers, higher cost for transport and installation of the equipment, cost of anti-fouling equipment and on-line cleaning devices and treatment plants. In most cases, the additional surface area to compensate for fouling is calculated by incorporating one or more fouling resistances into the basic design Equation 5.1. Basically, such approach is having low accuracy, as fouling is developing in time and even if it is quickly reaching asymptotic value, this value can be influenced by a number of factors not accounted by numerical values for fouling resistances quoted in the literature. As is discussed by Crittenden and Yang (2011), the development of a fouling resistance reduces the thermal performance of a heat exchanger and economic fouling resistance could be selected. One problem with the extra heat transfer surface in a heat exchanger as a way of compensating for fouling is that there is too much surface area when the exchanger is clean. Bypassing fluid around the exchanger to counter the increased heat transfer rate is leading to increase of fouling due to lower velocities through the exchanger. Energy consumptions and environmental impact in recuperative systems due to increase of fouling directly involve the additional fuel required for the heat energy to produce the temperature required by other process equipment. For example, the drop of the temperature on 1 °C at the end of the crude oil preheat train at a 100,000 bbl/d refinery leads to about $40,000 additional fuel costs and the release of about 750 t of CO_2/y. There are also significant production losses during shutdowns for cleaning of heat exchangers, which are difficult to estimate. After shutdowns there is also an additional cost after the production is restarted due to some period of out of specification product quality. Maintenance costs involve staff and other expenditure for cleaning of equipment from fouling, as also an economic and ecological cost for disposal of cleaning chemicals.

In compact heat exchangers (CHEs) with narrow channels and high heat transfer intensity, the accounting for possible fouling and its mitigation is of primary importance. Analysing the different methods available, Panchal and Knudsen (1998) emphasised the use of enhanced heat transfer surfaces as one of the important methods to mitigate water fouling. Heat transfer enhancement and mitigation of fouling renders further advantages to the use of CHEs in different applications in the process industries. Wang et al. (2012) have stressed that the reliable prediction of fouling in enhanced heat exchangers is of great importance for heat exchanger network optimisation, as well as for the design of coolers in an existing cooling water networks discussed by Picón-Núñez et al. (2012).

Nowadays, 'recommended good practice' for tubular heat exchangers suggested by TEMA (2007) requires the use of tabulated fouling resistances for different process fluids and water of different quality (the exempt see in Table 6.1) which should be used with purchaser assistance in a following way. 'The purchaser should attempt to select an optimal fouling resistance that will result in a minimum sum of fixed, shutdown and cleaning costs. The following tabulated values of fouling resistances allow for oversizing the heat exchanger so that it will meet performance requirements with reasonable intervals between shutdowns and cleaning. These values do not recognise the time related behaviour of fouling with regard to specific design and operational characteristics of particular heat exchangers.' It means that overall responsibility for correct accounting of fouling in heat exchanger design is of customer or engineering company carrying the project. About fouling factors in CHEs with intensified heat transfer the information even more scarce. For example, available data for plate heat exchangers (PHEs) are discussed in book by Wang et al. (2007) with reference on still the most certain information on fouling factors (see Table 6.1) published by Marriott (1971). All fouling resistances for PHEs in Table 6.1

Table 6.1 Approximate Data on Fouling Thermal Resistance

Fluid	Fouling Resistance in Shell and Tubes HE by TEMA (2007) (m^2 K/W)	Fouling Resistance in PHE According to Wang et al. (2007) (m^2 K/W)
Water:		
Demineralised or distilled	0.00009	0.000009
Towns (soft)	0.00018	0.000017
Towns (hard)	0.00053	0.000043
Cooling tower (treated)	0.00018–0.00035	0.000034
Sea (coastal) or estuary	0.00035	0.000043
Sea (ocean)	0.00048	0.000026
River, canal	0.00018–0.0007	0.000043
Engine jacket	0.00035	0.000052
Oils, lubricating	0.00018–0.00035	0.000017–0.000043
Oils, vegetable	0.00053	0.000017–0.000052
Solvents, organic	0.00018–0.00035	0.000009–0.000026
Steam	0.00009	0.000009
Process fluids, general	0.00018–0.00088	0.000009–0.000052

are about 10 times smaller than for shell and tube heat exchangers and this is fact well established in practice. But accurate prediction of fouling in CHEs requires more detailed knowledge of fouling process and procedures for accounting and mitigation of different types of fouling.

6.2 FORMS OF FOULING

As it is pointed out for heat transfer equipment in *Perry's Chemical Engineers Handbook* (2008), currently the following six different categories of fouling mechanisms are recognised in heat exchangers, including CHEs:

1. Corrosion
2. Biological
3. Chemical reaction
4. Freezing
5. Particulate
6. Precipitation (or crystallisation)

Of this Tubular Exchangers Manufacturers Association TEMA (2007) five, excluding freezing, recognises as major fouling mechanisms. The brief description of which and the main features of such fouling mitigation in CHE are summarised in the following, according to more detailed analysis of fouling in CHEs presented by Hesselgreaves (2001).

Corrosion fouling is formed when the heat transfer surface material can react chemically with components of the washing it fluid producing less thermally conductive layer of corrosion products. The fluid can also carry corrosion products from other parts of the system upstream that deposit on the heat transfer surface. The formation of fouling deposits can also lead to corrosion under them promoting electrochemical reactions. To avoid corrosion fouling construction materials resistant to corrosion in specific working media should be selected. Alternatively, corrosion inhibitors can be used. It should be noted that corrosion products could be significant in fouling whilst not endangering the exchanger from the point of view of leaks or pressure containment. The mitigation of corrosion fouling in CHEs can be achieved by using appropriate corrosion-resistant steels, which can be much more economical than in shell and tubes due to much smaller surface area for the same conditions and thinner walls. For example, PHEs have thickness of plate wall down to 0.5–0.3 mm and more than two times smaller heat transfer surface area. The corrosion products of material upstream suspended in fluid are depositing on heat transfer surface according to mechanism of particulate fouling.

Biological fouling is caused by the deposition of suspended in the fluid stream organisms on the warm heating heat transfer surfaces, where they attach, grow and reproduce. There are two subgroups, one of which is microbiofoulants such as bacterial slime and algae. Another subgroup is macrobiofoulants such as snails and barnacles. Bacteria are the organisms which are most likely to create problems in CHEs. They can also assist corrosion by, for example, reducing sulphate to hydrogen sulphide, which is corrosive to common stainless steels. Treatment with biocides is very helpful to prohibit biofouling grows, especially in closed systems. Chlorine and ozone are used as oxidising biocides, which kill the bacteria and oxidise the deposited film. To be fully effective they need high concentration that can be more easily achieved in a CHE with smaller volume and fluid inventory compared to shell and tube heat exchangers. The extensive research of biological fouling was made for the food industrial sector, especially dairy industry. Bansal and Chen (2006) have published comprehensive review of milk fouling in heat exchangers. The mechanisms of this complex phenomenon still not completely understood. The denaturation and aggregation reactions initiate just after milk is subjected to heating and fouling cannot be completely eliminated in heat exchangers. However, it can be controlled and mitigated by adjusting appropriate thermal and hydraulic conditions. Fouling can be reduced by increasing the flow rate and decreasing the temperature. The CHE and especially such type as PHE are proved preferable type of heat exchangers in such conditions. They have lower fouling formation compared with tubular heat exchangers because comparatively lower surface temperature and higher turbulence.

Chemical reaction fouling occurs with fluid in which chemical reaction taking place between its components creating the layer of reaction products on heat transfer surface. The surface material itself is not involved in the reaction. This form of fouling is usually associated with polymerisation reactions such as, for example, coking. When the deposit turns from a tar to a hard coke, its removal is very difficult. Different chemical reactions can cause fouling in organic mixtures. It includes auto-oxidation, polymerisation and thermal decomposition. Crude oil fouling mechanism is related to chemical reactions, but it also combined with the precipitation of asphaltenes and the involved into process inorganic components. The survey of research on chemical reaction fouling of organic fluids is presented by Watkinson and Wilson (1997) and more recent publications analysed by Crittenden and Yang (2011). Chemical reaction fouling is also taking place in heat exchanger-reactors. CHEs have a number of advantages that can help to minimise chemical reaction fouling due to good temperature control and low residence times.

Freezing or solidification fouling occur at cooling of the fluid below freezing point of that fluid or some of its components at the heat transfer surface, which causes solidification with deposition of solid phase on the surface. To mitigate this form of fouling a strict temperature control is required so that any such deposit can be melted if even started to form. One such example is the deposition of wax from crude oil in pipelines and heat exchangers. Air-to-air heat exchangers are often used with heating or cooling systems in buildings. Frosting inside such heat exchangers is common in cold regions and results in a significant decrease in the performance of the exchangers. The survey of research on frosting in air-to-air heat exchangers is presented by Nasr et al. (2014).

Particulate fouling is the deposition on the heat transfer surface of the particles held in suspension in the fluid flow. Small particles are usually harder to remove than larger ones, as the forces holding them together and to the surface can be greater. This type of deposit is easily removed mechanically or with back flashing. The problems with removal arise when it is joined by another fouling mechanism, especially such as tar formation. The rate of particulate fouling much depends on the fluid velocity. At higher velocities it can be significantly reduced. This type of fouling has a primary importance in heat exchangers working with gaseous streams which contain suspended particulate matter. The deposition on compact heat transfer surfaces of contaminants contained in the air affects the thermal performance of air cooling systems and can lead to component failures, expensive repairs and loss of service. The detailed analysis of publications on particulate fouling in transport air conditioning systems and the ways of it mitigation is presented by Wright et al. (2009). A detailed review of experimental works and modelling of particulate fouling is presented by Henry et al. (2012). It brings out the essential mechanisms occurring during different stages of deposit formation. It is found that deposit formation and clogging results from the competition between particle–fluid, particle–surface and particle–particle interactions.

Precipitation (or crystallisation) fouling occurs when the fluid of the stream contain some dissolved substance that with a change of conditions is precipitating forming crystals, either directly on the heat transfer surface or in the bulk of fluid with subsequent deposition on the surface. In case of water containing calcium or magnesium salts with inverse solubility, these salts are precipitating with increase of temperature on heating surfaces. This fouling mechanism is usually referred to as scaling. Opposite way waxes can precipitate out as temperature of fluid is reduced at the cooling surface. The rate of scaling depends on wall temperature, crystallisation kinetics and local flow structure and temperature distribution. Most commonly, crystallisation fouling is avoided by pre-treatment of the fluid stream before heat exchanger or by the addition of chemicals that reduce deposit formation. There are positive indications that the addition into fluid stream of appropriate chemicals can minimise scaling and waxing, but environmental regulations are limiting the application of such fouling mitigation. Another method of scaling prevention is electromagnetic descaling technology, which proved efficient for fouling prevention in water-carrying pipes to inhibit calcium carbonate deposition on the surfaces. The method can be effective for both inverse solubility salts and such as silica, exhibiting normal solubility characteristics. Another method of minimising scaling is based on surface treatment. The metallurgical surface modifications like ion implantation and magnetron sputtering can lead to significant more than 70% reduction in the growth of crystalline deposits and an even bigger reduction in bacteria adhesion. Many of these treatments act also to minimise corrosion and erosion, as was shown by Muller-Steinhagen and Zhao (1997).

6.3 FOULING DEPOSITION MECHANISMS

The majority of situations of fouling in heat exchangers involve different forms of fouling, which can interact and complement each other. For example, corrosion fouling can provide nucleation sites for precipitation fouling and particles for particulate fouling. This creates difficulties in prediction and modelling of the fouling process. One of the most complex and at the same time extensively researched area is crude oil fouling that can be considered as example to illustrate the complexity of the mechanisms that can be involved in fouling. Watkinson (2007) has pointed out that the potential causes of fouling in a crude oil heat exchanger could involve some or all of the following (1) suspended impurities, such as dirt, clay or rust, that attach to the surface; (2) insoluble gum formation in the bulk oil caused by presence of oxygen in storage tanks; (3) asphaltenes or resins that precipitate out due to changes in composition, temperature or pressure; (4) soluble sulphur species that react with the heat transfer surface, possibly forming iron sulphides believed to be important in starting the overall fouling process and (5) oil components that undergo thermal decomposition, thereby forming coke. It was suggested that in heavier, unstable crude oils, asphaltenes are the fouling species. The overall fouling process involves mass transport of species between fluid bulk and surface with complex reactions taking place. The effect of fluid shear stress on a deposit will also influence the overall fouling rate if it leads to convective or diffusive removal of deposits. Chemical reaction fouling generally involves the following multi-step process:

$$\text{Reactants (A)} \rightarrow \text{soluble precursors (B)} \rightarrow \text{insoluble foulant (C)}$$

All processes involved are influenced by flow structure in channels near heat transfer surface, with corresponding turbulent mixing and shear stress levels. Accordingly, the possible components of crude oil fouling mechanism on some intensified heat transfer surface are summarised in Figure 6.1. In this representation, the mechanism of specific fouling case, such as for example, predominantly precipitation or particulate fouling, can be simplified and some components are absent or can be neglected.

Fouling is dynamic process developing in time and depending on the specific fouling case and the nature of involved substances its time dependence can considerably vary. Figure 6.2 presents typical curves for change in time of the thickness

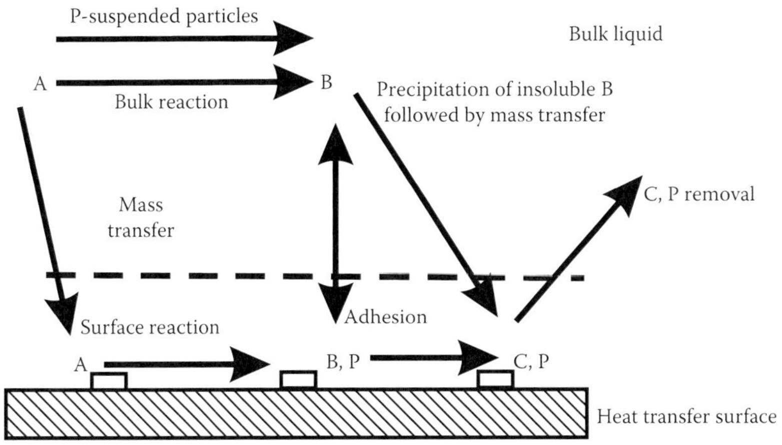

Figure 6.1 Multi-component fouling mechanism.

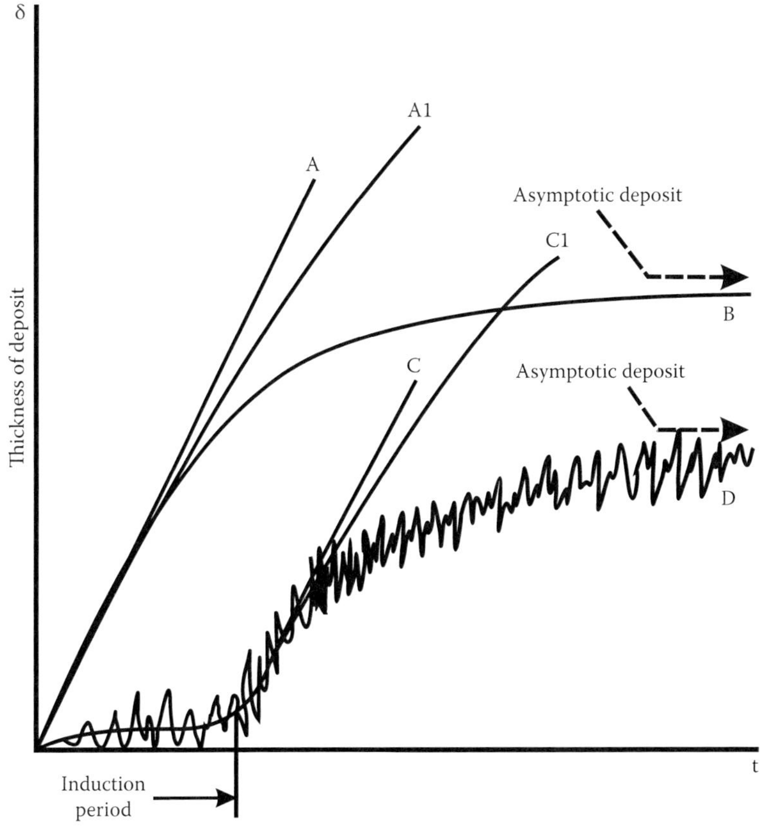

Figure 6.2 General shapes of typical fouling curves.

of fouling deposit (it can be mass of deposit or fouling thermal resistance which dependent from time in the same way). In practice, the smooth character of curves is rarely realised and they are looking similar to curve D due to stochastic nature of the real fouling process.

Curve A in Figure 6.2 corresponds to constant fouling rate when fouling deposit thickness and its thermal resistance increases linearly with time. Such fouling rates are rare in practice and mostly observed in laboratory studies for experiments not long enough to witness a decrease in fouling rate as the deposit grow up. Curve B shows an asymptotic behaviour of fouling in which fouling rate decreases with the time becoming zero when the asymptote is reached. Such behaviour directly follows from theoretical approach of Kern and Seaton (1959), in which the fouling process is presented as the simultaneous action of the fouling deposition and fouling removing processes. Net fouling rate φ can be determined as the difference between the fouling deposition rate φ_d and deposit removal rate φ_r created by the fluid shear forces on its boundary with fouling:

$$\varphi = \varphi_d - \varphi_r \tag{6.1}$$

The asymptotic value of fouling resistance is reached if net fouling rate is equal to zero that can happen when deposition and removal rates are equal. The gradual increase of deposit thickness is leading to the change of temperature on its boundary with washing it fluid as also change of strength of the deposited material. It can lead to both decrease of deposition rate and increase of removal rate with final their equality when the asymptotic fouling conditions are reached. The fouling resistance R_f as a function of time t would then be described by the following equation in which R_f^* and b^* are the asymptotic fouling resistance and time constant:

$$R_f = R_f^* \cdot \left[1 - \exp(-b^* \cdot t)\right] \tag{6.2}$$

At the initial stage when fouling is starting on the clean surface, the deposit formation can start after some period of time which is called induction period (also the terms initiation or delay period are used). In Figure 6.2, curves C and D show induction period after which fouling starts according to linear and asymptotic cases. Majority of fouling models are based on representation such as Equation 6.1 determining the rate of fouling as the difference between the deposition and the removal rates.

During *initiation period*, the overall heat transfer coefficient in heat exchanger remains equal to its value in clean heat exchanger. The presence of initiation period and its duration depends on such factors as type of fouling and nature of fouling deposit, surface material conditions (e.g., polishing or other special treatment) and surface temperature. The initiation period can last from seconds to several days and can be absent after cleaning of heat exchanger. Often no initiation period exists for particular fouling.

There are four major constituent processes which control the fouling formation and build up, namely (1) mass transport, (2) deposition, (3) removal and (4) ageing.

Mass transport is responsible for the delivery to heat transfer surface of fouling precursors which are suspended or dissolved in the bulk of fluid. It can be made by diffusive or convective mass transfer. The driving force of this process is the difference of precursor concentration in the bulk of fluid and at the surface. The rate of transportation can be determined by the following equation (6.3):

$$\frac{dm}{dt} = k_m \cdot (C_b - C_s) \tag{6.3}$$

where

m is the mass flux

C_b and C_s are the precursor mass concentrations in the bulk of fluid and at the surface

k_m is mass transfer coefficient

When $C_s > C_b$ and deposit is soluble, fouling substance can diffuse back into the bulk of fluid and this effect can be used for surface cleaning.

Deposition is taking place when the fouling precursors come to heat transfer surface and stick to it or react on it forming substances that stick to the surface. Some part of precursors or their reaction products can leave the surface. It is stochastic process that can be controlled by chemical reaction, diffusion or adhesion mechanisms.

Removal of some deposited material from outer boundary of fouling layer is caused by the action of fluid shear stress and mass transfer. The intensity of removing deposit from the deposit outer surface depends on the strength of deposited layer, shear stress and mass transfer magnitude.

Ageing of the deposit material occur after long enough time of its formation, especially at prolong exposure to high temperature. During ageing the strength of the deposit may increase with polymerisation, recrystallisation or other processes.

The main factors influencing fouling are surface temperature, fluid bulk temperature, flow velocity (through shear stress on the fouling layer surface) and fluid composition. To account for these factors, different fouling models were developed which are mostly applicable to specific types of fouling and predominant fouling mechanisms.

6.4 FOULING MODELS

According to Crittenden and Yang (2011) in the literature, there are four major categories of models developed based on different features of fouling process (1) reaction and transport, (2) induction, (3) ageing and (4) threshold. There is still a need for a full model to include all of these aspects for development of the general system applicable to different heat exchangers in a heat exchange network. But due to substantial differences in modelling approaches and complexity of the problem, it is still problem for a future development.

6.4.1 Reaction and Transport Models

For many types of fouling mechanisms, different reaction and transport models are published in the literature by a number of researchers. The background of models development and resulting equations exhibit considerable differences; nevertheless, all the models are verified with some set of experimental data showing rather good agreement with modelling results. It confirms the big variety of processes involved in different specific cases of fouling formation and difficulties in obtaining generalised models accounting for all the factors determining fouling on heat transfer surfaces.

The classic model expressing the net fouling rate as a result of competition between fouling deposition and fouling removal was introduced by Kern and Seaton (1959) and in general form expressed by Equation 6.1. Overall, this expression covers most of proposed fouling models, even those where removal term is absent, by putting in Equation 6.1 its value equal to zero ($\varphi_r = 0$). Such models usually correspond to non-asymptotic fouling behaviour shown by curves A and A1 or C and C1 in Figure 6.2.

For the deposition term in many models employ an Arrhenius-type expression for the rate of chemical reaction:

$$\frac{dR_f}{dt} = k_r \cdot \exp\left(-\frac{E}{RT_s}\right) \tag{6.4}$$

In this equation, E is the activation energy. Its value and the pre-exponential factor, k_r, can be determined from experimental data with constant velocity, fluid composition and equipment geometry, but at varying surface temperatures. Importance of mass transport effects in the deposition rate is accounted by combining mass transfer component with the Arrhenius expression in deposition term, following the earliest such model that was proposed by Crittenden and Kolaczkowski (1979a). Such models include in deposition term a transportation component expressed similar to Equation 6.3, as for example model proposed by Epstein (1983). To account for the inverse influence of flow velocity (or wall shear stress) some authors, such as for example Paterson and Fryer (1988) for skimmed milk fouling, included flow velocity or shear stress into denominator of the deposition rate expression.

The removal term φ_r in expression of Equation 6.1 is rather difficult to determine due to the fundamental drawback in identifying the physical phenomena involved. Suggested mechanisms here included diffusion of fouling species away from the surface, turbulent bursts sweeping particles from the surface and surface shear forces removing deposited material. Nowadays, it is more common to assume that an increased velocity will reduce fouling by increasing the shear stress at the heat transfer surface and removal term is usually expressed as directly proportional to shear stress in some power and the amount of deposited material. In this way with the increasing the amount (mass or thickness) of the deposit the removal term also increases and at some time becomes equal to deposition term and amount (thickness) of material deposited on the heat transfer surface stabilises, getting its asymptotic value.

6.4.2 Initiation Period Models

Most fouling models are focused on the fouling rate when thermal resistance of fouling deposit is increasing, not accounting for the beginning of the fouling process, which can start with some delay. According to Muller-Steinhagen (1999), this delay is very rare with particulate fouling mechanism, but for many cases of other fouling types the fouling deposition grows is preceded by an initiation (or induction period) in which fouling just starting to develop and no significant change of thermal resistance is observed with time increase. The initiation period could be from a few seconds to hours and up to month or so depending on fouling type, surface conditions, flow velocity, temperature and other parameters. A number of researchers have studied the initiation period phenomenon for different fouling mechanisms. For example, Belmar-Beiny et al. (1993) observed a much shorter induction period in a PHE than in a tubular heat exchanger for milk fouling. They attributed this to the higher turbulence in PHE.

Some researchers have developed empirical correlations for the length of initiation period for fouling from specific process streams. For example, Muller-Steinhagen (1998) presented following correlation for delay time t_d in minutes as function of surface temperature T_s at fouling from Kraft Black Liquor in the heat exchanger installed at Kraft process in the pulp and paper industry:

$$\frac{1}{t_d} = 2.444 \cdot 10^{20} \cdot \exp\left(\frac{-19{,}810}{T_s}\right) \tag{6.5}$$

Empirical estimation of the initiation time for fouling from calcium sulphate solutions during boiling heat transfer was proposed by Malayeri and Müller-Steinhagen (2007) in a frame of phenomenological model for the prediction of fouling thermal resistance.

The approach to include initiation period into the general phenomenological model of crystallisation fouling accounting for other periods of fouling grows was also proposed by Mwaba et al. (2006). On analysis of initiation period, they have concluded that there is a region where the fouling thermal resistance is zero for clean surface and then becomes even negative when nucleates start forming on the heat transfer surface in a random way. The surface became rough that leads to an increase in convective heat transfer in some part of initiation period of fouling formation and some relatively small increase in overall heat transfer coefficient that looks like some decrease in fouling thermal resistance. With following another regions of fouling grows and its asymptotic behaviour, the fouling curve is taking S – shape, like curve D in Figure 6.2. It was noticed that such shape is similar to that of the solution for differential logistic equation developed as a model for population grows in a finite environment, which form was proposed for correlation of experimental data for all period of time by continuous function of the form:

$$R_f = \frac{R_{f\infty}}{1 + \exp\left[-r\left(t - t_0\right)\right]} \tag{6.6}$$

where

$R_{f\infty}$ is asymptotic value of fouling thermal resistance
r and t_0 are parameters evaluated according to experimental data

Similar expression was also used by Nebot et al. (2007) for correlation of experimental data on fouling in power plant steam condenser including initiation period of fouling formation. However, while such approach allows convenient correlation of available experimental data gives no information about mechanism of initiation period or tools to predict the initiation time in conditions different from experimental.

The fouling mechanism in initiation period is very important in view of fouling mitigation. Having a good reliable knowledge about this phenomenon, it might be possible to prolong fouling delay to considerable times or even indefinitely, as pointed out Crittenden and Yang (2011). For correlation of experimental data in initiation period, Yang et al. (2012) proposed a model of fouling in initiation period based on assumption that active foulant species stick to the surface and gradually cover its fraction θ. The rate of surface coverage change $d\theta/dt$ is assumed being proportional to the fraction of free surface $(1 - \theta)$. Another assumption is that the already deposited fouling material on the surface acts a seed attracting more foulant and the fouling growth is directly proportional to first order of fractional coverage θ with a rate constant k. These assumptions lead to the following relationship:

$$\frac{d\theta}{dt} = k \cdot \theta \cdot \left(1 - \theta\right) \tag{6.7}$$

The fouling rate then can be expressed as $\theta(dR_f/dt)$ with (dR_f/dt) expressed corresponding to any model established for fouling growth period. Using this model different experimental data including the data for initiation periods have been correlated with good accuracy for systems including crude oil fouling, water scaling and whey protein fouling.

6.4.3 Ageing Models

Many of the fouling deposits are subjected to ageing with considerable time periods. It may increase the strength of the deposit through polymerisation, recrystallisation, dehydration, etc. Biological deposits can be poisoned by metal ions and washed away by the fluid flow. Currently, ageing is the least studied and understood step and is usually ignored in modelling attempts. To account phenomena in which fouling deposit undergoes further change to its composition and properties under the prolonged influence of high surface temperatures, Crittenden and Kolaczkowski (1979b) introduced two layers model of deposit. One layer could represent fresh foulant and the other aged deposit. More recently, this two-layer concept was further developed by Ishiyama et al. (2011) by couple heat transfer, deposition and ageing and extending it to cleaning algorithms for identification of optimal cleaning cycles with introduction of the cleaning super cycle based on cost model accounting for periodical chemical and/or mechanical cleaning procedures. The model and its applications for cleaning schedule of heat exchangers were further developed by this research group in paper of Pogiatzis et al. (2012).

6.5 THRESHOLD FOULING MECHANISM

Extensive experimental researches on pilot plants and in laboratories of fouling in heat exchangers used to preheat crude oil in crude oil distillation units have established that under some conditions exchangers may not foul. It is particularly interesting since if to find and maintain such favourable conditions in heat exchanger then the fouling problem is ultimately solved. To deal with this phenomenon, Ebert and Panchal (1997) initially introduced the concept of 'threshold fouling' models for crude oil fouling inside tubes of shell and tubes heat exchangers. Extending Kern and Seaton approach, they expressed fouling rate in their model as a difference between deposit formation and suppression terms as follows:

$$\frac{dR_f}{dt} = \alpha \cdot \mathrm{Re}^{\beta} \cdot \exp\left(\frac{E}{R \cdot T_f}\right) - \gamma \cdot \tau_w \tag{6.8}$$

where

α, β, γ and the activation energy E are dimensional parameters which vary for different crude oils and can be adjusted

Re is fluid Reynolds number

τ_w is the wall shear stress

The threshold fouling conditions could be determined by setting right part of Equation 6.8 to zero. For the same fluid and its thermophysical properties, wall shear stress τ_w is directly related to Reynolds number. Then when parameters α, β, γ and the activation energy E are determined, the threshold conditions can be presented in the form of curve depicted in Figure 6.3 in coordinates $T_f - \tau_w$. At film temperatures and wall shear stresses corresponding to points below this curve, the fouling on heat transfer surface will not start and it will remain clean during all time of heat exchanger operation. For film temperatures and shear stresses above this curve, the fouling thermal resistance can be determined by Equation 6.9.

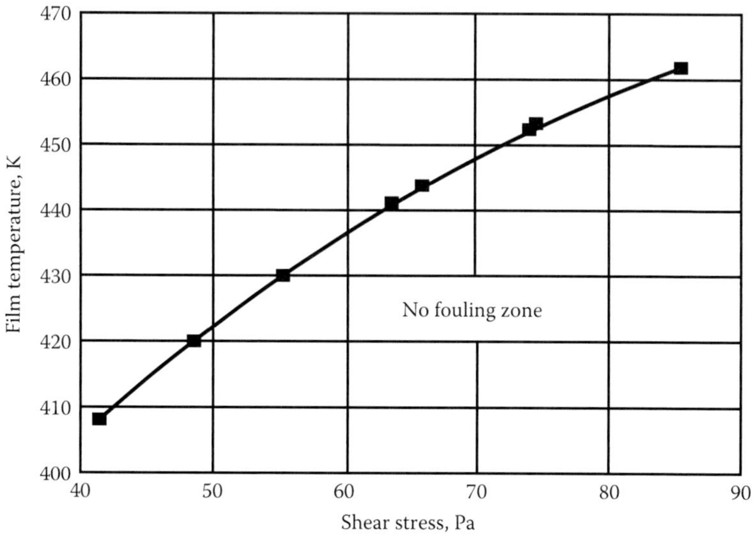

Figure 6.3 The curve of threshold fouling conditions for temperature T_s and shear stress τ_w for Australian light crude oil (after Nasr and Givi, 2006).

Panchal et al. (1999) modified Equation 6.9 with the aim to correlate a wider range of physical parameters and sets of experimental data proposed the following expression involving Prandtl number Pr:

$$\frac{dR_f}{dt} = \alpha \cdot \mathrm{Re}^{-0.66} \cdot \mathrm{Pr}^{-0.33} \exp\left(\frac{E}{R \cdot T_f}\right) - \gamma \cdot \tau_w \tag{6.9}$$

Polley et al. (2002) proposed another form of threshold model linking removal term to the rate of convective mass transfer between surface of fouling deposit and bulk of fluid. There is also introduced surface temperature instead of film temperature and fluid velocity u instead of wall shear stress:

$$\frac{dR_f}{dt} = \alpha \cdot \mathrm{Re}^{-0.8} \cdot \mathrm{Pr}^{-0.33} \exp\left(\frac{E}{R \cdot T_s}\right) - \gamma \cdot u^{0.8} \tag{6.10}$$

The same form of suppression term is used by Yeap et al. (2004) with modification of Epstein's model for chemical reaction fouling with first-order kinetics as deposition term and introducing empirical model parameters A, B, C and E:

$$\frac{dR_f}{dt} = \frac{A \cdot C_f \cdot u \cdot T_s^{2/3} \cdot \rho^{2/3} \cdot \mu^{-4/3}}{1 + B \cdot u^3 \cdot C_f^2 \cdot \rho^{-1/3} \cdot \mu^{-1/3} \cdot T_s^{2/3} \cdot \exp(E/(R \cdot T_s))} - C \cdot u^{0.8} \tag{6.11}$$

Another modification of deposition term in Equation 6.8 of Ebert and Panchal model was proposed by Nasr and Givi (2006) based on experimental data for Australian light crude oil:

$$\frac{dR_f}{dt} = \alpha \cdot \mathrm{Re}^{\beta} \cdot \exp\left(\frac{E}{R \cdot T_f}\right) - \gamma \cdot \mathrm{Re}^{0.4} \tag{6.12}$$

The earlier described fouling threshold models require estimation of the included empirical parameters according to available data sets on crude oil fouling. As there are a lot of different crude oil feed stocks and the properties of crude oils vary significantly depending on its type and place of oil field, these parameters require corrections to be used in specific operating conditions of heat exchangers. Nasr and Givi (2006) have compared the results of fouling rate calculations by different threshold models with their experimental data for Australian light crude oil and some other sets of experimental data available in the literature. Their model gave the best results for their experiments with only two results of 15 experiments with error 30% and 41% and other 13 less than 17%. For other data sets the average error of their model was about 60%, while for one data set Polley et al. (2002) model gave average error 28%. Similar comparison on the same data sets was made by Aminian and Shahhosseini (2009), who presented artificial neural network (ANN) model for prediction of fouling threshold. For other models, the results of comparison are similar to presented by Nasr and Givi (2006). ANN model gave better accuracy, absolute mean relative error was about 16%, while this error for most accurate threshold fouling model was 52%. While the black box character of ANN model makes difficult its application for understanding of the process mechanism, it can be used in on-site analysis and control of heat exchangers operation in crude oil distillation.

Beside problems with accuracy and estimation of fouling threshold models parameters for different crude oils, there are difficulties to use these models for channels with geometries other than round tubes. For complicated channel geometries, Polley et al. (2011) have proposed to use Ebert–Panchal model with deposition term expressed through film heat transfer coefficient h in a form:

$$\frac{dR_f}{dt} = \frac{A}{h} \cdot \exp\left(\frac{E}{R \cdot T_f}\right) - \gamma \cdot \tau_w \tag{6.13}$$

To determine wall shear stress, it was recommended to calculate the suppression term on an 'apparent wall shear' based upon the heat transfer coefficient. After calculating the heat transfer coefficient in shell side of tubular HE or in channel of any geometry, it requires to determine the flow conditions within a tube that yield the same value of heat transfer coefficient. The wall shear stress in tube with this flow condition can be used in the model. Such method was also recommended for the use in tubes with inserts.

To extend fouling model to tubes with inserts and other channels of complex geometries, Yang and Crittenden (2012) proposed another approach. They have developed a CFD-aided approach and incorporated shear stress distribution obtained with CFD simulation into the model. In this way fouling model developed for bare round tubes can be applied for use with more complex geometries. Accordingly Yeap et al. (2004) model was modified by incorporating surface shear stress τ_w into the removal term instead of fluid velocity u.

$$\frac{dR_f}{dt} = \frac{A_m \cdot C_f \cdot u \cdot T_s^{2/3} \cdot \rho^{2/3} \cdot \mu^{-4/3}}{1 + B_m \cdot u^3 \cdot C_f^2 \cdot \rho^{-1/3} \cdot \mu^{-1/3} \cdot T_s^{2/3} \cdot \exp(E/(R \cdot T_s))} - C \cdot \tau_w \tag{6.14}$$

For deposition term, it was introduced concept of 'equivalent velocity' which means the velocity in bare round tube of the same equivalent diameter and with the same wall shear stress like the average wall shear stress in the considered channel of complex geometry. This concept was successfully used for prediction of fouling rate in tubes with HiTRAN wire inserts (see Figure 3.19).

6.6 PRESSURE DROP ASSOCIATED WITH FOULING

Beside deterioration in heat transfer performance fouling is the cause of pressure drop increase in heat exchangers for the same flow rate of the heat exchanging media. The growing fouling deposits are leading to partial decrease of channel cross section and forming certain form of roughness on deposit surface. The local thickness of fouling deposit layer δ_f is directly linked to local thermal resistance of fouling R_f through fouling thermal conductivity λ_f by simple expression:

$$\delta_f = R_f \cdot \lambda_f \tag{6.15}$$

The values of fouling thermal conductivity for some deposits are presented in Table 6.2. This table is compiled from data published by Hesselgreaves (2002) based on data of different authors and table published by Muller-Steinhagen (1998).

Pressure drop in fouled channel of heat exchanger can be easily calculated in case when fouling deposition layer is evenly distributed over the heat transfer surface. The pressure drop can be determined with Equation 3.13 accounting for the increase of flow velocity due to partial blockage of cross section area by fouling and decrease of hydraulic diameter. The error arises with uncertainty of thermal conductivity values for cases of practical applications, roughness and uneven distribution of fouling deposit. It requires introduction of correction factor by comparing with available measurements of pressure drop data.

Models for the prediction of fouling rates and fouling deposit thickness generally apply to a local condition (temperature, shear stress and physical properties). In industrial heat exchangers, film deposit temperatures can significantly vary between entrance (where heated media is at its coolest) to exit (where heated media is at its highest temperature in heat exchanger). This means that the thickness of fouling deposit at zones close to entry can be much less than in zones close to exit

Table 6.2 Values of Thermal Conductivity of Some Deposits

Substance	λ_f (W/mK) Hesselgreaves (2002)	λ_f (W/mK) Muller–Steinhagen (1998)
Alumina	0.42	
Biofilm	0.2–0.7	0.7
Boiler scale	2.0	
Calcium carbonate, $CaCO_3$	2.19–2.94	0.9
Calcium sulphate, $CaSO_4$	0.74–2.34	0.8–2.2
Calcium phosphate	2.6	0.6
Carbon (solid)	1.6	
Carbon (fine)	0.07	
Carbon (lamp black)	0.147	
Calcite (boiler deposit)	0.93	0.9
Gypsum (boiler deposit)		1.3
Hematite (boiler deposit)	0.53	0.6
Hydrocarbon polymers	0.1–0.2	
Ice	2.25–3.95	
Magnesium carbonate	0.43	
Magnesium phosphate		2.3
Magnetic iron oxide	2.87	2.9
Milk components		0.5–0.7
Snow	0.465	
Sodium aluminium silicate		0.2–0.4
Water scale	0.6–1.4	
Wax	0.24	

from heat exchanger. These differences in deposit thickness invoke changes in shear stress at the deposit surface and also surface and deposit film temperatures. It can be accounted by calculation with one-dimensional mathematical model of heat exchanger, accounting the distribution of local process parameters along the heat transfer surface. Such method can significantly complicate the design. However, counting for uncertainties involved (in fouling models itself, fouling thermal conductivity values, surface roughness, etc.) such method also not guaranties high accuracy. In any case the results should be corrected on comparison with available pressure drop measurements on site.

6.7 FOULING ON ENHANCED HEAT TRANSFER SURFACES

As presented in Chapter 4, analysis of different methods of heat transfer intensification has shown that all these methods are based on considerable modifications in flow structure leading to the increase in intensity of heat, mass and momentum transfer. The rise of film heat transfer coefficients is accompanied by increase in convective diffusion between heat transfer surface and bulk of the fluid, increase and redistribution along the surface of shear stress. These phenomena are playing important role in the earlier discussed fouling mechanisms and fouling models that can lead to significant difference of fouling deposition rates on different enhanced surfaces and in smooth tubes.

6.7.1 Fouling in Tubes with Artificial Roughness

A number of researchers have investigated fouling in enhanced tubes with different forms of artificial roughness. The results of these researches and judgments about the intensity of fouling deposition compare to smooth tubes are fairly different. The majority of experiments on fouling in enhanced tubes with artificial roughness were conducted on water solutions with crystallisation, particulate or biological fouling types. In majority of cases, the asymptotic fouling behaviour was observed.

Analysing the published in literature experimental data, Panchal and Knudsen (1998) noted that for some forms of artificial roughness (such as spirally indented tubes) the performance decrease in 1-year period was about 10% compared to 15%–25% for plane tubes. On other data, the asymptotic fouling resistance for the longitudinal finned tubes was on 15%–30% greater than for plane tubes, but for spirally indented tube it was 25%–50% lower. The analysis of field data for fouling in condenser tubes (both enhanced and plane) with different cooling water sources has also shown different trends in fouling rates which were from similar up to 2.5 times higher in enhanced tubes. There was no satisfactory explanation for such differences in fouling rates, but consistent trend was observed: The higher the fouling propensity, the higher the fouling rate of the enhanced tube with respect to the plain tube.

Somerscales and Bergles (1997) in their survey of different publications came to conclusion that fouling in enhanced tubes is mitigated in cases of chemical reaction, particulate fouling and asymptotic precipitation by inverse solubility. While not asymptotic precipitation fouling is not mitigated in enhanced tubes. In analysis of laboratory and field tests results, it was observed cases of the same fouling rates with plane tubes, as also cases with higher and lower fouling rates depending on artificial roughness type and water quality.

Model of particulate fouling in enhanced tubes was proposed by Chamra and Webb (1994). The comparison with their experimental data published earlier (Chamra and Webb, 1993) has shown good accuracy of predicted results for two geometries of internally corrugated tubes as also for results of some other authors.

It was found that the tendencies for asymptotic behaviour of liquid particulate fouling on enhanced tubes with asymptotic fouling resistances higher than those in the plane tube. It was observed the increase of asymptotic fouling thermal resistances with increase of particles concentration and particle diameter and surface shear stress decrease.

The precipitation fouling from water with different salts content inside and outside of tubes with transverse turbulence promoters (see Figure 4.8) was studied by Kalinin et al. (1990). They have reported significant reduction of fouling rate compared to plane tubes and asymptotic character of fouling thermal resistance time dependence. The developed model for fouling thermal resistance on tubes outside surface is presented in the form of Equation 6.2. The results of experimental research with enhanced tube of outside diameter $D = 16$ mm inside annular channel formed by outer tube with inner diameter 25 mm were generalised by the following correlation:

$$R_f = \frac{2.6 \cdot 10^{-7} \cdot (P/D_e)^{0.2} \cdot C_s}{4 \cdot f \cdot w^{0.912}} \cdot \left[1 - \exp\left(0.45 \cdot 4 \cdot f \cdot t \cdot w^{-0.129}\right)\right] \tag{6.16}$$

where

C_s is salts concentration in (mg eq)/L
t is time in h
fouling thermal resistance R_f is in (m^2 K)/W

The investigated range of parameters is as follows: water velocity w varied from 0.05 to 1.2 m/s; indents pitch to equivalent diameter ratio $P/D_e = 0.4\ldots1.2$; indents depth to equivalent diameter ratio $h/D_e = 0.064\ldots0.086$; the wall temperature was varied from 50 °C to 107 °C and salts concentration was 20 (mg eq)/L.

It was observed that when with water velocity 0.15 m/s the overall heat transfer coefficient with plane tube was decreased about 5 times, with investigated enhanced tube such decrease was 1.5 times. With water velocity 1 m/s, the decrease of overall heat transfer coefficient was less than 15%. The fouling deposition inside investigated enhanced tubes (see Figure 4.8) was also about half of that inside plane tubes.

Kukulka et al. (2012) in experiments made with water from Lake Erie have observed the significant decrease of the fouling rate in stainless steel Vipertex EHT enhanced tubes with artificial roughness depicted in Figure 3.18. The experiments were made in tubes 1EHT with diameter 9.525 mm and 2EHT with diameter 15.9 mm. The fouling rate in enhanced tubes was about six times lower than in smooth stainless steel tube.

Based on preceding analysis of different data on fouling in enhanced tubes, it can be concluded that the fouling rate beside other parameters which are influencing the process in plane tubes depends on artificial roughness geometry. When thermal resistance of fouling and fouling rate in some enhanced tubes are even bigger than in plane tubes, some other exhibit big potential for fouling mitigation. It requires cautious attitude for enhanced tubes application for media having big fouling tendencies with estimation of possible fouling influence on overall performance of heat exchanger with such tubes.

6.7.2 Fouling in Tubes with Inserts

Enhanced tubes with artificial roughness have shown different fouling tendencies compared to smooth tubes in different media, from increase of fouling to it considerable mitigation, depending on the artificial roughness geometry. But there are developed special designs of in-tube inserts that have been shown to be effective not only in heat transfer enhancement but also in mitigating fouling, as is discussed by Somerscales and Bergles (1997). Krueger and Pouponnot (2009) have emphasised increasing interest in the use of such inserts in different applications of the oil industry as well as Bott (2001) in many cases of using water as cooling media. Ritchie et al. (2009) studied the characteristics of fluid flow in the tube with inserts and indicated that the flow near the wall resembled a turbulent profile even at a Reynolds number of 500. They also discussed some theoretical analysis of the fouling in tube with inserts.

Wire inserts of various forms are placed into the fouling stream to enhance heat transfer by increasing the mixing between the fluid in regions near to the wall and the fluid in the bulk flow. These devices are also mitigating the fouling. Krueger and Pouponnot (2009) have analysed the possibilities to enhance performance of shell and tube heat exchangers in refineries and chemical plants through the use of SPIRELF®, TURBOTAL® and FIXOTAL® systems of inserts placed inside tubes. SPIRELF and TURBOTAL are designed primarily at reducing fouling rate inside heat exchanger tubes by mechanical effect (rotation in case of TURBOTAL and vibration for SPIRELF). The enhancement of heat transfer is a favourable side effect for these devices. FIXOTAL is working as a turbulence promoter producing the swirl effect. It is fixed inside tube device enhancing heat transfer and also reducing fouling for such types of it as polymerisation, solidification of wax and water scaling.

Gough and Rogers (1984) from Cal Gavin Limited have patented the wire insert with registered trade mark hiTRAN® and method of its formation (example shown in Figure 3.19). As discussed by Droegemueller et al. (2013), this insert can be used to influence both temperature and shear stress on the tube wall and has been applied successfully in fouling applications in the oil industry. With this device the increase of film heat transfer coefficient on the tube side up to four times can be achieved. The wall shear stress can be increased by up to seven times at different flow regimes, that is, leading to substantial reduction in fouling rate inside tubes of heat exchanger.

6.7.3 Fouling in Channels of PHEs

Starting with the publication on fouling in PHEs by Marriott (1971) with examples of approximate recommended fouling thermal resistances presented in Table 6.1, the significant reduction of fouling deposition rate on the walls of PHE channels can be regarded as established fact. For design purposes, Cooper et al. (1980) recommended to use fouling thermal resistances not greater than one-fifth of the recommended in the literature for shell and tube heat exchangers.

The estimation for asymptotic value of fouling thermal resistance was proposed by Tovazhnyansky and Kapustenko (1984) based on Kern and Seaton (1959) fouling model and modified for PHE channel Reynolds analogy for heat and momentum transfer expressed at $Pr = 1$ by Equation 4.73. From that equation the relation between film heat transfer coefficient and wall shear stress can be expressed as follows:

$$h = \tau_w^{3/7} \cdot d_e^{-1/7} \cdot 0.065 \cdot \lambda \cdot \left(\frac{8}{\rho \cdot \nu^2} \right)^{3/7} = \tau_w^{3/7} \cdot d_e^{-1/7} \cdot A_{ph} \tag{6.17}$$

Respectively, wall shear stress is related to film heat transfer coefficient as:

$$\tau_w = h^{7/3} \cdot d_e^{1/3} \cdot A_{ph}^{7/3} \tag{6.18}$$

The coefficient A_{ph} is determined by physical properties of fluid and for the same fluid and temperature does not change.

According to the approach by Kern and Seaton (1959), the removal term in Equation 6.1 can be expressed through the deposit thickness δ as:

$$\varphi_r = b \cdot \delta \cdot \tau_w \tag{6.19}$$

where b is some parameter determined by properties of fluid and fouling deposit. When the asymptotic fouling deposition thickness δ^* is reached, the removal term φ_r becomes equal to deposition term φ_d and it is as follows:

$$\delta^* = \frac{\varphi_d^*}{b \cdot \tau_w} \tag{6.20}$$

From Equations 6.20 and 6.18 it follows that for the same fluid and fouling deposit properties, the asymptotic value of fouling thermal resistance is inversely proportional to film heat transfer coefficient in power 7/3 and to channel equivalent diameter in power 1/3:

$$R_f^* = \frac{\delta^*}{\lambda_f} \sim \frac{1}{h^{7/3} \cdot d_e^{1/3}} \tag{6.21}$$

The experience of designing and operating plate heat exchangers suggests that the film heat transfer coefficients are usually two, three and more times higher than in shell and tubes heat exchangers and even with the smaller equivalent diameters they have much smaller thermal resistance of fouling deposits. This conclusion explains the values presented in Table 6.1 and is relevant for those types of fouling which are described by Kern and Seaton model (sedimentation of suspended solids, scaling, etc.). However, in case of oversized heat exchange surface area (incorrect design or excessive surface reserve), the aforementioned lowering of the fouling thermal resistance may be cut down.

The significant reduction of fouling in welded PHE COMPABLOCK working for heating of crude oil in crude oil preheat train was reported by Tamakloe et al. (2013). They have proposed method for modelling and integration of the fouling behaviours in these heat exchangers. It was shown to be good instrument for their subsequent design since the fluid shear stresses generated in these exchangers can be high to minimise fouling. By proper selection of the passes arrangement, it was possible to considerably improve the performance of heat exchangers.

6.7.4 Cooling Water Fouling in Channels of PHEs

Crystallisation fouling on heat exchanger surfaces can create chronic operational problems in a broad range of processing applications that include cooling water systems, desalination, steam generation, etc. Energy losses, additional power consumption and the costs of cleaning all make this practical operational problem a significant challenge in the progression towards sustainable development. In the systems of water cooling circuits at industrial enterprises, there are also a lot of suspended solid particles in water formed by corrosion products at the pipelines and also particles removed from fouling deposits in the elements of heat transfer equipment. The process of fouling formation with such media involves mechanism of particulate fouling.

The comprehensive study of particulate water fouling in PHE channels was reported by Karabelas et al. (1997). The experiments were made for PHE channels formed by commercial plates with corrugation inclination angle 60° and 30°. The asymptotic behaviour with a time of fouling thermal resistance was observed, as well as strong influence of flow velocity. The experimental data of that paper on asymptotic value of fouling thermal resistance are plotted against wall shear stress in Figure 6.4. The Reynolds number Re is calculated for velocities presented in paper, water properties at temperature 40 °C and equivalent diameter $D_e = 0.005$ mm. The corrugation aspect ratio is assumed as $\gamma = 0.581$. The friction factor of total pressure losses (due to friction on the wall and form drag) calculated using Equation 4.65 and the share of friction losses ψ estimated by Equation 4.75. The influence of wall shear stress on asymptotic thermal resistance of fouling is distinctly seen. Data are correlated by Equation 6.22 with $B^* = 3.5 \cdot 10^{-4}$ K·s/m (solid line in Figure 6.4). For wall shear stress higher than 40 Pa, thermal resistance becomes very small (less than 10^{-5} m^2 K/W) and data unstable.

$$R_f^* = \frac{B^*}{\tau_w} \tag{6.22}$$

The process of calcium sulphate fouling was investigated by Bansal et al. (2000) in a PHE made from commercial plates with chevron corrugations. The corrugations inclination angle was $\beta = 60°$. To investigate velocity effect, there are three experimental runs with velocities 0.183, 0.352 and 0.667 m/s. All other conditions were the same. Two points in Figure 6.4 correspond to fouling thermal resistances at the end of runs (after 7,000 min), velocities 0.352 and 0.667 m/s. The test run at lowest velocity finished after only 2,000 min and do not permit to estimate R_f^*. The wall shear stresses are estimated using Equations 4.65 and 4.75, as described earlier. The water temperature 61 °C, $D_e = 0.005$ mm and $\gamma = 0.581$. Data correlated by Equation 6.22 with $B^* = 2.05 \cdot 10^{-4}$ K·s/m.

The process of calcium carbonate deposit formation on the surface of inner tube in annulus channel was investigated experimentally by Quan et al. (2008). There are two test runs at velocities 0.6 and 1.2 m/s with all other conditions the same.

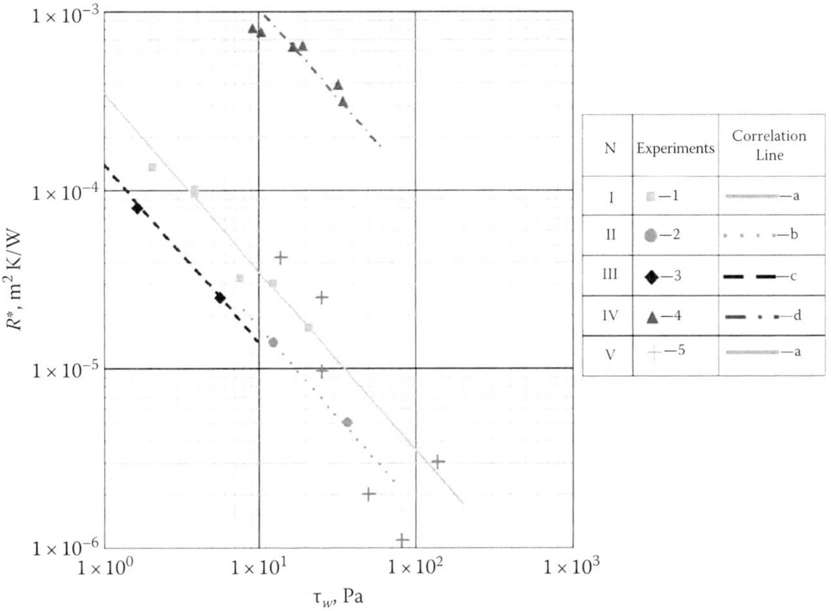

Figure 6.4 The influence of wall shear stress on asymptotic value of fouling thermal resistance: 1 and line b – annular channel data of Quan et al. (2008); 2 and line d – PHE channel $\beta = 60°$ data of Chernyshov (2002); 4 and line c – PHE channel $\beta = 60°$ data of Bansal et al. (2000); 3 and line a – PHE channel $\beta = 30°$; 5 – PHE channel $\beta = 60°$ data of Karabelas et al. (1997).

Estimating wall shear stress as for smooth annular duct, these data are also presented in Figure 6.4. They are correlated by Equation 6.22 with $B^* = 1.45 \cdot 10^{-4}$ K·s/m.

All aforementioned tests were performed in laboratory conditions on small PHEs or experimental models of channel. The experimental study of precipitation fouling in industrial PHE for fresh water heating was reported by Chernyshov (2002). AlfaLaval M10B PHE of 78 plates was installed in District Heating (DH) system of city Tula in Russia. This DH system had nowadays obsolete 'open circuit' of hot water supply, where hot water is just taken from radiator circuit. It required to heat fresh water up to the temperatures of radiator circuit, which are much higher than temperatures of hot tap water in modern 'closed radiator circuit' schemes. Because of high salts content in fresh water (780 mg/L) and necessity to heat it to temperatures 60 °C–63 °C and more, heat exchanger experienced severe fouling. The fouling depends strongly from temperature of the heated fresh water and hence from the position along the plate length. The typical fouling deposit distribution after asymptotic fouling thicknesses were achieved can be seen on photo in Figure 6.4b. The data on average asymptotic fouling thermal resistance, reported by Chernyshov (2002) at three different flow velocities (from 0.26 to 0.57 m/s), are presented in Figure 6.4. The fouling thermal resistance is calculated from data on fouling thickness and the value of fouling thermal conductivity which is 1.03 mK/W in the cited work. The wall shear stresses are estimated using Equations 4.65 and 4.75 for water temperature 61 °C, D_e = 0.005 mm and $\gamma = 0.581$. Data are correlated by Equation 6.22 with $B^* = 115 \cdot 10^{-4}$ K·s/m. Slightly lower data for smallest velocity can be explained by relatively big fouling thickness (up to 0.8 mm) compared to plate spacing (2.5 mm). It leads to decrease of real cross section area for water flow, higher velocity and higher shear stress, than calculated for clean channel cross section.

The earlier presented data are obtained at four different conditions of water fouling. For the same level of wall shear stress, the asymptotic values of fouling thermal resistance are much different. They depend on different factors: concentration, nature and size distribution of suspended particles in study by Karabelas et al. (1997); soluble salts concentration and different salts content and temperatures of water and the channel walls. But for the same water and temperature conditions, the dependence of asymptotic fouling thermal resistance from wall shear stress is described by the same relation and is inversely proportional to wall shear stress. It also confirm that the removal term φ_r^* when asymptotic fouling thickness is reached is proportional to wall shear stress in power $m = 1$, as is described by Equation 6.19.

The determination of deposition term φ_d^*, at the conditions when asymptotic fouling reached, is a very complicated task. A lot of parameters influence it. But for the big enterprise with certain water circuit, the quality of the cooling water is the same for all coolers. In that case, when water asymptotic fouling in one heat exchanger can be determined, we can calculate the coefficient B^* in Equation 6.22 and to use this value to determine cooling water asymptotic fouling thermal resistance for PHEs calculated for this enterprise on all other cooling positions.

Special attention should be paid when scaling can be expected as predominant fouling mechanism. In Figure 6.5, one can see two sides of the plate, which has worked about 1 y at DH system of Kyiv (Ukraine) for tap water heating. Side (a) was in contact with DH water. Here only particulate fouling takes place. The plate has some deposits which are almost evenly distributed along its length, with some increase towards the entrance of hot medium. It can be attributed to higher strength of the deposits at higher temperatures. Side (b) of the plate was in contact with fresh water heated from 5 – 10 °C to 57 °C for domestic hot water supply. On this side the plate surface near the entrance of cold fresh water is clean, the deposits start and grow significantly towards the hot water exit with increase of water and wall temperatures. It can be judged by examining the pictures of fouling deposit distribution on small plate areas along the plate presented in Figure 6.6. We can conclude that for scaling fouling there exist some threshold conditions on water temperature and salt content, after which fouling deposition starts. So when the scaling is possible for cooling water of enterprise circuit, the outlet temperature should be kept below certain level depending on wall shear stress in PHE and cooling water quality of that enterprise.

When the conditions below fouling threshold cannot be reached and asymptotic value of fouling thermal resistance is too higher, the periodical cleaning of heat exchanger is required. To establish the cleaning schedule, the correct prediction of fouling deposit grows with time is required.

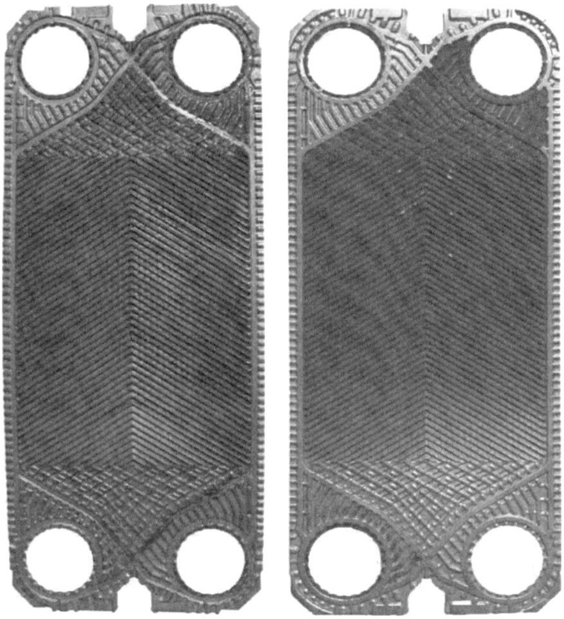

Figure 6.5 Two sides of the plate after 1 year in operation for tap water heating in DH system (fresh water coming from the bottom, heating DH radiator water from the top).

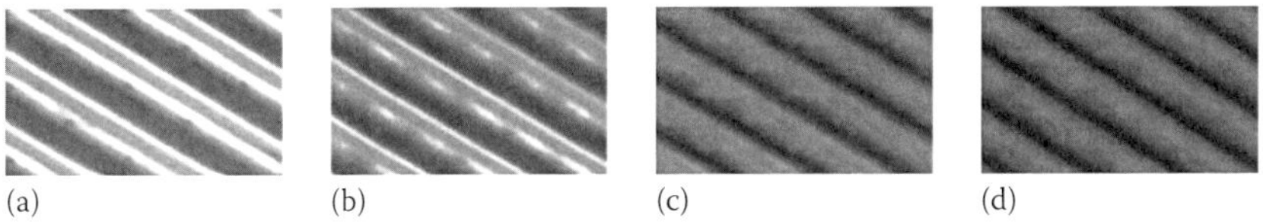

Figure 6.6 Photos of four parts of plate surface a, b, c and d – from heated water inlet to its exit from the PHE channel.

The development of fouling thermal resistance in time for known deposit thermal conductivity λ_f can be described by equation, which is directly following from Equation 6.1:

$$\frac{dR_f}{dt} = \frac{\varphi_d}{\lambda_f} - b \cdot \tau_w \cdot R_f \tag{6.23}$$

When φ_d, b, λ_f and τ_w not depend on the time of deposit formation t and deposit thickness δ, the integration of linear ordinary differential Equation 6.23 causes no problem. For approximate solution at time t, we can use time averaged values of these parameters on time period $0 - t$. Then, the fouling thermal resistance at time t after integrating Equation 6.23 is given as

$$R_f(t) = \frac{B}{\tau_w} \cdot \left[1 - \exp\left(-\frac{\varphi_d}{B} \cdot \tau_w \cdot t\right)\right] \tag{6.24}$$

where $B = \varphi_d/(b \cdot \lambda_f)$.

To estimate coefficient B, we can take its value at asymptotic fouling conditions $B = B^*$. As it was observed on examining fouling deposit distribution along the plate, the water fouling process in PHEs exhibits the threshold behaviour discussed earlier in Section 6.4. Yang and Crittenden (2012) have proposed the application of their model represented by Equation 6.14 also for the tubes with heat transfer enhancement. The deposition term in their model is expressed as

$$\frac{\varphi_d}{\lambda_f} = \frac{A_m \cdot C_f \cdot u \cdot T_s^{2/3} \cdot \rho^{2/3} \cdot \mu^{-4/3}}{1 + B_m \cdot u^3 \cdot C_f^2 \cdot \rho^{-1/3} \cdot \mu^{-1/3} \cdot T_s^{2/3} \cdot \exp(E/(R \cdot T_s))} \tag{6.25}$$

where

T_s is the surface temperature, K
ρ is the fluid density, kg/m^3
μ is the fluid dynamic viscosity, $Pa \cdot s$
$R = 8.314$ J/(mol·K) is the universal gas constant
C_f is the fanning friction factor; and u is the average flow velocity, m/s

The parameter values that give the best fittings with experimental data for investigated crude oil are

$$E = 52{,}100 \text{ J/mol}; A_m = 7.93 \cdot 10^{-10} \text{ kg}^{2/3} \text{ K}^{1/3} \text{ m}^{5/3}/\text{kW/s}^{1/3}/\text{h} \tag{6.26}$$

$$B_m = 1.8 \cdot 10^{-5} \text{ m}^{13/3} \text{ kg}^{2/3} \text{ s}^{8/3}/\text{K}^{2/3}$$

For intensified heat transfer, the velocity u is defined as equivalent velocity, which is velocity in a bare tube that gives the same wall shear stress as in a tube with heat transfer enhancement of the same internal diameter operating at a different average fluid velocity. Let us assume that this is justified also for PHE channel and bare tube of the same equivalent diameter *de*.

For the developed turbulent flow in a straight bare tube with internal diameter *de*, the Fanning friction factor can be calculated by Blasius Equation:

$$C_f = 0.0791 \cdot \left(\frac{u \cdot \rho \cdot de}{\mu}\right)^{-0.25} \tag{6.27}$$

Wall shear stress in a straight bare tube is given as

$$\tau_w = \frac{C_f \cdot \rho \cdot u^2}{2} \tag{6.28}$$

For known shear stress in channel with enhanced heat transfer, the equivalent velocity in bare tube can be calculated from Equation 6.28, substituting C_f from Equation 6.27:

$$u = \left(\frac{\tau_w \cdot de^{0.25} \cdot 2}{\mu^{0.25} \cdot \rho^{0.75} \cdot 0.0791} \right)^{\frac{1}{2-0.25}} \tag{6.29}$$

The product of C_f and u from Equation 6.28 is given as

$$P_{cu} = C_f \cdot u = \frac{2 \cdot \tau_w^{1-\frac{1}{1.75}}}{\rho} \left(\frac{de^{0.25} \cdot 2}{\mu^{0.25} \cdot \rho^{0.75} \cdot 0.0791} \right)^{-\frac{1}{1.75}} \tag{6.30}$$

Eliminating C_f and u from Equation 6.25, using Equations 6.28 and 6.30 we obtain expression for deposition term through wall shear stress:

$$\frac{\varphi_d}{\lambda_f} = \frac{A_m \cdot P_{cu} \cdot T_s^{2/3} \cdot \rho^{2/3} \cdot \mu^{-4/3}}{1 + B_m \cdot P_{cu} \cdot 2 \cdot \tau_w \cdot \rho^{-4/3} \cdot \mu^{-1/3} \cdot T_s^{2/3} \cdot \exp(E/(R \cdot T_s))} \tag{6.31}$$

Using Equation 6.31 and the value of B^* which was obtained by analysis of asymptotic fouling thermal resistance data, fouling thermal resistance at time t can be calculated by Equation 6.24. As the properties of crude oil and water considerably different, the empirical parameters A_m, B_m and E should be corrected to fit experimental data for water fouling.

To fit experimental data for precipitation of calcium sulphate in PHE channels, presented by Bansal et al. (2000) only one parameter A_m is adjusted. For the value $A_m = 6.5 \cdot 10^{-12}$ kg$^{2/3}$ K$^{1/3}$ m$^{5/3}$/W/s$^{1/3}$/h, the computed thermal resistances of fouling at different flow velocities are presented by curves in Figure 6.7 in comparison with experimental data taken from a graph at paper by Bansal et al. (2000). The lower values at initial period can be explained as it was done by the authors of cited paper. The initial delay period may be required for the formation of the deposit on the heat transfer surface.

The calcium carbonate scaling in annular smooth channel between enclosing tube of 22 mm inner diameter and inside tube of 16 mm outer diameter was investigated by Quan et al. (2008). These data for two different velocities are fitted by the model at $A_m = 1.5 \cdot 10^{-10}$ kg$^{2/3}$ K$^{1/3}$ m$^{5/3}$/W/s$^{1/3}$/h (see Figure 6.7). Other model parameters B_m and E not changed.

The particulate fouling in PHE channels from water with suspended calcium carbonate particles was investigated by Karabelas et al. (1997). Figure 6.8 presents the comparison of these data with prediction by model (solid lines calculated for T_s = 37 °C). All parameters in Equation 6.31 are the same as in calculations for Figure 6.7. It let to conclude that the values of parameters in Equation 6.31 for similar salt and impurities content have close numerical values (Figure 6.9).

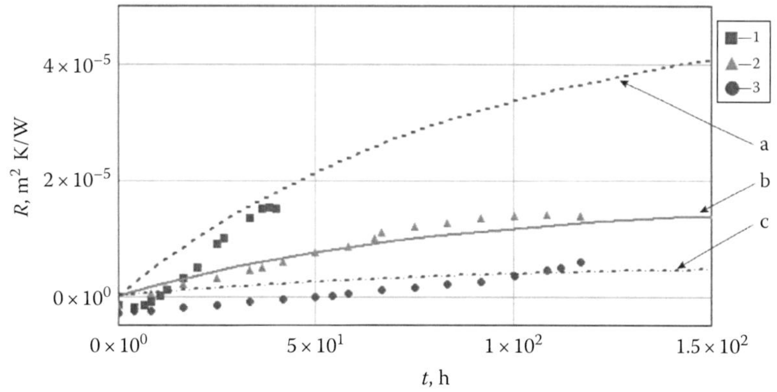

Figure 6.7 Fouling thermal resistance at PHE channel with β = 60° under calcium sulphate precipitation: 1 and line a – w = 0.183 m/s; 2 and line b – w = 0.352 m/s; 3 and line c – w = 0.667 m/s (from experimental data of Bansal et al., 2000).

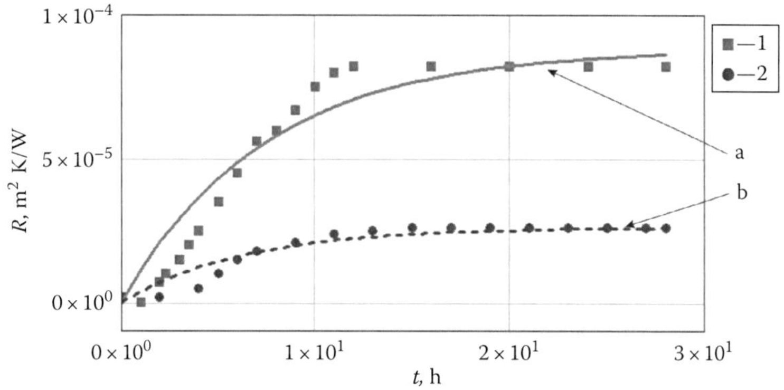

Figure 6.8 Fouling thermal resistance at calcium carbonate scaling in annular channel: 1 and line a – w = 0.6 m/s; 2 and line b – w = 1.2 m/s (from experimental data of Quan et al., 2008).

Equation 6.31, beside wall shear stress and physical properties of the fluid, accounts for the effect of surface temperature T_s. We can guess that the coefficient B depends on the surface temperature in the same way. For two surface temperatures T_{s1} and $T_{s2,}$ when all other conditions the same, let us assume:

$$\frac{B(T_{s1})}{B(T_{s2})} = \frac{\varphi_d(T_{s1})}{\varphi_d(T_{s2})} \tag{6.32}$$

The data for calcium carbonate scaling at three different wall temperatures, when velocity and salt concentration are the same, are presented in paper by Quan et al. (2008).

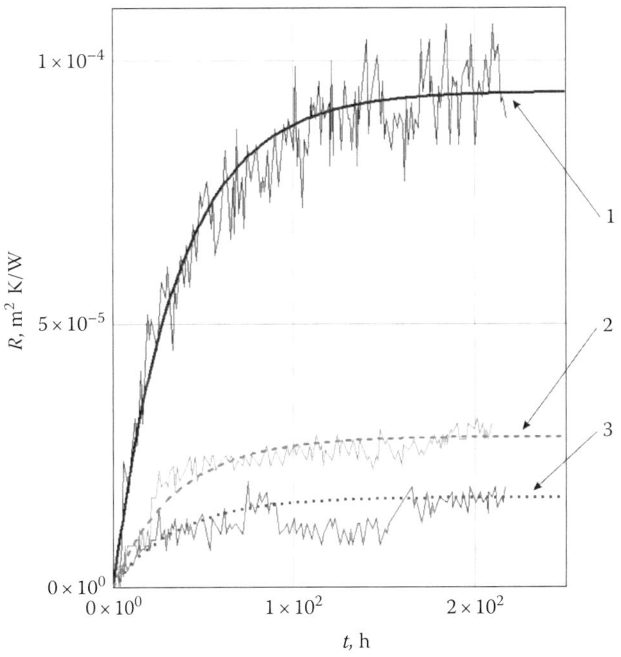

Figure 6.9 Particulate fouling thermal resistance at PHE channel with β = 30°: 1 and line a – w = 0.5 m/s; 2 and line b – w = 1.0 m/s; 3 and line c – w = 1.35 m/s (from experimental data of Karabelas et al., 1997).

These data are shown in Figure 6.10. The lines on the graph are calculated by Equation 6.24 using Equations 6.31 and 6.32. The value of coefficient B is taken from previous estimation for data of this paper, which were taken at T_s = 51 °C. The prediction of the asymptotic fouling thermal resistance is fairly good. The description by the model of time dependence is reasonable, with some overestimation of fouling thermal resistance.

Analysing the results, we can conclude that presented mathematical model is capable to predict fouling thermal resistance for precipitation and particulate fouling at different flow velocities and surface temperatures. The model can be used for PHEs with enhanced heat transfer and also for straight channels without heat transfer intensification. It predicts the effects of wall shear stress and surface temperatures, but the influence of salts concentrations and solid particles content and sizes is not accounted.

For correct predictions by the model, the reliable data about its parameters are needed. The data about these parameters obtained for the straight tubes or channels can be used for prediction of fouling thermal resistance in PHEs working with the same cooling water.

The big industrial enterprises usually have quite a big numbers of heat exchangers, which are using water of the centralised cooling water circuit of the enterprise.

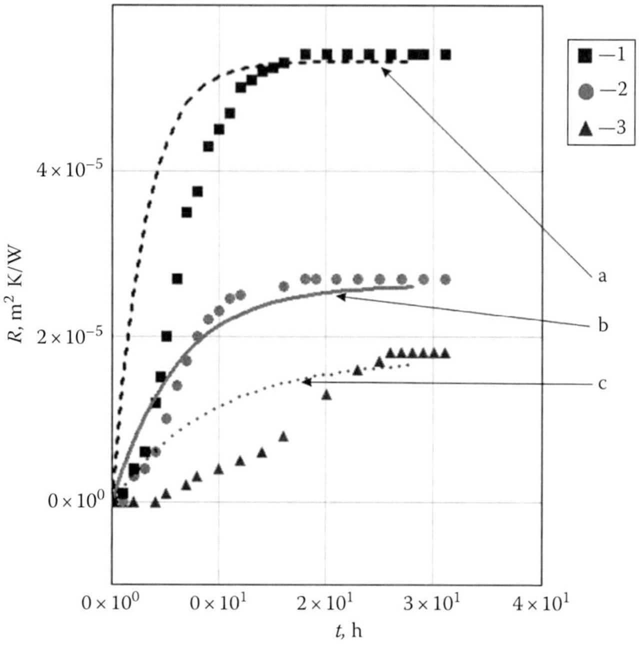

Figure 6.10 Fouling thermal resistance at calcium carbonate scaling in annular channel w = 1.2 m/s: 1 and line a – T_s = 64 °C; 2 and line b – T_s = 51 °C; 3 and line c – T_s = 44 °C (from experimental data of Quan et al., 2008).

The salts and solid particles contents in this water are the same for all heat exchangers. Therefore, by monitoring water side fouling development in one heat exchanger (in PHE or inside tubes of tubular heat exchanger) we can estimate parameters B and A_m in proposed mathematical model here. It will enable to account for cooling water fouling in all PHEs of this enterprise. The threshold values of wall shear stress and surface temperature can be calculated. After that in a design and selection of PHEs for the enterprise the wall shear stress should be kept higher than threshold value or maximal if to reach threshold is not possible. The wall surface temperature must be kept lower threshold value, or minimal. When the conditions to prevent fouling completely cannot be reached, the asymptotic fouling thermal resistance should be calculated by the model and included into design of the PHE.

For flows inside tubes and straight channels the calculation of wall shear stress is not problem, it can be made by formulas available in open literature. For PHEs the correlations for friction factor are usually proprietary by PHEs manufacturers. The important feature of the fouling prediction approach proposed here is that the wall shear stress is calculated based on Equations for friction factor on a main corrugated field of the PHE channel. To use these Equations, engineer must know only the geometrical characteristics of the main corrugated field of the plates – the corrugation angle to longitudinal plate axis β and corrugation aspect ratio γ, which can be measured on real plates.

In present form, the proposed mathematical model can give accurate results only having some reference point for cooling water concerned. It does not account for salt content and solid particles size and concentrations effects on fouling thermal resistance. To estimate fouling thermal resistance just on a data of water purity and chemical content, more experimental data are needed. Important that the data obtained for smooth channels can be used directly for PHEs, utilising proposed approach.

REFERENCES

Aminian J., Shahhosseini S. 2009. Neuro-based formulation to predict fouling threshold in crude preheaters. *International Communications in Heat and Mass Transfer* 36: 525–531.

Bansal B., Chen X.D. 2006. A critical review of milk fouling in heat exchangers. *Comprehensive Reviews in Food Science and Food Safety* 5: 27–33.

Bansal B., Muller-Steinhagen H., Chen X.D. 2000. Performance of plate heat exchangers during calcium sulphate fouling – Investigation with an in-line filter. *Chemical Engineering and Processing* 39: 507–519.

Belmar-Beiny M.T., Gotham S.M., Peterson W.R., Fryer P.J., Pritchard A.M. 1993. The effect of Reynolds number and fluid temperature in whey protein fouling. *Journal of Food Engineering* 19: 119–139.

Bott T.R. 2001. Potential physical methods for the control of biofouling in water system. *Transactions of IChemE* Part A 79: 484–490.

Chamra L.M., Webb R.L. 1993. Effect of particle size distribution on particulate fouling in enhanced tubes. *Journal of Enhanced Heat Transfer* 1(1): 65–75.

Chamra L.M., Webb R.L. 1994. Modeling liquid-side particulate fouling in enhanced tubes. *International Journal of Heat and Mass Transfer* 37(4): 571–579.

Chernyshov D.V. 2002. Prognosis of scaling in plate water heaters for increasing reliability of their operation. PhD thesis, VAC RF May 23, 2003, Tula, Russian Federation, 2002. (In Russian.) www.disserCat.com (accessed 25 September 2014).

Cooper A., Suitor J.W., Usher J.D. 1980. Cooling water fouling in plate heat exchangers. *Heat Transfer Engineering* 1(3): 50–55.

Crittenden B., Yang M. 2011. Technical review of fouling and its impact on heat transfer. Report on project FP7-SME-2010-1 262205/INTHEAT. intheat.dcs.uni-pannon.hu/wp-content/uploads/2011/11/D1.2.pdf (accessed 10 October 2014).

Crittenden B.D., Kolaczkowski S.T. 1979a. Mass transfer and kinetics in hydrocarbon fouling. In: *Proceedings of Conference on Fouling -Science or Art? Institute of Corrections Science and Technology/Institute of Chemical Engineering*, London, UK, 169–187.

Crittenden B.D., Kolaczkowski S.T. 1979b. Energy savings through the accurate prediction of heat transfer fouling resistances. In: O'Callaghan P.W. (Ed.) *Energy for Industry*. Pergamon, Oxford, UK, 257–266.

Droegemueller P., Osley W., Philipp D. 2013. Increased shear, reduced wall temperatures, use of hiTRAN wire matrix inserts in systems subject to fouling. In: *Proceedings of International Conference on Heat Exchanger Fouling and Cleaning*, June 9–14, 2013, Budapest, Hungary, 372–378.

Ebert W.A., Panchal C.B. 1997. Analysis of exxon crude slip stream coking data. In: Panchal C.B. et al. (Eds.) *Fouling Mitigation of Industrial Heat-Exchange Equipment*. Begell House, New York, USA, 451–460.

Epstein N. 1983. Fouling of heat exchangers. In: Taborek J., Hewitt G.F., Afgan N. (Eds.) *Heat Exchangers: Theory and Practice*. Hemisphere, Washington, DC, USA, 795–815.

Gough M.J., Rogers J.V. 1984. Insert for placement in a vessel and method of forming the insert. US Patent 4,481,154.

Green D.W., Perry R.H. 2008. *Perry's Chemical Engineers Handbook*, 8th edn. McGraw-Hill, New York, USA.

Henry C., Minier J-P., Lefèvre G. 2012. Towards a description of particulate fouling: From single particle deposition to clogging. *Advances in Colloid and Interface Science* 185–186: 34–76.

Hesselgreaves J.E. 2001. *Compact Heat Exchangers Selection, Design and Operation*. Pergamon, Oxford, UK.

Hesselgreaves J.E. 2002. An approach to fouling allowances in the design of compact heat exchangers. *Applied Thermal Engineering* 22(7): 755–762.

Ishiyama E.M., Paterson W.R., Wilson D.I. 2011. Optimum cleaning cycles for heat transfer equipment undergoing fouling and ageing. *Chemical Engineering Science* 66(4): 604–612.

Kalinin E.K., Dreitser G.A., Yarkho S.A. 1990. *Intensification of Heat Transfer in Channels (Intensifikatsiya teploobmena v kanalah)*. Mashinostroenie, Moscow, Russian Federation. (In Russian.)

Karabelas A.J., Yiantsios S.G., Thonon B., Grillot J.M. 1997. Liquid side fouling of heat exchangers. An integrated R&D approach for conventional and novel designs. *Applied Thermal Engineering* 7(8–10): 727–737.

Kern D.Q., Seaton R.E. 1959. A theoretical analysis of thermal surface fouling. *British Chemical Engineering* 4: 258–262.

Krueger A.W., Pouponnot F. 2009. Heat exchanger performance enhancement through the use of tube inserts in refineries and chemical plants – Successful application examples: SPIRELF, TURBOTAL and FIXOTAL systems. In: *Proceedings of Eurotherm Conference on Fouling and Cleaning in Heat Exchangers*, June 14–19, 2009, Schladming, Austria, 400–406.

Kukulka D.J., Smith R., Zaepfel J. 2012. Development and evaluation of Vipertex enhanced heat transfer tubes for use in fouling conditions. *Theoretical Foundations of Chemical Engineering* 46(6): 627–633.

Malayeri M.R., Muller-Steinhagen H. 2007. Initiation of $CASO_4$ scale formation on heat transfer surface under pool boiling conditions. *Heat Transfer Engineering* 28: 240–247.

Marriott J. 1971. Where and how to use plate heat exchangers. *Chemical Engineering* 78(8): 127–134.

Muller-Steinhagen H., Zhao Q. 1997. Investigation of low fouling surface alloys made by ion implantation technology. *Chemical Engineering Science* 52: 3321–3332.

Muller-Steinhagen H.M. 1998. Mitigation of process heat exchanger fouling: An integral approach. *Transactions of IChemE* 76(Part A): 97–107.

Muller-Steinhagen H.M. 1999. Cooling water fouling in heat exchangers. *Advances in Heat Transfer* 33: 415–495.

Mwaba M.G., Golriz M.R., Gu J. 2006. A semi-empirical correlation for crystallization fouling on heat exchange surfaces. *Applied Thermal Engineering* 26: 440–447.

Nasr M.R., Fauchoux M., Besant R.W., Simonson C.J. 2014. A review of frosting in air-to-air energy exchangers. *Renewable and Sustainable Energy Reviews* 30: 538–554.

Nasr M.R.J., Givi M.M. 2006. Modeling of crude oil fouling in preheat exchangers of refinery distillation units. *Applied Thermal Engineering* 26: 1572–1577.

Nebot E., Casanueva J.F., Casanueva T., Sales D. 2007. Model for fouling deposition on power plant steam condensers cooled with seawater: Effect of water velocity and tube material. *International Journal of Heat and Mass Transfer* 50: 3351–3358.

Panchal C.B., Knudsen J.G. 1998. Mitigation of water fouling: Technology status and challenges. *Advances in Heat Transfer* 31: 431–474.

Panchal C.B., Kuru W.C., Liao C.F., Ebert W.A., Palen J.W. 1999. Threshold conditions for crude oil fouling. In Bott T.R., Melo L.F., Panchal C.B., Somerscales E.F.C. (Eds.) *Understanding Heat Exchanger Fouling and Its Mitigation*. Begell House, New York, USA, 273–279.

Paterson W. R., Fryer P. J. 1988. A reaction engineering approach to the analysis of fouling. *Chemical Engineering Science* 43(7): 1714–1717.

Picón-Núñez M., Polley G.T., Canizalez-Dávalos L., Tamakloe E.K. 2012. Design of coolers for use in an existing cooling water network. *Applied Thermal Engineering* 43: 51–59.

Pogiatzis T., Ishiyama E.M., Paterson W.R., Vassiliadis V.S., Wilson D.I. 2012. Identifying optimal cleaning cycles for heat exchangers subject to fouling and ageing. *Applied Energy* 89: 60–66.

Polley G.T., Tamakloe E., Picón Nuñez M. 2011. Models for Chemical Reaction Fouling. In: *11th AIChE Spring Meeting*, vol. Abstract No. 94c, Chicago, Illinois, USA.

Polley G.T., Wilson D.I., Yeap B.L., Pugh S.J. 2002. Evaluation of laboratory crude oil threshold fouling data for application to refinery preheat trains. *Applied Thermal Engineering* 22: 777–788.

Quan Z., Chen Y., Ma C. 2008. Experimental study of fouling on heat transfer surface during forced convective heat transfer. *Chinese Journal of Chemical Engineering* 16(4): 535–540.

Ritchie J.M., Droegemueller P., Simmons M.J.H. 2009. hiTRAN® wire matrix inserts in fouling applications. *Heat Transfer Engineering* 30: 876–884.

Somerscales E.F.C., Bergles A.E. 1997. Enhancement of heat transfer and fouling mitigation. *Advances in Heat Transfer* 30: 197–253.

Tamakloe E.K., Polley G.T., Picón-Núñez M. 2013. Design of Compabloc exchangers to mitigate refinery fouling. *Applied Thermal Engineering* 60(1): 441–448.

TEMA. 2007. *Standards of the Tubular Exchanger Manufacturers Association*, 9th edn. TEMA Inc., New York, USA.

Tovazhnyansky L.L., Kapustenko P.A. 1984. Intensification of heat and mass transfer in channels of plate condensers. *Chemical Engineering Communications* 31(6): 351–366.

Wang L., Sunden B., Manglik R.M. 2007. *PHEs. Design, Applications and Performance.* WIT Press, Southampton, UK.

Wang Y., Smith R., Jin-Kuk K. 2012. Heat exchanger network retrofit optimization involving heat transfer enhancement. *Applied Thermal Engineering* 43: 7–13.

Watkinson A.P. 2007. Deposition from crude oils in heat exchangers. *Heat Transfer Engineering* 28: 177–184.

Watkinson A.P., Wilson D.I. 1997. Chemical reaction fouling: A review. *Experimental Thermal and Fluid Science* 14: 361–374.

Wright S., Andrews G., Sabir H. 2009. A review of heat exchanger fouling in the context of aircraft air-conditioning systems, and the potential for electrostatic filtering. *Applied Thermal Engineering* 29: 2596–2609.

Yang M., Crittenden B. 2012. Fouling thresholds in bare tubes and tubes fitted with inserts. *Applied Energy* 89: 67–73.

Yang M., Young A., Niyetkaliyev A., Crittenden B. 2012. Modelling fouling induction periods. *International Journal of Thermal Sciences* 51: 175–183.

Yeap B.L., Wilson D.I., Polley G.T., Pugh S.J. 2004. Mitigation of crude oil refinery heat exchanger fouling through retrofits based on thermo-hydraulic fouling models. *Transactions of IChemE* 82: 53–71.

CHAPTER 7

Integration of Intensified Compact Heat Exchangers in a Heat Exchanger Network

The efficient use of heat from a variety of low-grade heat sources described in Chapter 2 requires their efficient integration into the heat supply, distribution and usage system including all other existing energy sources and heat consumers. Such system can include the number of streams to be heated and streams that have heat energy that can be extracted as they are cooled. These streams can exchange heat energy with the use of heat exchangers that as a system of devices transforming heat energy of hotter streams into the heat energy of cooler streams represent a heat exchanger network (HEN). With the same system of streams, the structure of HEN and heat transfer areas of constituting it heat exchangers can vary to a great extent with the significant differences in total heat energy exchanged (recuperated), as well as heat exchangers numbers, heat transfer areas and costs. The design of HEN represents the problem of finding the structure and parameters constituting heat exchangers that correspond to a certain optimum objective function. The widely adopted approaches for HEN design can be divided into two groups, namely, mathematical programming and Pinch Analysis as a basic methodology of Process Integration.

7.1 PROCESS INTEGRATION FOR THE SYNTHESIS OF ENERGY-EFFICIENT HEN

The powerful tool to solve the problem of developing complex energy-efficient systems for heat utilisation is Process Integration, the resent state of which is comprehensively described in the *Handbook of Process Integration* edited by Klemeš (2013). It started as mainly Heat Integration (HI) and Pinch Analysis, stimulated by the energy crisis of the 1970s. Introduced initially as a method for generation of energy-efficient HENs by Linnhoff and Flower (1978), Process Integration was described in detail in the User Guide by Linnhoff et al. (1982) with a second edition by Kemp (2007). The role of Process Integration in preserving sustainable development of modern process industry, its potential for reduction of harmful environmental effects, is discussed by Klemeš et al. (2010).

The application of the Process Integration methodology for HEN synthesis is briefly described below in sections 7.1 – 7.5.

The Pinch Analysis methodology (see, e.g., Linnhoff et al., 1982) became popular owing to its simplicity and efficient management of complex energy utilisation systems. The method has evolved into a complete suite of tools for heat recovery and design techniques for energy efficiency, including guidelines for energy-efficient retrofit and integration of energy-intensive processes. The main strategy of Pinch-based Process Integration is to determine targets on system energy performance prior to the main design activity. From the optimisation viewpoint, using a thermodynamic background, it gives the opportunity to approach directly the vicinity of the optimal solution, giving estimation for the value of the objective function at the global optimum. This strategy yields important guidelines for the design. The energy performance of the system is evaluated by its need of external energy sources (hot utilities [HU]) and means to extract unused heat energy (cold utilities [CU]).

According to the Second Law of thermodynamics, heat is flowing from higher-temperature to lower-temperature bodies with the increase of real system and its surrounding entropy. When two streams are exchanging heat energy, the change of entropy and loss of the useful part of energy (exergy) is smaller when the final temperature difference between the streams is minimal. At the same time in a heat exchanger according to Equation 3.9, the smaller the temperature difference between the streams, the bigger the required heat transfer area. So making the temperature difference smaller saves heat energy for the expense of increase in heat transfer surface area and the corresponding capital cost of the heat exchanger. In HEN design, the minimum allowed temperature difference (ΔT_{min}) in the system of heat exchanging streams is the lower bound on any temperature difference to be encountered in any heat exchanger in the network. The value of ΔT_{min} is a design parameter determined by exploring the trade-offs between saving heat energy by heat recovery and the larger heat transfer area requirement. Any given pair of hot and cold process streams may exchange as much heat as allowed by their temperatures and the minimum temperature difference.

Consider the two-stream example shown in Figure 7.1a. The amount of heat recovery is 10 MW, which is achieved by allowing ΔT_{min} = 20 °C. If ΔT_{min} = 10 °C,

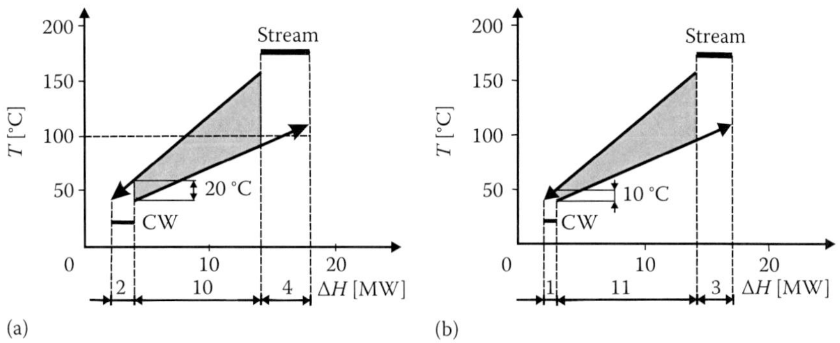

Figure 7.1 Thermodynamic limits on heat recovery for two cases: (a) ΔT_{min} = 20 °C; (b) ΔT_{min} = 10 °C.

as in Figure 7.1b, then it is possible to save one more megawatt of heat energy. To obtain the heat recovery targets for a practical HEN design problem, this principle is extended to handle multiple streams.

The analysis starts by combining all hot streams and all cold streams into two Composite Curves (CCs) (Linnhoff et al., 1982). An example of these curves is shown in Figure 7.2. For each process there are two curves: one for the hot streams (Hot Composite Curve [HCC]) and another for the cold streams (Cold Composite Curve [CCC]). Each CC consists of a temperature–enthalpy (T–H) profile, representing the overall heat availability in the process (the HCC) and the overall heat demands of the process (the CCC). The procedure of constructing composite on the data of streams' target temperatures and heat capacities is well described in books on process integration as, for example, by Linnhoff et al. (1982).

CCs can be moved along the ΔH axis (in horizontal direction on the graph). Usually the HCC position is fixed and the CCC is shifted. This means varying the amount of heat recovery and at the same time the amount of required utilities for heating and cooling. Where the curves overlap, heat can be recuperated between the hot and cold streams. More overlap means more heat recovery and smaller utility requirements, and vice versa. As the overlap increases, the temperature differences between the overlapping curve segments decrease. The closest approach between the curves is termed the *Pinch Point* (or simply the Pinch); it is also known as the *Heat Recovery Pinch*.

On shifting the CCC the larger is the value of ΔT_{min}, the smaller is the possible heat recovery. Specifying the minimum hot utility, the minimum cold utility or the minimum temperature difference fixes the relative position of the two CCs and hence the maximum possible amount of heat recovery. The identified heat recovery targets are relative to the specified value of ΔT_{min}. If that value is increased, then the

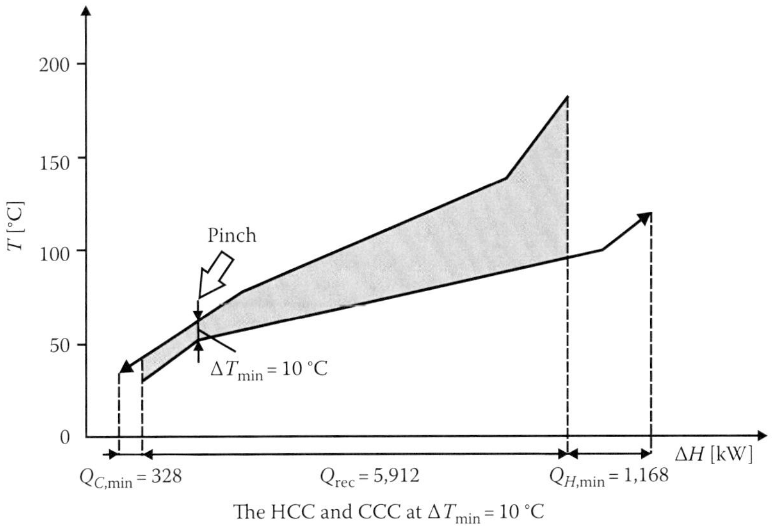

Figure 7.2 Example of CCs at ΔT_{min} = 10 °C.

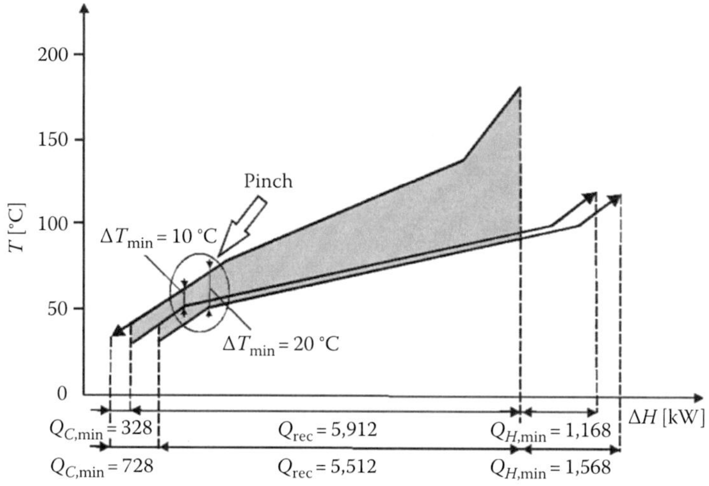

Figure 7.3 Variation of heat recovery targets with ΔT_{min}.

minimum utility requirements also increase and the potential for maximum heat recovery drops, as illustrated in Figure 7.3.

The appropriate value for ΔT_{min} is determined by economic trade-offs. Increasing ΔT_{min} results in larger utility demands and increased energy costs; choosing a higher value reflects the need to reduce the heat transfer area and its corresponding investment cost. Conversely, if ΔT_{min} is reduced, then utility costs go down but investment costs go up. This trade-off is illustrated in Figure 7.4.

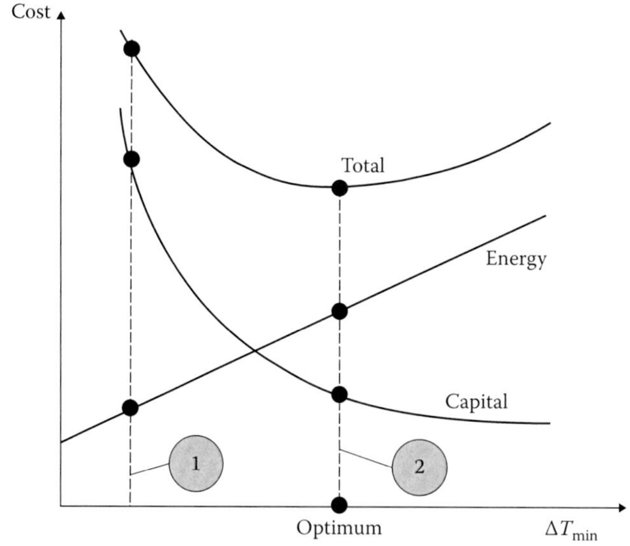

Figure 7.4 Trade-off between investment and energy cost as a function of ΔT_{min}.

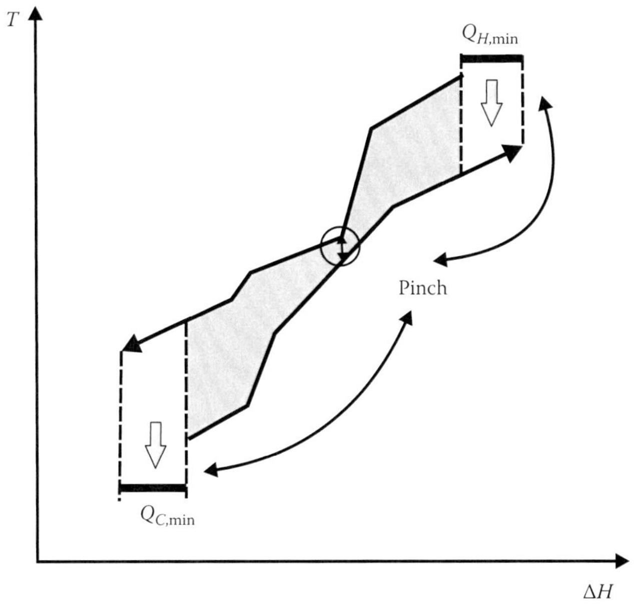

Figure 7.5 Limits for process heat recovery set by the Pinch.

The heat recovery Pinch has important implications for the designing of HEN. As illustrated in Figure 7.5, the Pinch sets the absolute limits for heat recovery within the process. The Pinch Point also divides the heat recovery problem into a net heat sink above the Pinch Point and a net heat source below it. The heat sink above the pinch is in balance with the minimum hot utility ($Q_{H,\min}$) and the heat source below the Pinch is in balance with the minimum cold utility ($Q_{C,\min}$), while no heat is transferred across the Pinch via utilities or via process-to-process heat transfer.

The heat from below to above the Pinch cannot be transferred, as this is thermodynamically infeasible. But it is feasible to transfer heat from hot streams above the Pinch to cold streams below the Pinch. Cold streams below the Pinch are feasible to heat by hot streams above the Pinch or even by a hot utility. Similar, all the hot streams could be cooled by cold streams below the Pinch or by a cold utility. But applying such arrangements would cause utility use to exceed the minimum, as identified by the Pinch Analysis. This is a fundamental relationship in the design of heat recovery systems.

Pinch for heat recovery problems implies the following three main conditions, which must hold to achieve the minimum energy targets for a process:

1. Heat must not be transferred across the Pinch.
2. No external cooling above the Pinch.
3. No external heating below the Pinch.

These rules are applied explicitly in the context of HEN synthesis by the Pinch Design Method. Other HEN synthesis methods achieving the minimum utility demands also conform to the Pinch rules (though sometimes only implicitly).

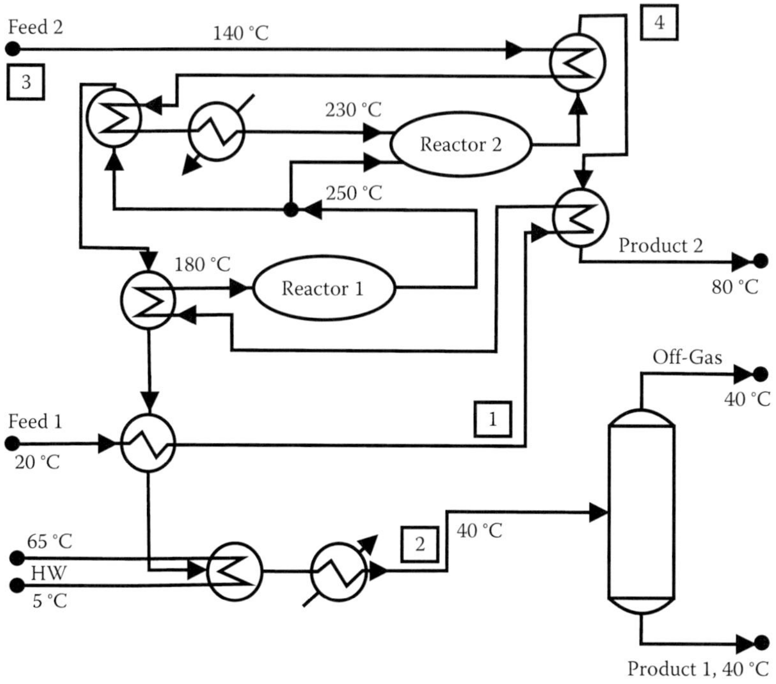

Figure 7.6 The example of HEN representation by general process flowsheet.

The Process Integration is developed in the mature branch of science with a vast set of rules and procedures described in detail in the literature cited at the beginning of this Chapter (most of them are out of scope of this book). It encompasses some conventions about graphical representations, designations and terminology.

One of such conventions is graphical representation of HEN in a form of the *Grid Diagram.*

The representation of HENs by general process flowsheets, as depicted in Figure 7.6, is not convenient for the purpose of heat exchanger placement according to the Pinch Analysis methodology. The reason is that this representation makes it difficult to answer a number of important questions: 'Where is the Pinch situated?' 'What is the amount of heat recuperated in the system?' 'How much cooling and heating from external utilities are needed?'

Shown in Figure 7.7 is an example of so-called conventional HEN flowsheet used in earlier works on HEN design; see, for example, a paper by Grossmann (1985), which offers a small improvement. It shows only heat transfer operations and is based on a simple convention: cold streams run horizontally and hot streams vertically. The pinch location in this representation can be marked for simple cases, but it is still difficult to see it distinctly. This representation also makes it difficult to express the proper sequencing of heat exchangers and to designate clearly the network temperatures. Moreover, changing the positions of certain matches corresponding to some heat exchangers often results in complicated path representations.

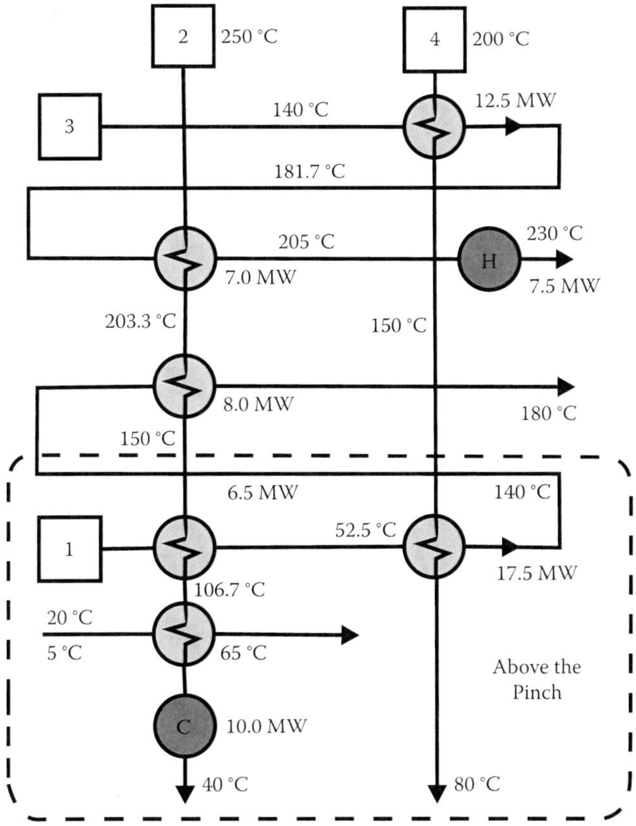

Figure 7.7 Example of conventional HEN flowsheet.

In Figure 7.8 the HEN is represented graphically by the *Grid Diagram* according to the methodology of Process Integration. It eliminates the problems described earlier for other forms of graphical representation and is a convenient and efficient tool for HEN description. The grid diagram has several advantages: the representation of streams and heat exchangers is distinct and clear, and it is also more convenient for designating temperatures. Pinch location (and its implications) is distinctly visible, as shown in Figure 7.9. Similar to the conventional HEN flowsheet, only heat transfer operations are shown in the grid diagram. The direction of temperature increases in the grid is from left to right, which is in line with CC diagrams of the Pinch Analysis. The (re)sequencing of heat exchangers on the Grid Diagram is straightforward.

The procedure for designing a HEN follows several simple steps:

1. Specification of the heat recovery problem
2. Identification of the heat recovery targets and the Heat Recovery Pinch
3. Synthesis
4. Evolution of the HEN topology

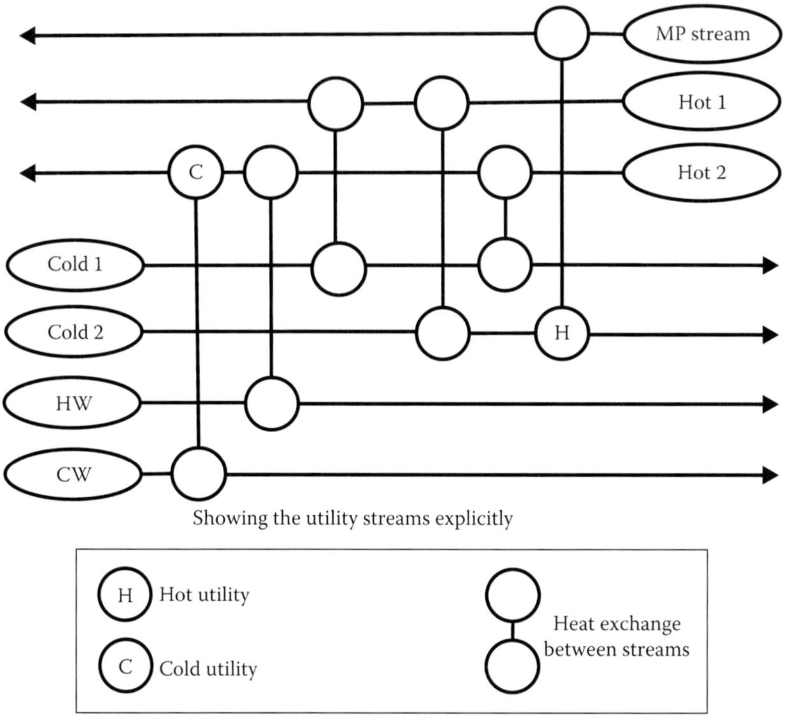

Figure 7.8 Example of Grid Diagram representation of HEN.

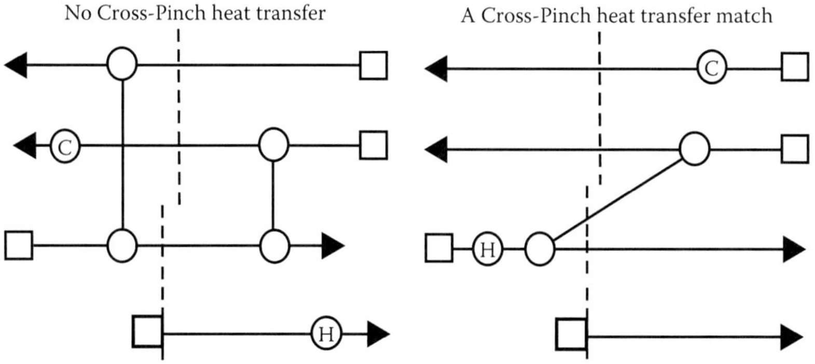

Figure 7.9 The Grid Diagram and implications of the Pinch.

The engineering practice suggests starting the network design from the Pinch (the most restricted part of the design owing to temperature differences approaching ΔT_{min}) and then to place heat exchanger matches while moving away from the Pinch. When placing matches, several rules have to be followed in order to obtain a network that minimises utility use: (1) no exchanger may have a temperature difference smaller than ΔT_{min}, (2) no process-to-process heat transfer may occur across the pinch and (3) no inappropriate use of utilities should occur.

Following the set of other guidelines and heuristic rules described in the cited literature on Process Integration, the designer comes to a certain option of the HEN structure, which is fulfilling the targets set when identifying heat recovery targets. Any HEN obtained using the previously mentioned design guidelines are optimal with respect to its energy requirements, but it is usually away from the total cost optimum. It requires the step of network evolution which involves solving of the Mixed Integer Non-Linear Programming for a limited number of options close to the identified network structure.

7.2 SUPERSTRUCTURE APPROACH FOR ENERGY-EFFICIENT HEN DESIGN

As discussed earlier, the Pinch Design Method is based on a sequential strategy for the conceptual design of HENs. It first develops an understanding of the thermodynamic limitations imposed by a set of process streams, and then it exploits this knowledge to design a highly energy-efficient HEN. However, another approach has also been developed: the superstructure approach to HEN synthesis, which relies on developing a reducible structure (the superstructure) of the network under consideration. An example of the spaghetti-type HEN superstructure fragments typically generated by such methods (Yee and Grossmann 1990) is shown in Figure 7.10.

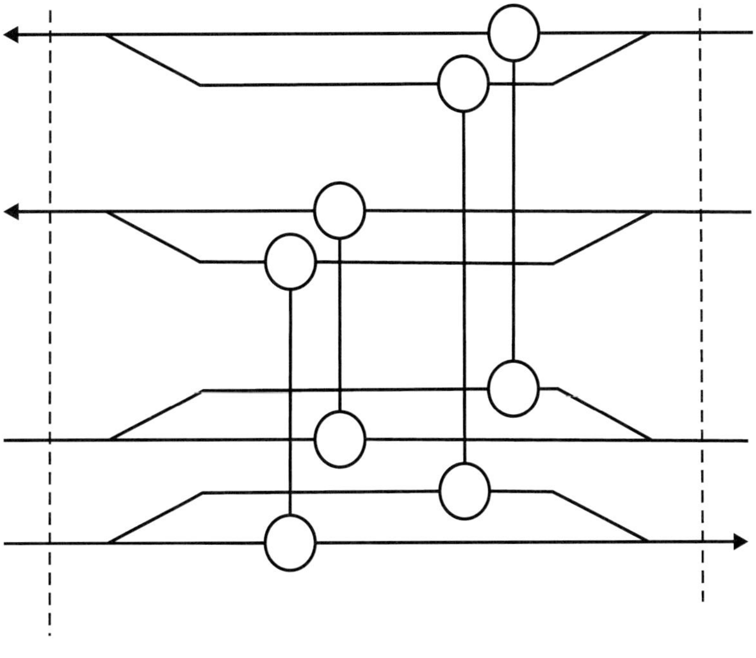

Figure 7.10 Spaghetti superstructure fragment (after Klemeš et al., 2010).

Typically, this kind of superstructure is developed in stages or blocks, each of which is a group of several consecutive enthalpy intervals. Within each block, each hot process stream is split into a number of branches corresponding to the number of the cold streams presented in the block, and all cold streams are split similarly. This is followed by matching each hot branch with each cold branch. Once developed, the superstructure is subjected to optimisation. The set of decision variables includes the existence of the different stream split branches and heat exchangers, the heat duties of the exchangers and the split fractions or flow rates of the split streams. The objective function involves mainly the total annualised cost of the network, although the function may be augmented by some penalty terms for dealing with difficult constraints. Because the optimisation procedure makes structural as well as operating decisions about the network being designed, it is called a *structure–parameter* optimisation. Depending on what assumptions are adopted, it is possible to obtain both Mixed-Integer Linear Programming (MILP) and Mixed-Integer Nonlinear Programming (MINLP) formulations. Linear formulations are usually derived by assuming isothermal mixing of the split branches and then using piecewise linearisation on the heat exchanger capital cost functions.

With the superstructure approach, it is possible to include other heat exchange options – for example, direct heat transfer units (i.e., mixing) and different heat exchanger types (e.g., double pipe, plate–fin, PHE and shell and tube). For example, Soršak and Kravanja (2004) presented a method incorporating different heat exchanger types into a superstructure block for each potential heat exchange match.

7.3 HYBRID APPROACH FOR HEN DESIGN

The superstructure methodology offers some advantages in the synthesis of HENs. Among these advantages are (1) the capacity to evaluate a large number of structural and operating alternatives simultaneously, (2) the possibility of automating (to a high degree) the synthesis procedure and (3) the ability to deal efficiently with many additional issues, such as different heat exchanger types and additional constraints.

However, these advantages also give rise to certain weaknesses. First, the superstructure approaches, in general, cannot eliminate the inherent non-linearity of the problem. Consequently they resort to linearisation and simplifying assumptions, such as allowing only isothermal mixing of split streams. Second, the transparency and visualisation of the synthesis procedure are almost completely lost, excluding the engineer from the process. Third, the final network is merely given as an answer to the initial problem, and it is difficult to assess how good a solution it represents or whether a better solution is possible. Fourth, the difficulties of computation and interpreting the result grow dramatically with problem size; this is a consequence of the large number of discrete alternatives to be evaluated. Finally, the resulting networks often contain subnetworks that exhibit a spaghetti structure: a cluster of parallel branches on several hot and cold streams with multiple exchangers between them. Because of how superstructures are constructed, these subnetworks often cannot be eliminated by the solvers.

All these considerations highlight the fundamental trade-off between applying techniques that are based on thermodynamic insights, such as the Pinch Design Method, and relying on a superstructure approach. It would therefore be useful to find a combination of both approaches and, if there is, to see whether it offers any advantages. One such middle way is the class of hybrid synthesis methods described next. A method of this class first applies Pinch Analysis to obtain a picture of the thermodynamic limitations of the problem; but then, instead of continuing on to direct synthesis, it builds a reduced superstructure. At this point the method follows the route of the classical superstructure approaches, including structure–parameter optimisation and topology simplifications. The cycle of optimisation and simplification is usually repeated several times before the final optimal network is obtained. All resulting networks feature a high degree of heat recovery, though rarely the maximum possible. A key component of this technique is avoiding the addition of unnecessary features to the superstructure, and this is an area where Pinch Analysis can prove helpful.

7.4 HEN DESIGN WITH THE COMPACT AND ENHANCED HEAT EXCHANGERS

Initially the Process Integration methodology had possibility for accounting different types of heat exchangers, including some types of compact heat exchangers, in HEN design. Linnhoff et al. (1982) have presented the description of PHEs and recommendations on their heat transfer coefficients for use in Process Integration. As the method needs just initial values of film heat transfer coefficients, the use of any kind of heat exchanger can be accounted for with proper guesses on heat transfer coefficient values at heat exchanging streams and proper heat exchanger cost function. The recent developments in designing HENs with heat transfer intensification are discussed by Smith et al. (2013).

Earlier works on Process Integration assumed that all exchangers in HEN obey a single cost law that in practice meant that their materials of construction, pressure rating and design type must be assumed uniform. Hall et al. (1990) proposed an approach which allows capital cost targets to take account of differences in exchanger specification. They have demonstrated that variations in exchanger specification can considerably alter the capital/energy trade-off. It can change resulting in the design structure of the global cost-optimal network. Without correct initialisation of the design to account for these differences, the obtained network will be far from global cost optimum. In case of compact heat exchangers with enhanced heat transfer, the correct initialisation of the design is a complicated task, as the final structure of the network and parameters of constituent heat exchangers are much dependent on the optimality of heat exchanger design and its correspondence to specific process conditions at which heat exchanger is used, as it is discussed in Chapter 5. In this case reliable data on performance of enhanced heat exchangers are needed to obtain trustworthy design of HEN. It requires the use of adequate enough mathematical models of compact heat exchangers, which as a rule are non-linear. In this sense it makes necessary to use the MINLP hybrid approach for HEN design.

The superstructure MINLP approach has no limitations on inclusion of different types of heat exchangers and can operate with compact, enhanced or traditional shell and tube units. The problem consists in the availability of accurate enough and at the same time quick computer calculations for heat exchanger models and cost functions.

The networks obtained by the different synthesis methods have distinct features, which influence their total cost and their properties of operation and control. Because the Pinch Design Method incorporates the tick-off heuristic rules, the networks synthesised by this method tend to have simple topologies with few stream splits and feature a minimum number of units (Linnhoff et al., 1982). Both the tick-off rule and the Pinch principle dictate that utility exchangers be placed last, so they are usually located immediately before the target temperatures of the streams. However, the tick-off rule may also result in many process streams not having utility exchangers assigned to them, which may reduce control efficiency. The Pinch Design Method may reduce network flexibility because it relies on the pinch decomposition of the problem and so, to a large degree, fixes the network behaviour.

Both the pure superstructure approach and the hybrid approach tend to produce more complex topologies. Their distinctive feature is the greater number of heat exchangers and stream splits, a result of how the initial superstructure is built. Spaghetti-type subnetworks also present a significant challenge to control.

With a lot of energy-intensive industries requiring improvement in energy efficiency, the problem of HEN's retrofit became of paramount importance. The survey of developments in the retrofit of HENs was presented by Smith et al. (2010), who proposed a modified network pinch approach combining in a single step structural modifications and cost optimisation to avoid missing cost-effective design solutions. One of the most efficient techniques to increase energy saving with retrofitting HENs is the use of heat transfer intensification. Mostly HEN retrofit problems with intensified heat transfer are considered without network topology modification. Wang et al. (2012) proposed an approach for HEN retrofit which employs detailed models of heat exchangers combined with five heuristic rules to identify heat exchangers suitable for implementing enhanced heat transfer techniques. To avoid the disadvantage of computations with non-linearity, Pan et al. (2012) developed a new MILP-based method for HEN retrofit with intensified heat transfer using iterative procedure. This method is capable of effectively dealing with the computational difficulties of non-linearity for HENs with constant structure. Its extension for considering exchanger details was later proposed by Pan et al. (2013a). But these approaches cannot account for network topology modification and usually fail to increase energy recovery in case of the network pinch problem when the minimum temperature difference does not change during the retrofit. The method to overcome this obstacle was proposed by Pan et al. (2013b) who proposed a new development which allows removing existing exchangers, increasing heat transfer coefficients in exchangers, but remains in the same locations of the existing matches. It enables to avoid substantial capital investments in practice as well as safety and maintenance problems associated with structural changes (exchanger moving, changing streams' piping) in the HEN.

While MILP-based methods of HEN design can avoid difficulties of the MINLP approach in solving computational problems of creating complicated structures with big numbers of streams and heat exchange matches, the higher accuracy of accounting all features of compact and enhanced heat exchangers is possible only with detailed models accounting for the non-linearity of the real processes. Such models must be robust, computationally efficient and accurate enough to reflect the heat transfer and pressure drop performance of actual equipment. The development of such detailed models for shell-and-tube heat exchangers with heat transfer enhancement by tube inserts is discussed by Jiang et al. (2014). The development and application of detailed models for HEN design with optimisation of heat transfer surface of compact plate heat exchangers is discussed by Arsenyeva et al. (2013).

7.5 ESTIMATION OF ENHANCED HEAT TRANSFER AREA TARGETS

Optimising heat transfer equipment used in industrial processes by using compact heat exchangers with heat transfer enhancement can significantly increase the heat recovery of the whole process. It is especially important when the heat transfer surface is manufactured from sophisticated and costly corrosion-resistant alloys. This is often needed when utilising low-grade heat from industry, geothermal sources and others. In this case the use of compact heat exchangers allows to reduce the cost of equipment and the environmental footprint of industrial process units and minimises energy consumption and carbon dioxide emissions. But the principles of the construction and correlations for the design of enhanced heat exchangers generally substantially differ from those for conventional shell-and-tube Heat Exchangers (HEs).

In Chapter 5, on example of Plate Heat Exchangers (PHEs) as one of the representatives of compact heat exchangers with enhanced heat transfer, it was discussed that for specified process conditions, the optimal geometrical parameters of the compact heat exchanger can be selected. PHEs of different constructions (plate and frame, brazed, welded) are one of the most efficient types of compact heat exchangers suitable for low-grade heat utilisation. They have already proved their efficiency and reliability in different industrial applications.

There are generalised correlations for PHE heat transfer and hydraulic performance for a wide range of geometrical parameters of plates and corrugations on their surface (see Chapters 4 and 5). It enables to create detailed models of PHE and calculate the optimal geometrical parameters of plates and PHE configuration for given process conditions. The inclusion of such model into HEN design software gives the possibility to obtain optimal HEN topology and the target for heat transfer surface for total HEN and at different matches that can be achieved with the optimised geometrical parameters of PHEs. For practical implementation of the described methodology, it is essential to have a software tool which enables to design the plate with the optimal geometrical parameters for the unit used for specified process conditions and to estimate the heat transfer area and cost targets for PHEs. In a paper by Arsenyeva et al. (2013), they discussed the development of such software in a form of a dynamic-link library (DLL) module, which can be included in a general software package of HEN design. For software module development, the principle of optimising the plate itself and its corrugation geometrical parameters for the most common 'chevron' corrugation type has been used. In the DLL module software developed, the problem of finding the best plate geometry for specified process conditions is solved by estimating the optimum for continuous non-linear objective function in a space of discrete-continuous optimising variables. Based on the developed PHE model and DLL software module, the optimisation algorithm is developed using the MINLP method with inequality constraints described by Pan et al. (2012). The objective function is PHE heat transfer area. Plate spacing, plate length, the corrugation inclination angle to plate axis and corrugation pitch to height ratio are optimised variables. The resulting optimal solution can be considered as a target for PHE heat transfer area, when the geometrical parameters of the plate and its corrugation strictly correspond to calculated optimal values.

The developed DLL module software can be used also for multiple calculations when optimising HEN. In that case the procedure of optimal HEN development is stepwise, using the retrofit strategy described by Smith (2005). In the first step the initial HEN structure is developed with conventional tubular HEs using the methodology of pinch analysis and process integration. After that the matches and streams, where conditions are suitable for the use of PHEs, are identified. For these positions, the obtained values of heat transfer coefficients and cost functions of PHEs are taken. With these values, new optimisation steps should be made until the final optimal solution is obtained. The use of enhanced PHEs widens the space of possible HEN options and can lead to the better global optimum. When performing HEN retrofit on the basis of the concept of the Network Pinch, once structural modifications have been made, the developed software can be utilised for a detailed capital–energy trade-off using NLP optimisation. It can help when estimating a new heat transfer area to be added to existing matches in the form of additional PHE or by changing the existing heat exchanger to PHE with the enhanced heat capacity.

After the optimal solution for HEN design is obtained, at the stage of final equipment selection, the design of PHEs should be made by the manufacturer of that equipment according to those specified in HEN design process requirements. The preliminary results obtained during the HEN design can be used as a reference to what can be achieved with this kind of heat transfer equipment.

REFERENCES

Arsenyeva O.P., Smith R., Bulatov I., Tovazhnyanskyy L.L., Kapustenko P.O., Khavin G.L. 2013. Estimation of enhanced heat transfer area targets in process industries. *Computer Aided Chemical Engineering* 32: 355–360.

Grossmann I.E. 1985. Mixed-integer programming approach for the synthesis of integrated process flowsheets. *Computers & Chemical Engineering* 19(5): 463–482.

Hall S.G., Ahmad S., Smith R. 1990. Capital cost targets for heat exchanger networks comprising mixed materials of construction, pressure ratings and exchanger types. *Computers & Chemical Engineering* 14(3): 319–335.

Jiang N., Shelley J.B., Smith R. 2014. New models for conventional and heat exchangers enhanced with tube inserts for heat exchanger network retrofit. *Applied Thermal Engineering* 70: 944–956.

Kemp J.C. 2007. *Pinch Analysis and Process Integration: A User Guide on Process Integration for the Efficient Use of Energy*, 2nd edn. Elsevier, Oxford, UK.

Klemeš J. (Ed.). 2013. *Handbook of Process Integration (PI): Minimisation of Energy and Water Use, Waste and Emissions*. Elsevier, Amsterdam, the Netherlands.

Klemeš J., Friedler F., Bulatov I., Varbanov P. 2010. *Sustainability in the Process Industry: Integration and Optimization*. McGraw-Hill, New York, USA.

Linnhoff B., Flower J.R. 1978. Synthesis of heat exchanger networks: I. Systematic generation of energy optimal networks. *AIChE Journal* 24(4): 633–642.

Linnhoff B., Townsend D.W., Boland D., Hewitt G.F., Thomas B.E.A., Guy A.R., Marsland R.H. 1982. *A User Guide on Process Integration for the Efficient Use of Energy*. IChemE, Rugby, UK [last revised edition published in 1994].

Pan M., Bulatov I., Smith R., Kim J.K. 2012. Novel MILP-based iterative method for the retrofit of heat exchanger networks with intensified heat transfer. *Computers & Chemical Engineering* 42: 263–276.

Pan M., Smith R., Bulatov I. 2013a. A novel optimization approach of improving energy recovery in retrofitting heat exchanger network with exchanger details. *Energy* 57: 188–200.

Pan M., Smith R., Bulatov I. 2013b. Intensifying heat transfer for retrofitting heat exchanger networks with topology modifications. *Computer Aided Chemical Engineering* 32: 307–312.

Smith R. 2005. Chemical process design and integration. Chichester, UK: Wiley.

Smith R., Jobson M., Chen L. 2010. Recent development in the retrofit of heat exchanger networks. *Applied Thermal Engineering* 30: 2281–2289.

Smith R., Pan M., Bulatov I. 2013. Heat transfer enhancement in heat exchanger networks. In: Klemeš J. (Ed.) *Handbook of Process Integration (PI): Minimisation of Energy and Water Use, Waste and Emissions*. Elsevier, Amsterdam, the Netherlands, 966–1037.

Soršak A., Kravanja Z. 2004. MINLP retrofit of heat exchanger networks comprising different exchanger types. *Computers & Chemical Engineering* 28: 235–251.

Wang Y., Pan M., Bulatov I., Smith R., Kim J.K. 2012. Application of intensified heat transfer for the retrofit of heat exchanger network. *Applied Energy* 89: 45–59.

Yee T. F., Grossmann I. E. 1990. Simultaneous optimization models for heat integration — II. Heat exchanger network synthesis. *Computers & Chemical Engineering* 14(10): 1165–1184.

CHAPTER 8

Economical Consideration

8.1 ENERGY–CAPITAL TRADE-OFF

In practice, any decision about low-grade heat utilisation and equipment selection for this purpose is made considering the economical aspect of the problem. The crucial role in taking such decision is playing the estimation of the trade-off between the payment received for the energy beneficially utilised after the project implementation and the cost of investment required for the design of utilisation system, purchasing, installation and commissioning of the equipment.

As it is discussed in the previous chapter, the energy–capital trade-off is an important feature of any heat exchanger network (HEN) design. When utilising low-grade heat with the use of compact enhanced heat exchangers, all methods like Process Integration, superstructure and hybrid approach can be used. To estimate capital cost, all the methods require cost functions of heat exchangers involved in HEN design, and for energy–capital trade-off, the reliable estimation of utility cost is needed.

When utilising low-grade heat, the utility cost can have two meanings. One is the purchasing cost that is the same as when designing HEN of any industrial enterprise. It is the cost of utility including cost of fuel, generation of this utility and other costs paid by the enterprise for the delivery of the utility to the process under consideration. For the case of low-grade heat utilisation, it is also the price in which the resulting utilised energy can be sold. The preferable situation is when the utilised energy is used on the same enterprise or on another property of the same owner. In this case, the cost of utilities can be taken as the same as for the whole enterprise under consideration or that produced at generation facilities that belong to the owner of the enterprise, as the use of the utilised energy will save the utilities. Another situation arises when the utilised energy should be supplied over the fence to some external consumers. The most frequent external consumers of the utilised low-grade heat energy are District Heating (DH) or District Energy networks or the power grid in case of electricity. Their purchasing price for energy can be much different from what they are selling energy to the same enterprise. This price should be considered as energy cost in energy–capital trade-off on designing HEN for waste heat utilisation.

The relationship between different organisations involved in the process of generation, distribution and use of energy is frequently playing a crucial role in the realisation of the projects of low-grade heat utilisation. Thollander et al. (2010) in an example of DH in Sweden analysed the factors which are influencing collaborations between DH and industry. After reviewing the previous works in this field, they come up with the final conclusion that with all complexity of technological and economical aspects, the establishing collaboration between the industry and DH has more to do with the individuals and organisations involved in the relationship than with the technology used. The importance of energy collaboration between different companies for improvement of energy efficiency is emphasised in a paper by Hackl et al. (2011). They have analysed Sweden's largest chemical cluster using the total site analysis method. The methodology enables investigation of opportunities to deliver waste heat from one process to another using a common utility system and can also provide the basis for heat integration within a larger geographical area, which in addition to the industrial sites also includes building complexes, offices and residences. The accuracy of the method is limited by assumption of steady-state operation of the processes within the cluster and preposition that all the processes are running simultaneously. Therefore, changes in loads and temperatures depending on, for example, the production rate, maintenance or the product changes are not considered, as well as part-time operations of DH generating equipment.

DH nowadays is well established, especially in northern countries, like Sweden, where it covers more than 40% of the total heat demand for apartment blocks, detached houses and buildings (Holmgren and Gebremedhin, 2004). There is an extensive experience of using the heat supply from various sources, especially low-grade heat from industry (Holmgren, 2006). This experience can be used for development and retrofit of new and existing District Energy systems. The industry sector plays a dual role, being in some cases a supplier of waste heat and in others a consumer. Of the total energy supply of 60 TWh for production of DH, which heats buildings and tap water in 2004, 6 TWh originated from industrial excess heat (Klugman et al., 2009) and DH consumption within the industry sector was 4.6 TWh (Holmgren and Gebremedhin, 2004). The prospects to use low-grade thermal energy from the process industry in the United Kingdom have been analysed by Ammar et al. (2012).

The economical and technical and environment feasibility of integration of waste heat from pharmaceutical industry with a DH network was evaluated by Ajah et al. (2007, 2008). The possibility to upgrade low-grade heat by chemical and mechanical heat pumps is evaluated. The economical aspect of integration of waste heat from industries and waste incineration into the DH network is discussed by Holmgren (2006). The waste heat from industry covered approximately 15% of the overall heat demand. But this work is not accounted for the part load performance of combined heat and power (CHP) and boiler.

Svensson et al. (2008) and Jönsson et al. (2008) evaluated the usage of excess heat in pulp mill to supply DH system. They evaluated the options with integration of new energy-efficient technologies, such as process-integrated evaporation and drying identified by Axelsson et al. (2006) for increasing the surplus energy from the pulp mill. This surplus heat can be either integrated within the process or exported. They studied the techno-economic impact of integration of excess heat with DH system. However, as shown by Kapil et al. (2012), their model does not account for the part load performance of CHP and boilers. Heat load and electricity production in the DH system vary with the time of the day, seasons, etc. Therefore, the performance of the DH system is dependent on the part load performance of generation equipment.

Kapil et al. (2012) proposed new process integration methodology accounting the fact that integration of waste heat from the process site with the DH network reduces the heat produced by the DH itself. This decrease in energy production reduces the part load on the CHP and boilers, which operate under partial load. The economic feasibility of the integrated system is to a big extent determined by the part load performance in DH systems, which provides an opportunity to produce heat and electricity by the combination of boiler, CHP, etc. But there are interactions between power production, heat consumption and part load performance depending on the relative cost of electricity and heat. The electricity can be either imported or exported to the grid, depending on the deficit or surplus of electricity produced by DH. The heat supplied from the DH system can include the waste heat from the industry, which is not available during shutdowns. Therefore, additional capacity of the DH supply should be provided to maintain the continuous supply of heat during such periods. The CHP and the boiler in the DH system can operate at part load, depending on heat demand. Therefore, for the optimisation of operating cost of DH, the part load performance of CHP and boilers should be evaluated.

The proposed approach consists in application of process integration total site analysis to the industrial site with modification of source and sink composite curve according to DH requirement for temperature in supply line. It allows identifying of the industrial waste heat potential, which can be utilised by DH system. The next step is application of mathematical modelling accounting for variation of DH heat demand and part load performance of energy generating equipment. With the use of linear models, the optimisation problem with operating energy cost as objective function is modelled as a mixed integer linear programming problem. The model can accommodate the variability of total site profiles with different seasons.

The economical aspects of the model application are illustrated by case study. The authors have used total site data of a case study for industrial site utility systems and associated site-wide energy use presented earlier in literature. The case study discussed is concerned with the analysis of a DH system of a city with a population of about 130,000 inhabitants. The amount of low-grade heat available for DH at a temperature higher than 105 °C is 62.11 MW. The temperature of 105 °C corresponds to a temperature of supply of 90 °C and minimum temperature of approach 15 °C. The calculated optimised heat required to be supplied from generation equipment has decreased by over 62 MW. From an environmental point of view, it means considerable reduction in CO_2 emissions, but the economical consideration is not favourable in this particular case. The revenue for the DH system is generated by selling electricity to the grid and heat to DH consumers and from the consumption of waste at its incineration, while the consumption of fuel (oil and coal) is the expenditure of the system. The integration of waste heat with DH decreases the thermal load. This, in turn, reduces the part load on CHP units, which leads to the decrease in the annual electricity production. Electricity revenue drops from 56.5 M$/y to 49.26 M$/y. There is also a decrease in the utilisation of the waste fuel to produce steam from the boiler. Therefore, with integration of the waste heat from the industry, overall annual operating energy revenue from the DH system decreases from 202.53 M$/y to 193.26 M$/y. It shows that for already existing optimised DH system, having installed generating capacities, the detailed analysis should be made for taking decisions on industry waste heat integration. Beneficial for the environment by the reduction of CO_2 emissions, renovation cannot be cost-effective and does not create problem with municipal waste utilisation.

The necessity to minimise fossil fuel consumption and environmental hazards from CO_2 and greenhouse gas emissions moved forward the utilisation of energy from renewable sources. Lund et al. (2010) analysed different scenarios of energy system development in Denmark up to 2060 with the aim to obtain finally 100% renewable energy system. The research, based on Geographic Information System, has shown that a substantial reduction in fuel demands and CO_2 emissions as well as cost can be achieved by converting to DH from 46% now in Denmark up to 70%. They identified the main perspective sources of renewable energy in that area as biomass, heat pumps for upgrading low-grade heat, waste incineration and use of waste heat from industry, with prospects for solar energy, geothermal heating and biogas production.

The renewable resources are usually available on smaller scale and distributed over a given area. Their availability (with the exception of biomass and geothermal energy) is usually well below maximal values and varies significantly with time and location. The energy demands (heating, cooling and power) also vary significantly with time of the day and period of the year. As a result, optimising energy conversion systems using renewable resources is more complex than when using just fossil fuels. One option to smooth the inherent variations and satisfy the demands at high efficiency is to combine the supply and demand streams of the individual users in larger systems as the Total Sites using process integration methodology and accounting economical aspect of the problem with regard to the full system.

The importance of economical consideration for any system designed to include low-grade heat as important source of energy put forward the problem of accurate enough estimation of capital investments and energy cost. These factors can have a major impact both on project profitability and the choice of technical solution. This leads to a commitment of substantial amounts of money and manpower over an extended time. Various methods often provide different results. These differences can be rather significant towards the real cost of a project and selection of the right arrangements. It is therefore important to use a proper estimate that generates enough confidence to choose the right alternative. Another source of uncertainty during the economic analysis of design and retrofit projects is the future energy price value as part of operating costs. This value can significantly affect the project viability and profitability.

8.2 CAPITAL COST ESTIMATION

Both retrofit and grass root design projects of HEN require a reliable enough cost estimate in different stages of the project to be able to make the right decision on structure and heat transfer area of the HEN. The approximate guidelines for the requirements on accuracy of cost estimation on different stages of the project development according to Perry's *Chemical Engineers' Handbook* (2008) are presented in Table 8.1. At present days, some companies are using front-end loading process for project definition and execution. In that case three categories are used:

- Conceptual with required accuracy ±40%
- Feasibility with accuracy ±25%
- Definition with accuracy ±10%

There are several methods available for estimating the cost of heat exchangers for HENs. Most of them are based on FOB (free on board) cost estimates. Taal et al. (2003) presented analysis of different heat exchanger cost estimate approaches as well as trends in energy prices. They are also giving some guidelines for the time spent in the work concerning estimation, as presented in Table 8.2.

Heat exchangers are normally the most significant piece of equipment determining the capital cost for a HEN retrofit and also grass root projects. Taal et al. (2003) have analysed several different methods for cost estimations of heat exchangers that can be found in the relevant literature. Most of the methods are concerned with the estimation of the cost for tubular heat exchangers of different TEMA (tubular exchanger

Table 8.1 Types of Capital Cost Estimates, after Green and Perry (2008)

Type of Estimate	Usual Basis	Probable Accuracy (%)
Order of magnitude	Previous similar cost information	−30 to +50
Study (factored estimate)	Knowledge of flowsheet and major equipment	−25 to +30
Preliminary (initial budget)	Sufficient data for budget	−20 to +25
Definitive (project control)	Detailed data	−10 to +15
Detailed (firm, contractor)	Complete drawings and specifications	−5 to +10

Table 8.2 Typical Ranges for Spending on Cost Estimate in USD, according to Taal et al. (2003)

	Cost of Project		
Type of estimate	<USD 2 M	USD 2 M to 10 M	USD 10 M to 100 M
Order of magnitude	3,000	6,000	13,000
Study	20,000	40,000	80,000
Preliminary	50,000	80,000	130,000
Definitive	80,000	160,000	320,000
Detailed	200,000	520,000	1,000,000

manufacturers association) designs. The comparisons of the resulting estimations for prices have shown considerable differences in the results obtained by different methods. From nine compared methods, only four gave the results with differences between them in the range less than 15%. The cost of the same heat exchanger by other five methods could be from 50% higher to 60% smaller. Such discrepancies can be explained not just by the quality of the methods proposed but also by the price differences and pricing policies of the producers which data were used for method development, the differences in price quotations made for established companies and new for the market, different price levels in different countries and other market eventualities. Counting for such uncertainties in cost estimation for traditional widely used shell-and-tube heat exchangers, what can be expected for less common compact heat exchangers?

The literature data for the cost of compact heat exchangers are rather limited, especially when these data are required for multiple calculations in the process of designing HENs. There are some equations published for estimating the cost of plate heat exchangers (PHEs). Hall et al. (1990) presented equation for calculation in HEN design procedure of the installed cost in USD of plate-and-frame PHE with stainless steel plates and carbon steel frame as function of equivalent heat transfer area F_e:

$$IC_{PHE} = 1{,}950 \cdot F_e^{0.78} \tag{8.1}$$

The cost was calculated for plant life equal to 6 y and capital interest of 16% per annum.

Vatavuk (1995) has proposed equation for estimation of the purchasing cost in USD of plate-and-frame PHE with stainless steel plates in the range of surfaces F measured in square feet from 50 to 9,000 ft^2 (4.65 m$^2 < F <$ 836 m^2):

$$C_{PHE} = 231 \cdot F^{0.639} \tag{8.2}$$

This is obtained for prices at different time and different money values. To compare at the same time conditions in 2012 when more current prices are available, calculations with this equation are made with correction accounting for the Chemical Engineering Plant Cost Index (CEPCI) (see Green and Perry, 2008). In 2012, CEPCI was equal to 584.6 and in 1994 its value was 368.1. Correspondingly, the results obtained with Equation 8.2 are multiplied by 1.588.

The analysis made by authors of this book with a large number of quotations for PHEs with heat transfer surface up to 370 m^2 (mainly from Alfa Laval) estimated for money value in 2012 has shown that calculation of the price by indexed Equation 8.2 can give an error of ±50% and even more. The calculations by this equation mostly correspond to PHEs from AISI 304 stainless steel (according to American Iron and Steel Institute grades) of 0.5 mm thickness and nitrile gaskets with frame for working pressure below 10 bar, but sometimes also overestimate the lowest price, mostly for smaller heat transfer area ($F <$ 20 m^2) and when plates with thickness of plate metal 0.4 mm are used.

It can be explained by the fact that not only heat transfer surface area is influencing the price of PHE with stainless steel plates, but also the material of the gaskets, the working pressure and its corresponding strength of frame, the thickness of the metal from which the plates are stamped, the construction made by a specific manufacturer and the pricing policy in the market. There is also an intrinsic feature of plate-and-frame PHE construction, which is shown in Figure 3.21, which does not permit in obtaining the generalised simple relation between cost and heat transfer surface area. The cost of frame is usually made from carbon steel and can take considerable, or even largest, share in the cost of PHE with small number of plates. The PHE consisting of three or five plates with heat transfer area close to zero costs practically the same as its frame. With the increase of plate numbers, the cost of frame increases also in some steps, as well as the cost of work for assembling the plates, but not so considerably as its increase with the plate size. All further discussions about PHE cost prediction are assuming that the plate number in one PHE is not less than 20. The cost of the plates itself is not directly proportional to their size, as the cost share of pressing tool and equipment depreciation also significantly varies.

For the utilisation of low-grade heat, more expensive materials of plates and gaskets can be required, such as stainless steel of the American Iron and Steel Institute AISI 316 grade, gaskets of EPDM (ethylene propylene diene monomer) rubber and frames for working pressure 16 bar that are more reliable in many applications. It can increase the estimated purchasing price by 10%–15%. The reliability is a crucial factor in low-grade heat utilisation and energy-saving projects, especially of long payback periods and innovative. One failure on some pioneering installation can significantly damage the whole idea and dramatically postpone the time of its market realisation.

In certain cases of industrial low-grade heat utilisation, it is required to use heat transfer surface expensive metals, such as titanium and metal alloys of high grades, to estimate prices of PHE with plates of such materials is a complicated task. As a rough approximation to estimate the order of magnitude, the results obtained by Equation 8.2 can be multiplied by φ_m, accounting for cost of heat transfer surface material. The following can be recommended: for titanium and alloy 254 (SMO), $\varphi_m = 2$; for alloy A276, $\varphi_m = 3$; and for stainless steel AISI 316, $\varphi_m = 1.10$. After an order of magnitude and study stages (see Table 8.1), the quotations should be required from the PHE manufacturing company for PHEs on conditions according to preliminary estimated with HEN design software flowsheet with the price C_Q for heat transfer area F_Q in quotation. After that, the cost laws for positions requiring expensive PHEs can be changed with approximately estimated heat transfer area:

$$C_{PHEi} = C_{Qi} \cdot \left(\frac{F_i}{F_{Qi}} \right)^{0.8} \quad (8.3)$$

These cost functions should be used with HEN design software for preliminary and definitive project stages (see Table 8.1). For the detailed stage of project, the quotations for PHEs according to optimised final flowsheet has to be obtained for calculating the final budget on detailed stage of the project. It should be mentioned that for PHEs, as for all compact enhanced heat exchangers, the heat transfer surface area for the same duty can vary significantly depending on the intensification technique employed and quality of production by specific manufacturer.

The prices for PHEs of other types other than plate and frame should be estimated by considerably different relations. The introduction of Brazed PHEs (BPHEs) (see Figure 3.42) allowed significant price reduction for the range of heat transfer surface areas up to 80 m^2, for which BPHEs are produced. These heat exchangers are made by brazing in standardised sizes with the step of heat transfer surface area of about 8–10 plates for one plate dimension. There are available lists of prices for such heat exchangers from different manufacturing companies. These data from Alfa Laval price list in Euro of year 2012 are correlated by following two equations for different ranges of heat transfer surface area (the surface areas are approximately estimated according to size, construction and number of plates).

For the range $0.13\ m^2 < F < 5.8\ m^2$, the following equation is correlating price data in euro within error ±20% (excluding smallest type of BPHEs, for which the error is up to 30%):

$$C_{BPHE} = 540 \cdot F^{0.8} \tag{8.4}$$

For bigger plate sizes in the range $2\ m^2 < F < 86\ m^2$, the data in euro are correlated within maximal error ±17% by

$$C_{BPHE} = 805 \cdot F^{0.74} \tag{8.5}$$

The difference of prices calculated by both equations in the range $2\ m^2 < F < 5.8\ m^2$ is explained by the fact that same heat transfer area can be obtained by BPHE with large number of smaller plates (smaller diameter of portholes also) and with small number of bigger plates (bigger diameter of portholes).

For fusion-bonded all stainless steel PHE AlfaNova depicted in Figure 3.43, there is also a list of prices by Alfa Laval from 2012 in €. These prices in € in the range $0.6\ m^2 < F < 86\ m^2$ are correlated by the following equation within maximal error ±20%:

$$C_{FPHE} = 1{,}280 \cdot F^{0.8} \tag{8.6}$$

The comparison of the prices calculated by the presented equations for brazed (BPHE), fusion-bonded (FPHE) and plate-and-frame PHEs with the same AISI 316 heat transfer area ($F < 86\ m^2$) is presented in Figure 8.1. The BPHE is from 45% to 20% cheaper than plate-and-frame PHE of the same heat transfer area. The BPHE with plates of smaller sizes with heat transfer area less than 5.8 m^2 (Equation 8.4) can cost even 70%–55% (or less than half) of the plate-and-frame PHE. The higher working pressure and temperature make BPHE more attractive for streams with little fouling tendencies where fouling deposits can be removed with cleaning by chemical solutions. Besides no access to heat transfer surface, their application is also limited to media not aggressive to copper, as, for example, ammonia. Whole stainless steel fusion-bonded PHE can work in such media, but it can be up to 70% more expensive, especially at bigger heat transfer areas.

To compare the levels of prices with conventional shell-and-tube heat exchangers, Figure 8.1 shows the lines for these heat exchangers of different materials for

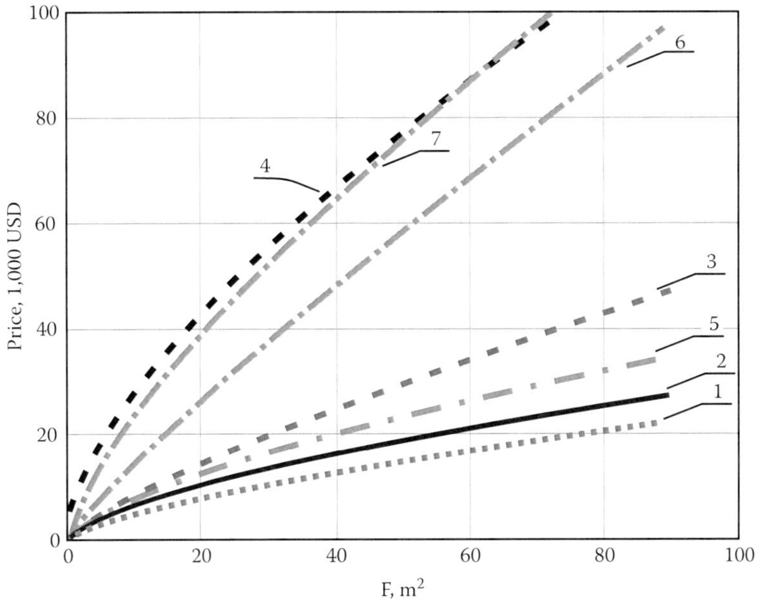

Figure 8.1 Estimated prices for heat exchangers of different types: line 1, BPHE, Equation 8.5; line 2, plate-and-frame PHE, Equation 8.2; line 3, AlfaNova FPHE, Equation 8.6; line 4, Compabloc™ welded PHE, Equation 8.7; lines 5, 6 and 7, shell-and-tube heat exchangers (with fixed tube sheet After Sinnott 2005.); line 5, all carbon steel; line 6, stainless steel tubes; line 7, all stainless steel. The exchange rate 1 € = 1.05 USD.

tubes and shells taken from the graph in a book by Sinnott (2005). The prices are estimated in a 2012 level accounting for CEPCI index in 2004 equal to 444.2. The working pressure of shell-and-tube heat exchangers is up to 10 bar. The prices of all carbon steel shell-and-tube heat exchangers (line 5 in Figure 8.1) are higher by 10%–25% than for plate-and-frame PHE with stainless steel plates. The cost advantage of PHE is even much bigger counting that this enhanced heat exchanger requires much less heat transfer area than shell and tube for the same duty. For BPHE, the cost advantage is even more significantly higher as it is shown in the earlier presented discussion on its comparison with plate and frame. All stainless steel FPHE is more costly than all carbon steel shell-and-tube heat exchanger of the same heat transfer area, but almost half the price of shell-and-tube heat exchanger with stainless steel tubes of the same heat transfer area (line 6 in Figure 8.1).

The welded Compabloc™ PHE (see Figure 3.31) can have bigger heat transfer area, up to 320 m². It is more expensive compared to other PHE types but has higher working parameters. The price of this heat exchanger can be estimated in euro by Equation 8.7 obtained by correlating the limited number of quotations with maximal error ±18%:

$$C_{CBL} = 4,280 + 4,690 \cdot F^{0.7} \tag{8.7}$$

The comparison in Figure 8.1 shows that the price of Compabloc PHE (line 4 in Figure 8.1) is of the same level as the price of all stainless steel shell-and-tube heat exchanger of the same heat transfer area (line 7 in Figure 8.1). Counting that this enhanced PHE requires much less heat transfer area than shell and tube for the same duty, Compabloc can have smaller cost even compared to shell-and-tube heat exchanger with carbon steel shell and stainless steel tubes.

The presented equations for purchasing cost calculation give only rough preliminary information for estimation of the capital investment. The real price for heat exchangers complying to the acting quality standards is determined to a certain extent also by the market situation in a particular country and industry, marketing policy of the manufacturers and other factors. For example, compact PHEs are now the predominant type of heat transfer equipment in DH which requires a big quantity of PHEs. The prices of PHEs in this sector can be considerably lower than average, especially during the big investment projects. Reliable data for detailed capital cost estimation can be obtained only with the quotation from manufacturers.

The different types of compact heat exchangers described in Chapter 3 have different purchasing cost levels depending on their design and technology of manufacturing. The discussed price levels for the different types of PHEs show economical advantage of compact heat exchangers over traditional shell and tube and also give guidelines for possible costs of the other types of compact heat exchangers, as to have application areas, they must be cost competitive with PHE.

8.3 ENERGY PRICES

From the start of energy usage by mankind, the fossil fuels were the main source of energy. Based on one of the popular theories, the recent amount of oxygen in the air was obtained by way of photosynthesis of plants with CO_2 involvement. Now, the carbon resulting in this process for millions of years is stored in the form of fossil fuels deep under the earth implementing the nature efforts to sequester CO_2 and to maintain concentration of the oxygen in the air sufficient to prolong the life of different living creatures. Digging fossil fuels out and oxidising them, we try to reverse this process and to make earth uninhabitable in a long term. The first sign is global warming with the increase of CO_2 concentration in the air, not considering seemly more distant in time decrease of oxygen concentration. It is our price for energy in global sense. Foreseeing this situation, nature has buried sequestrated CO_2 in very hidden places that to dig it out would be difficult and require a lot of money to be more understandable for the people. Therefore, the only hope for sustainable development of future generations is the increase of energy prices until other reliable, safe and not expensive sources of energy will be found. It will lead to energy price reduction and make fossil fuel excavation uneconomical.

The general trend in energy prices fully supports the scenario favourable for sustainability. Nowadays, fossil fuels are not the only sources for energy generation, but nuclear energy and renewables still cannot limit the growing demand for oil, coal and gas to satisfy the ever-increasing need of energy. According to the IEA (2013)

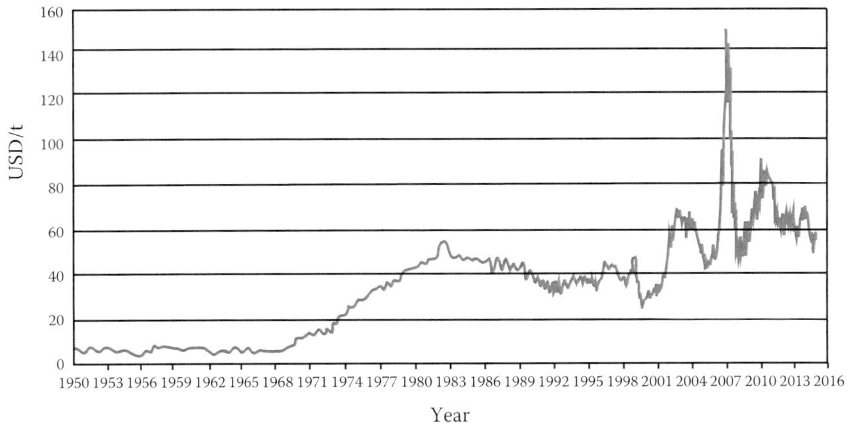

Figure 8.2 The history of coal prices in the United States since 1950 (before 2001 data of US energy information administration EIA, 2015 and after 2001 from InfoMine Inc., 2015a).

prediction for the period up to 2035, the demand will grow for all forms of energy. Even with contemporary emphasis on the utilisation of low-carbon energy sources (renewables and nuclear), they will meet around 40% of the growth in primary energy demand. Nearly half of the net increase in electricity generation will come from renewables. But still today's share of fossil fuels in the global mix of energy sources, at 82%, is the same as it was 25 years ago. The strong rise of renewables will only reduce this to around 75% in 2035, meaning an increase in fossil fuel consumption by 25% as compared to that in 2011. Such increase of demand with limited supply inevitably means there is a general trend for the increase in prices for energy utilities, which is of course subjected to fluctuations due to political and market eventualities, as well as technological innovations in exploration and development. This is illustrated by the history of prices for coal, the fossil fuel with the biggest available reserves, depicted in Figure 8.2. Oil prices are also going the same general trend, but with much bigger fluctuations, as more subjected to political and market conditions, as can be seen in Figure 8.3 showing data after the 1970s energy crisis which highly stimulated energy efficiency by dramatic increase in oil prices which dropped later in 1980s when the crises was finished.

The price of oil has significant influence on energy utility prices, especially in oil refineries with big HENs. But in overall energy production, the share of oil as fuel is limited to about 5%, as can be seen in Figure 8.4.

With overall tendency to price increase, some primary fuels can have big price fluctuations, as, for example, the price of natural gas in the United States (see Figure 8.5). Nevertheless, the total price on energy mix still has general tendency to increase, as can be judged by the history of average electricity prices in the United States from 2002 depicted in Figure 8.6. Different mixes of energy sources

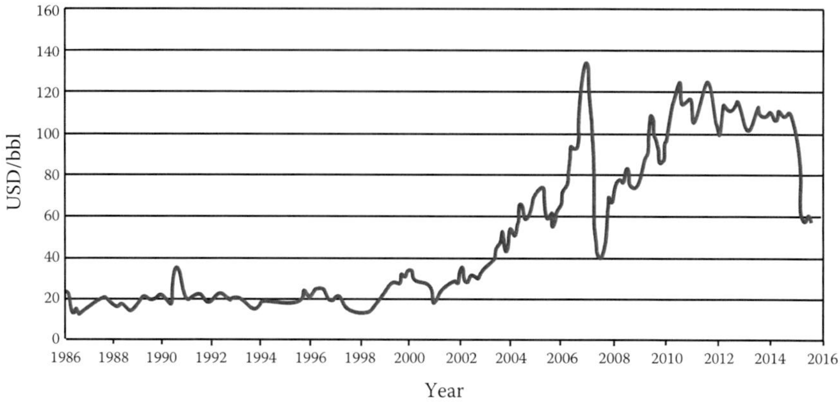

Figure 8.3 The history of oil prices from 1986 (before 2001 from Avalon energy, 2014 and after 2001 from InfoMine Inc., 2015b).

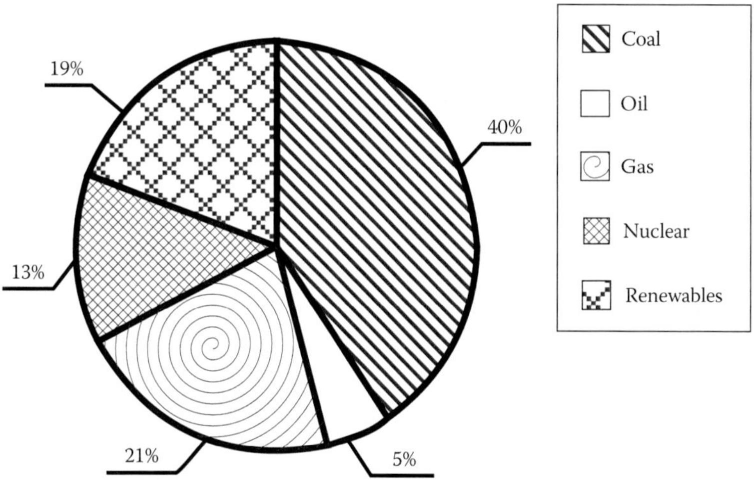

Figure 8.4 World energy generation by fuel type in 2009 (before 2001 from Inmesol, 2014 and after 2001 from InfoMine Inc., 2015b).

create substantial differences in energy utility prices, as can be judged from electricity prices in different countries. For example, from the data in Oilprice.com (2010), in 2010 the highest industry electricity price of 0.22 USD/kWh (without taxes) was that in Italy that is more than four times higher than in Norway and Russia. Besides, the price is also different for residencies, industries and businesses of different sizes.

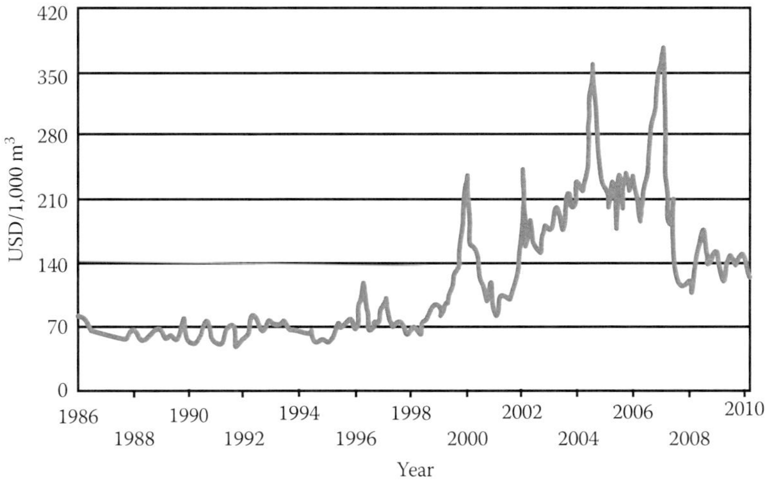

Figure 8.5 Natural gas prices in the United States from 1986 (after Avalon, 2014).

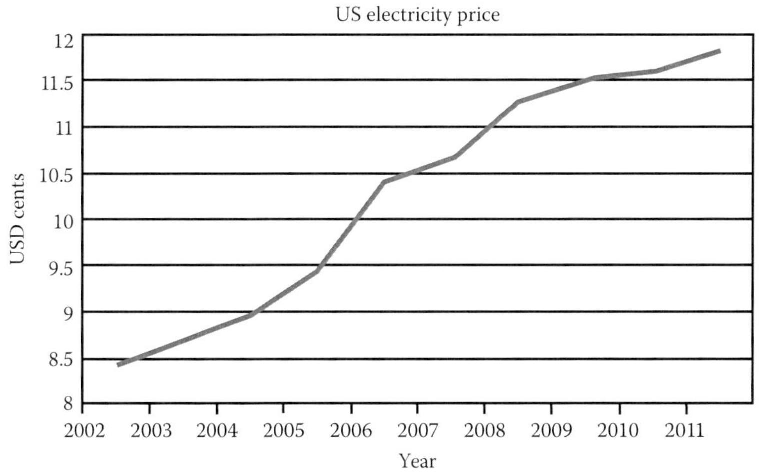

Figure 8.6 Electricity price in the United States, USD cents (after Savenialabs, 2011).

The earlier-presented analysis leads to a conclusion that the capital investment costs can be predicted with accuracy sufficient for different stages of the project with comparatively small differences for different industries and territories. On the contrary, the cost of energy utilities can be much influenced by the geographical place of project implementation, type of customer for project implementation (residential, industry, size of business) and energy market situation. It must be accounted for the final consumer of utilised energy, is it for customer own use or it is for sale over the fence. The best option is to take acting energy utility prices with calculations for different scenarios, giving priority to the scenario of price increase with time.

REFERENCES

Ajah A.N., Mesbah A., Grievink J., Herder P.M., Falcao P.W., Wennekes S. 2008. On the robustness, effectiveness and reliability of chemical and mechanical heat pumps for low-temperature heat source district heating: A comparative simulation-based analysis and evaluation. *Energy* 33: 908–929.

Ajah A.N., Patil A.C., Herder P.M., Grievink J. 2007. Integrated conceptual design of a robust and reliable waste-heat district heating system. *Applied Thermal Engineering* 27: 1158–1164.

Ammar Y., Joyce S., Norman R., Wang Y., Roskilly A.P. 2012. Low grade thermal energy sources and uses from the process industry in the UK. *Applied Energy* 89: 3–20.

Axelsson E., Olsson M. R., Berntsson T. 2006. Heat integration opportunities in average Scandinavian kraft pulp mills: pinch analyses of model mills, *Nordic Pulp and Paper Research Journal*, 21(4): 466–75.

Avalon energy services LLC. 2014. Crude Oil and Natural Gas Get a Divorce. avalonenergy.us/blog/?p=85 (accessed 20 March 2014).

EIA. 2015. US energy information administration. Coal Prices, 1949–2011. www.eia.gov/totalenergy/data/annual/showtext.cfm?t=ptb0709 (accessed 15 April 2015).

Green D.W., Perry R.H. 2008. *Perry's Chemical Engineers Handbook*, 8th edn. McGraw-Hill, New York, USA.

Hackl R., Andersson E., Harvey S. 2011. Targeting for energy efficiency and improved energy collaboration between different companies using total site analysis (TSA). *Energy* 36: 4609–4615.

Hall S.G., Ahmad S., Smith R. 1990. Capital cost targets for heat exchanger networks comprising mixed materials of construction, pressure ratings and exchanger types. *Computers & Chemical Engineering* 14(3): 319–335.

Holmgren K. 2006. Role of a district-heating network as a user of waste-heat supply from various sources – The case of Goteborg. *Applied Energy* 83: 1351–1367.

Holmgren K., Gebremedhin A. 2004. Modelling a district heating system: Introduction of waste incineration, policy instruments and co-operation with an industry. *Energy Policy* 32: 1807–1817.

IEA. 2013. World energy outlook 2013. www.worldenergyoutlook.org/pressmedia/recentpresentations/London November12.pdf (accessed 20 March 2014).

InfoMine Inc., 2015a. Historical Coal Prices and Price Chart. www.infomine.com/investment/metal-prices/crude-oil/all/ (accessed 15 April 2015).

InfoMine Inc., 2015b. Historical Crude Oil Prices and Price Chart. www.infomine.com/investment/metal-prices/crude-oil/all/ (accessed 15 April 2015).

Inmesol. 2013. World overview of the energy sector. www.inmesol.com (accessed 20 November 2014).

Jönsson J., Svensson I.L., Berntsson T., Moshfegh B. 2008. Excess heat from kraft pulp mills: Trade-offs between internal and external use in the case of Sweden part 2: Results for future energy market scenarios. *Energy Policy* 36(11): 4186–4197.

Kapil A., Bulatov I., Smith R., Kim J.-K. 2012. Process integration of low grade heat in process industry with district heating networks. *Energy* 44: 11–19.

Klugman S., Karlsson M., Moshfegh B. 2009. A Swedish integrated pulp and paper mill – Energy optimization and local heat cooperation. *Energy Policy* 37: 2514–2524.

Lund H., Möller B., Mathiesen B.V., Dyrelund A. 2010. The role of district heating in future renewable energy systems. *Energy* 35: 1381–1390.

Oilprice.com. 2010. How different energy sources create electricity price differences between countries. oilprice.com/Energy (accessed 20 November 2014).

Savenialabs. 2011. U.S. electricity price. www.savenialabs.com (accessed 20 November 2014).

Sinnott R.K. 2005. *Chemical Engineering Design*, vol. 6, 4th edn of Coulson & Richardson's Chemical Engineering Series. Elsevier, Oxford, UK.

Svensson I.L., Jönsson J., Berntsson T., Moshfegh B. 2008. Excess heat from kraft pulp mills: Trade-offs between internal and external use in the case of Sweden part 1: Methodology. *Energy Policy* 36(11): 4178–4185.

Taal M., Bulatov I., Klemeš J., Stehlík P. 2003. Cost estimation and energy price forecasts for economic evaluation of retrofit projects. *Applied Thermal Engineering* 23(14): 1819–1835.

Thollander P., Svensson I.L., Trygg L. 2010. Analyzing variables for district heating collaborations between energy utilities and industries. *Energy* 35: 3649–3656.

Vatavuk, W.M. 1995. A potpourri of equipment prices. *Chemical Engineering* 102(8): 68–73.

CHAPTER 9

Industrial Examples

9.1 FOOD INDUSTRY: INTEGRATION OF A HEAT PUMP INTO THE HEAT SUPPLY SYSTEM OF A CHEESE PRODUCTION PLANT

The steep rise of energy prices in recent years, as well as the necessity to curb carbon dioxide emissions, has intensified the need for solutions that are increasing energy efficiency in the food industry. The most efficient methods and tools for utilisation of low-grade heat, by elevating its thermal level, are compression heat pumps, which are rather well described in the literature (see, e.g., the book by Reay and Macmichael, 1988).

Many food processing factories, especially in Ukraine, use ammonia refrigerating plants to obtain low temperatures for production processes and product storage.

When in operation, such plants produce a considerable amount of low potential heat, which usually cannot be used and is rejected to the atmosphere through a water-cooling tower or air-cooled heat exchangers. The technique to utilise this heat for the process and other needs of the factory, using high-pressure ammonia heat pumps, was described by Korfitsen and Kristensen (1998). But without a detailed analysis of process streams, its potential, efficient recuperation and heat exchange on different levels of temperature are difficult to obtain for an optimal solution. An advanced methodology for the achievement of optimal energy efficiency is Process Integration. This methodology has considerable potential in the food processing industry, as was demonstrated in several publications by different authors (for survey see e.g. Klemeš and Perry, 2007).

Kapustenko et al. (2008) investigated the possibility for using heat at the condenser side of the refrigeration cycle. The two options are considered. The first one is limited to the utilisation of ammonia superheating and partly of ammonia condensation heat, which require only minor retrofit changes and a low investment cost. The second one utilises the additional compression for part of the ammonia stream from the refrigeration cycle. Both cases are explored using process integration methodology, which enables optimisation of energy consumption and integration of heat pumps and other heat machines into the heat exchanger network (HEN). Assuming that the evaporator side of the refrigeration unit remains unchanged, the investment cost of retrofit for both options is considerably lower compared to installation of a separate heat pump and thus enables achievement of a short payback period. This approach is discussed in Chapter 9 in more detail.

9.1.1 System Description

A short description of the cooling cycle of the cheese production plant in operation initially is under consideration. The stream of ammonia vapour ($T = -6$ °C; $P = 2$ bar) enters the compression unit, which consists of eight compressors with a total power of 1,874 kW; five to six compressors work in the regular mode. There is also the compression of cooling agent vapour to 12 bar in the compressor unit. Gaseous ammonia, with temperature of 130 °C and pressure of 12 bar, flows from the compressors through oil traps (OT-1 and OT-2) to the condensation unit (Figure 9.1). The block of condensers consists of seven tubular water spray condensers. The heat load of each condenser is about 400 kW. The condensation temperature is 28 °C. The temperature of moist air is 18 °C. Cooling and condensation of ammonia vapour take place due to four air-cooling ventilators with the power 4 × 2.2 kW. The flow rate of air is 58 m^3/h. The heat transfer area of the condenser is 88 m^2. The flow rate of circulation water is 33 m^3/h. In a winter period, condensation of ammonia takes place due to air cooling and the flow rate of water drops.

Liquid ammonia with a temperature at 28 °C and pressure of 12 bar flows to a linear receiver. There is cooling of the ammonia condensate at 3 °C in the pipelines and the linear receiver. Further, a liquid cooling agent passes through an expansion device, where ammonia changes pressure from condensation pressure to evaporation pressure and travels to the circulation receivers. Ammonia from the first receiver ($T = -6$ °C; $P = 2$ bar) flows to the refrigeration chambers of the factory. From the

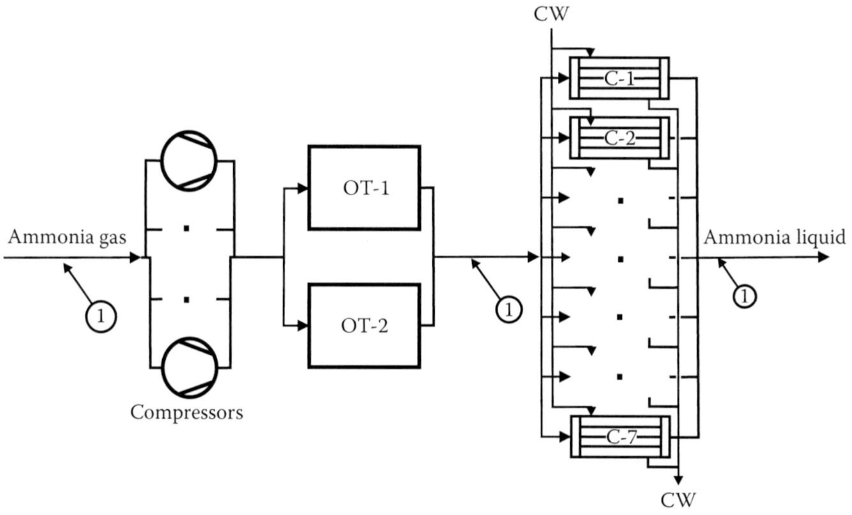

Figure 9.1 Principal flowsheet of existing ammonia unit.

second receiver, ammonia ($T = -10$ °C; $P = 2$ bar) flows to the panel evaporators of the workshops. There is evaporation of ammonia caused by the cooling water and brine, which is used in the technological process. Ammonia vapours flow through the liquid separators and enter the compressor unit of the enterprise cooling cycle. The principal flowsheet of the existing process for an ammonia unit is presented on Figure 9.1.

9.1.2 Data Extraction

A grid diagram of the existing process stream system of the ammonia unit and the streams now available for integration in the system of heat supply are shown in Figure 9.2. There is no evidence of any heat recuperation. The technological stream information, extracted in the process of inspection, is provided in Table 9.1.

9.1.3 Heat Integration

Based on these data, the Composite Curves for an existing process and its heat system are constructed (Figure 9.3), on which the necessary utility loads are clearly visible for the system of process streams under inspection.

The analysis of Composite Curves identifies the possibility for the integration of streams in the boiler house, in particular the stream of the make-up water to the demineraliser. The power necessary for this stream can be fully resourced from the ammonia streams (Figure 9.4), but the Composite Curves also show the availability of a large amount of low-grade heat, which is presently taken from the system by the

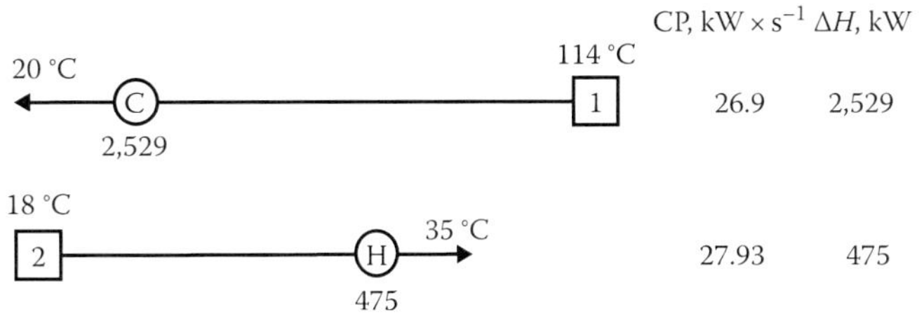

Figure 9.2 Grid Diagram for existing stream system in a boiler house and ammonia unit.

Table 9.1 Stream Data for an Existing Process

No.	Name of Stream	Type	TS, °C	TT, °C	G, t/h	C, kJ/(kg·°C)	r, kJ/kg	CP, kW/°C	ΔH, kW
1.1	Cooling of gaseous ammonia	Hot	114	28	6.22	3.25		5.615	482.91
1.2	Condensation of ammonia	Hot	28	28	6.22		1,146		1,980.03
1.3	Cooling of liquid ammonia	Hot	28	20	6.22	4.75		8.207	65.66
2	Make-up water to demineraliser	Cold	18	35	24.00	4.19		27.933	474.87

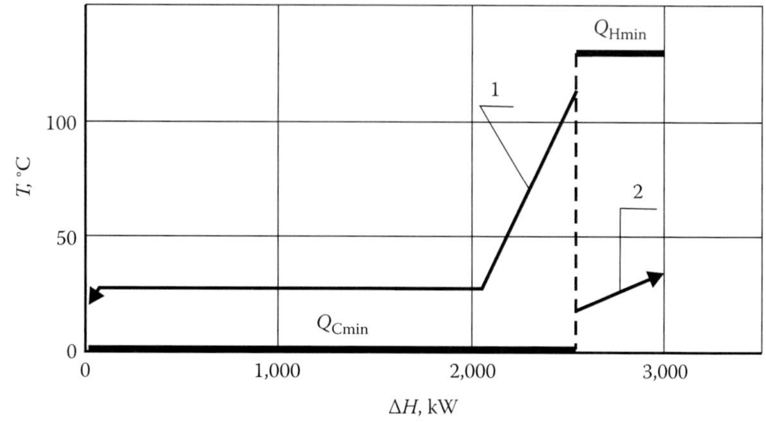

Figure 9.3 Composite Curves for existing process streams in a boiler house and compression unit: (1) Hot Composite Curve and (2) Cold Composite Curve. Q_{Hmin} = 475 kW, hot utility; Q_{Cmin} = 2,529 kW, cold utility.

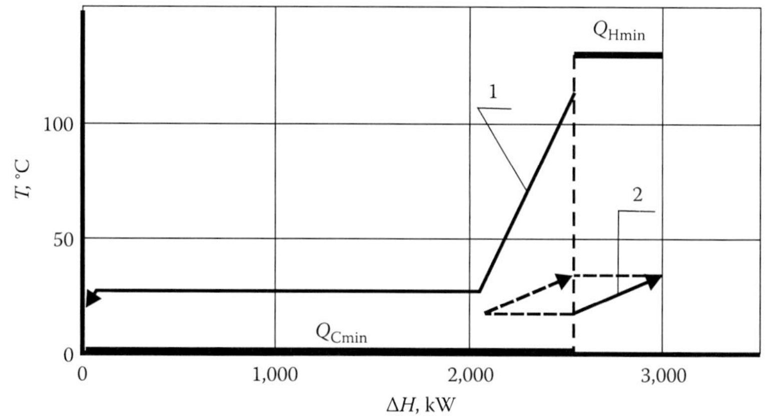

Figure 9.4 The possibility of Process Integration.

cooling water. This heat can be used for heating the make-up water to the demineraliser as also the stream of hot water for the enterprise process system.

By adding a new stream into the Heat Integration, a new table of streams information is composed (Table 9.2).

For the integration of the heat exchanger system, ΔT_{min} = 2 °C is chosen. Such difference of temperatures between the heat transfer agents can be achieved by using modern, highly effective, compact plate heat exchangers. Composite Curves constructed for stream data from Table 9.2 and ΔT_{min} = 2 °C also show the possibility for complete elimination of hot utilities from the process stream system under consideration and additional heating of the delivery water. In the system considered choosing ΔT_{min} = 2 °C, the possible duty of recuperation makes a value of 0.75 MW,

Table 9.2 Integrated Process Stream Data (without Additional Compression)

No.	Name of Stream	Type	TS, °C	TT, °C	G, t/h	C, kJ/(kg·°C)	r, kJ/kg	CP, kW/°C	ΔH, kW
1.1	Cooling of gaseous ammonia	Hot	114	28	6.22	3.250		5.615	482.91
1.2	Condensation of ammonia	Hot	28	28	6.22		1,146		1,980.03
1.3	Cooling of liquid ammonia	Hot	28	20	6.22	4.750		8.207	65.66
2	Make-up water to demineraliser	Cold	18	35	24.00	4.190		27.933	474.87
3	Water to process	Cold	18	70	4.53	4.190		5.270	274.00

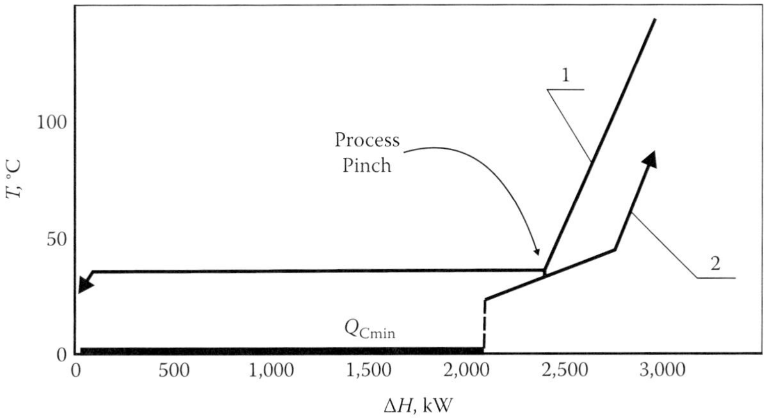

Figure 9.5 Composite Curves for boiler house and compression unit process streams after integration (without additional compression): (1) Hot Composite Curve and (2) Cold Composite Curve. Q_{Hmin} = 0 kW, hot utility; Q_{Cmin} = 1,780 kW, cold utility.

from which 0.47 MW is used for heating of the make-up water to the demineraliser and 0.27 MW for the heating of the delivery hot water.

The grid diagram of the HEN, which complies with the Composite Curves in Figure 9.5, is shown in Figure 9.6. To achieve the energy targets shown by these Composite Curves, five heat exchangers are required (Table 9.3). The flowsheet of the HEN, to satisfy the grid diagram (Figure 9.6), is presented in Figure 9.7.

9.1.4 Heat Integration with Additional Compression

By conducting further analysis of the Composite Curves for the proposed HEN (Figure 9.5), it can be seen that the localisation of the Pinch is on the hot temperature of 28 °C—the temperature of ammonia condensation. This temperature limits the amount of network delivery of hot water, which may be used for heating and the hot water supply. If we can increase the temperature of condensation, at least for some parts of the ammonia stream, it will be possible to increase the flow rate for the

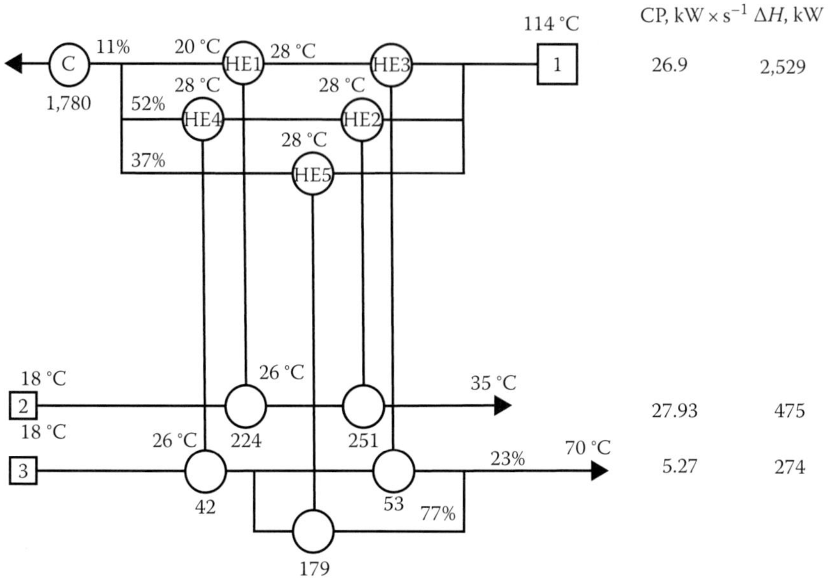

Figure 9.6 Grid Diagram for the integrated system of process streams in boiler house and ammonia unit. HE, heat exchanger; C, cooler; CP, stream heat capacity; ΔH, stream enthalpy.

Table 9.3 Evaluative Features of Heat Exchangers for the Heat Exchanger Network without Additional Compression

	Hot Stream			Cold Stream					
A	**No.**	**T_{in}, °C**	**T_{out}, °C**	**No.**	**T_{in}, °C**	**T_{out}, °C**	**ΔT_{min}, °C**	**Q, kW**	**S, m²**
T1	1	28	20	2	18	26	2	224	10.22
T2	1	114	28	2	26	35	2	251	17.28
T3	1	114	28	3	26	70	2	53	8.68
T4	1	28	28	3	18	26	2	42	7.04
T5	1	114	28	3	26	70	2	179	21.60
Total									64.82

delivery of hot water for the needs of the enterprise. It is possible to increase the temperature of ammonia condensation by applying additional compression to the part of the stream. Parameters for the stream are found by the use of specially developed software for integration of process streams.

The results of this analysis show that for the practical supply for the enterprise of heat and hot water, that is, for heating of delivery water with a flow rate of 20 t/h at 70 °C, an additional compression is needed of 2.3 t/h of gaseous ammonia from 12 bar to 26 bar and a temperature of condensation of 60 °C. Research results are presented in Table 9.4.

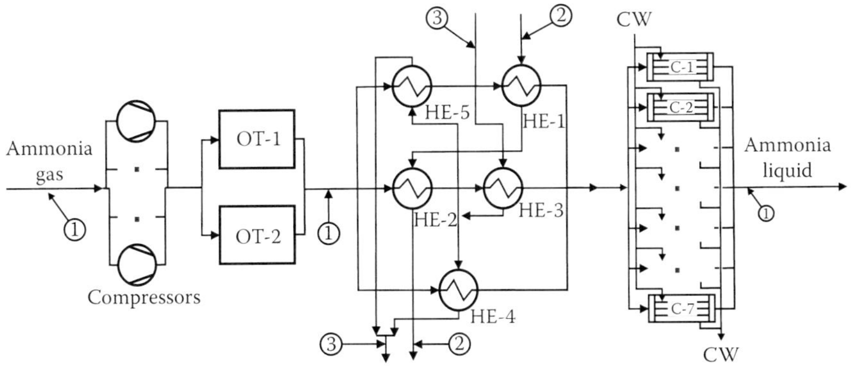

Figure 9.7 Flowsheet of integrated HEN (without additional compression): OT-1, 2, oil trap; CW, cooling water; C-1–C-7, condensers; HE-1–HE-5, heat exchangers.

Table 9.4 Stream Data for the Integrated Process with an Additional Compression

No.	Name of Stream	Type	TS, °C	TT, °C	*G*, t/h	*C*, kJ/(kg·°C)	*r*, kJ/kg	CP, kW/°C	Δ*H*, kW
1.1	Cooling of gaseous ammonia	Hot	114	28	3.914	3.250		3.533	303.88
1.2	Condensation of ammonia	Hot	28	28	3.914		1,146.0		1,245.96
1.3	Cooling of liquid ammonia	Hot	28	20	3.914	4.750		5.164	41.31
2.1	Cooling of gaseous ammonia	Hot	146	60	2.306	4.275		2.738	235.50
2.2	Condensation of ammonia	Hot	60	60	2.306		986.2		631.72
2.3	Cooling of liquid ammonia	Hot	60	20	2.306	4.935		3.161	126.45
3	Water to make-up demineraliser	Cold	18	35	24.000	4.190		27.933	474.87
4	Water to process	Cold	18	70	19.933	4.190		23.200	1,206.40

The Composite Curves constructed for the streams' data from Table 9.4 for an already chosen $\Delta T_{min} = 2$ °C show that the task of designing the HEN, in this case, represents a Two-Pinch problem (Figure 9.8). The Composite Curves show that in the system with additional compression of ammonia, the recovery produces $Q_{Rec} = 1.7$ MW, which is 2.3 times more than in the system without additional compression (Figure 9.5).

The heat duty which must be taken from the system of hot streams, in this case equal to $\Delta H_{Hot} = 2.585$ MW, is bigger than the power presently taken of 56 kW. This power will be spent on additional compression.

In order to synthesise the Grid Diagram of the HEN to satisfy the Composite Curves in Figure 9.8, the process streams participating in Heat Integration, with localisation of Near-Pinch and Process Pinch, are presented in Figure 9.9. These two

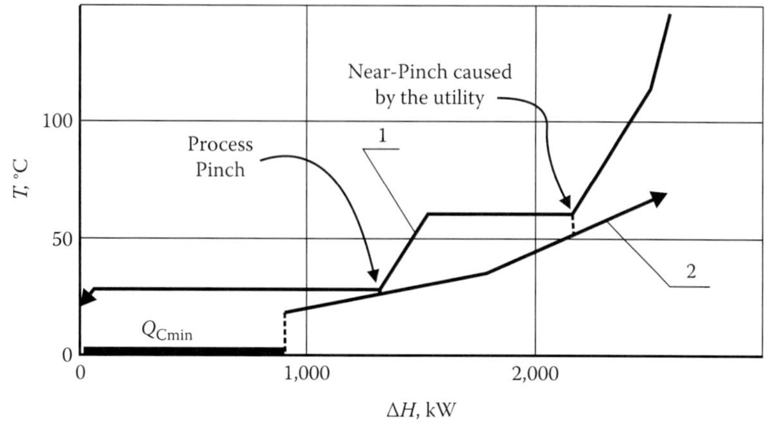

Figure 9.8 Composite Curves for the integrated process streams with additional compression: (1) Hot Composite Curve and (2) Cold Composite Curve. Q_{Hmin} = 0 kW, hot utility; Q_{Cmin} = 903 kW, cold utility.

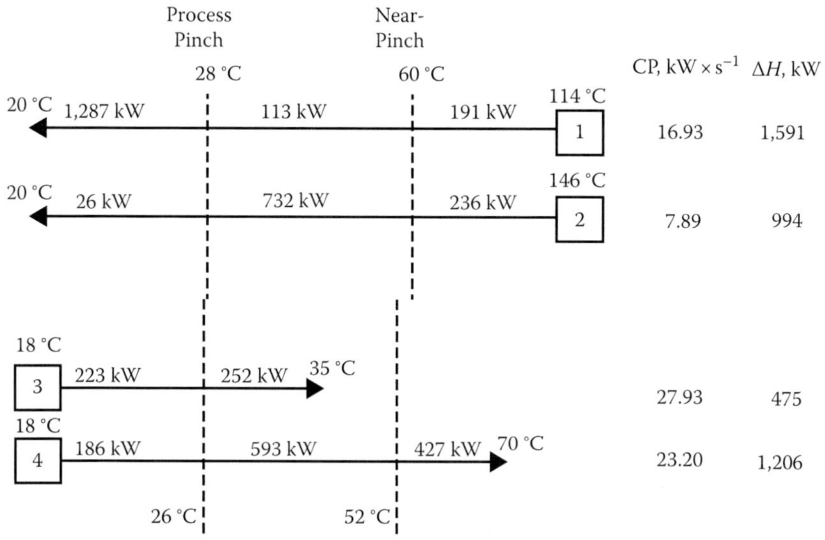

Figure 9.9 Process streams participating in Heat Integration with additional compression. Stroke lines show the division on Pinches: CP, stream heat capacity; ΔH, stream enthalpy.

Pinches divide the system of streams into three subsystems; in each of them we construct a HEN, which forbids the transfer of energy through both Pinches.

As a result refer to the Grid Diagram of the HEN, represented in Figure 9.10. Here a few loops can be noticed, but to break them with the purpose of simplification of the system, it seems inadvisable technically and economically, as there are different phase states in heat exchangers and also different flow rates of heat transfer agents in all loops.

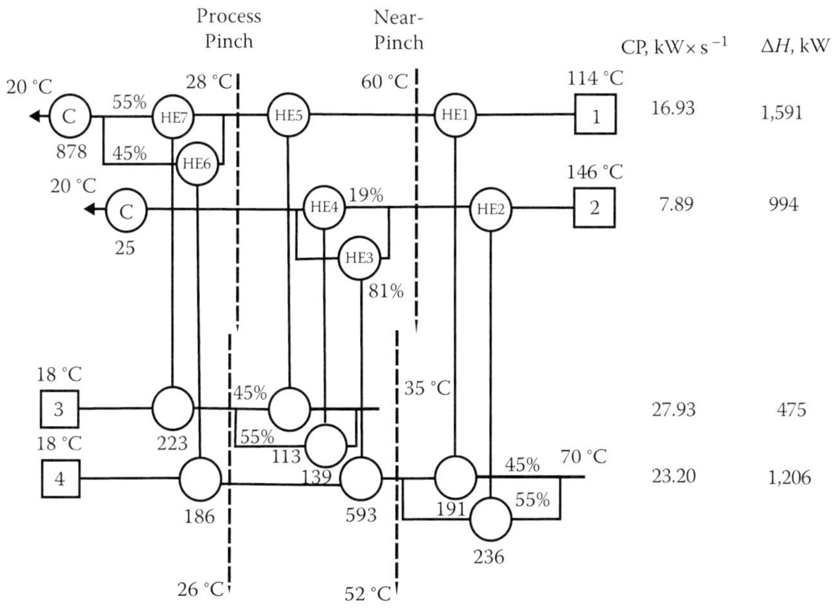

Figure 9.10 Grid Diagram for integrated HEN with additional compression. Stroke lines show the division on Pinches: HE, heat exchangers; C, coolers; CP, stream heat capacity; ΔH, stream enthalpy.

In this system, there are seven recuperative heat exchangers with a total heat transfer area of about 69 m^2, which is only 5 m^2 more than in the network pictured in Figure 9.7. Some technical parameters of the equipment are presented in Table 9.5. The flowsheet of the HEN in the case of an additional compression of ammonia is shown in Figure 9.11.

Table 9.5 Some Technical Parameters of Heat Exchangers for the Heat Exchanger Network with Additional Compression

	Hot Stream			Cold Stream					
No.	No.	T_{in}, °C	T_{out}, °C	No.	T_{in}, °C	T_{out}, °C	ΔT_{min}, °C	Q, kW	S, m^2
T1	1	114	60	4	52	70	8	191	11.44
T2	2	146	60	4	52	70	8	236	7.00
T3	2	60	28	4	26	52	2	593	9.38
T4	2	60	28	3	26	35	2	139	3.87
T5	1	60	28	3	26	35	2	113	21.12
T6	1	28	28	4	18	26	2	186	7.28
T7	1	28	28	3	18	26	2	223	9.24
Total									69.33

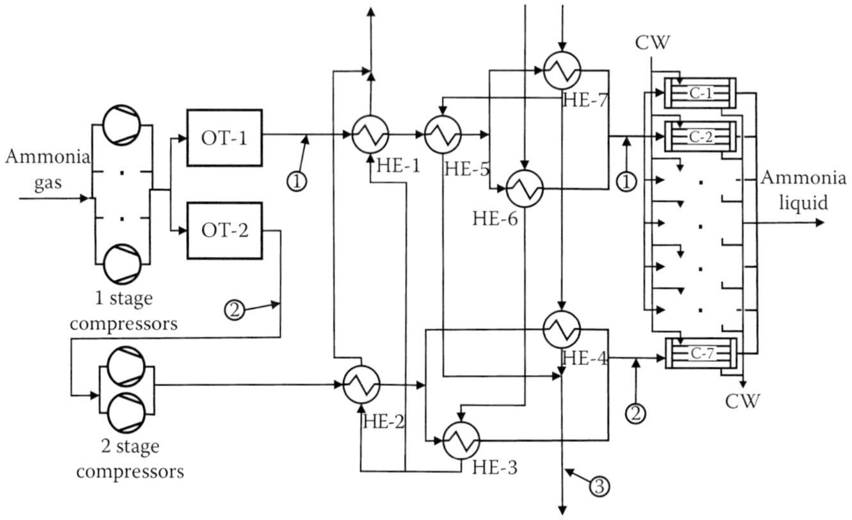

Figure 9.11 Flowsheet of the HEN of boiler house and ammonia unit after integration with additional compression: OT-1, 2, oil trap; CW, cooling water; C-1–C-7, condensers; HE-1–HE-7, heat exchangers.

9.1.5 Economic Efficiency

The cost of hot utility for the inspected cheese production plant is 50 $/kW/y and the cost of cold utility is 7$/kW/y (prices for year 2004 in Ukraine). Applying Heat Integration without additional compression can increase recuperation of heat energy of 750 kW. Evaluative total cost for HEN installation (Figure 9.7) are $35,500. The payback period of this retrofit option is about 10 months. Additional compression allows increased recuperation of heat energy in the system. However, it should also spend 60 kW of electric power for ammonia compression. In this case recuperation is increased to 930 kW in comparison with integration without additional compression. The total capital cost for the heat exchange network (Figure 9.10) and compressor installation is about $75,500. And the payback period of a retrofit project, with additional compression of the ammonia stream, is also about 10 months.

9.1.6 Conclusion

The energy-saving potential of the ammonia cooling cycle is rather large as is shown in this study. Application of results of this study can considerably improve heat integration on enterprises which use ammonia as a refrigerant. The use of compact and energy-efficient plate heat exchangers enabled to considerably increase heat recuperation by lowering the minimal temperature difference down to 2 °C. Another advantage of compact Plate Heat Exchangers (PHEs) is considerable reduction of internal volume that allows a significant decrease of the charge of ammonia in the refrigeration system.

The study has wide industrial application because ammonia cooling cycles are used on most Ukrainian, Russian, Byelorussian and Kazakh food enterprises. However, inspection and Heat Integration of enterprises of different industries should not be made without careful consideration of all site features.

9.2 CHEMICAL INDUSTRY: THE USE OF INTENSIFIED HEAT EXCHANGERS TO IMPROVE ENERGY EFFICIENCY IN PHOSPHORIC ACID PRODUCTION

9.2.1 Process Description

The chemistry of phosphoric acid production from apatite concentrate by the wet method is sufficiently developed and rather well described in the literature – see, e.g., European Fertilizer Manufacturers Association [EFMA] (2000) Best Available Techniques and paper by Abu-Eishah and Abu-Jabal (2007). According to these literature sources, the typical process flowsheet of phosphoric acid production by the dehydrate method is presented in Figure 9.12. This process enables one to obtain weak acid with a P_2O_5 concentration of 26%–32% which should be followed by an evaporation unit to obtain acid with higher P_2O_5 concentration (see Figure 9.13).

Most of the technology in the phosphorus chemical industries is outdated and unable to 'solve' the problems of waste minimisation and full utilisation of material and energy resources. New research and innovation strategy for the phosphoric

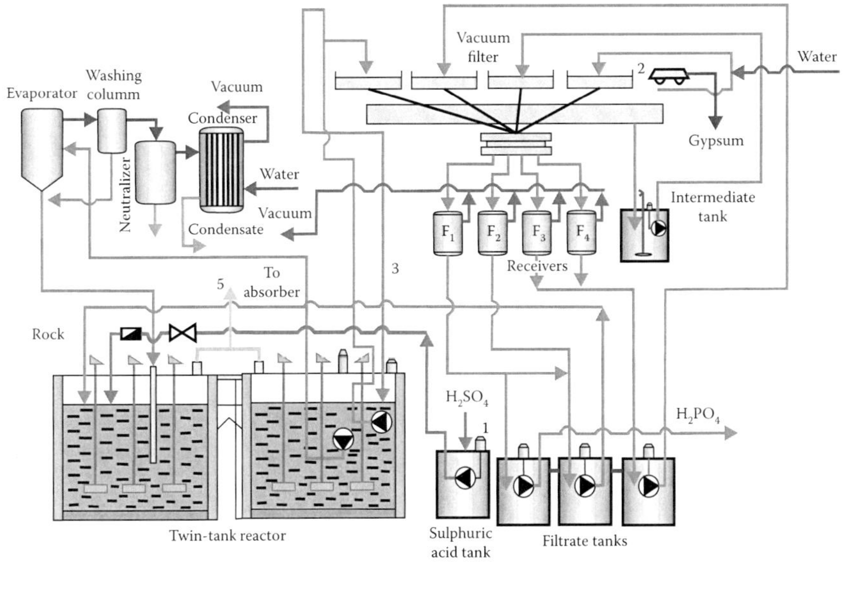

Figure 9.12 The typical process flowsheet of phosphoric acid production by the dehydration method.

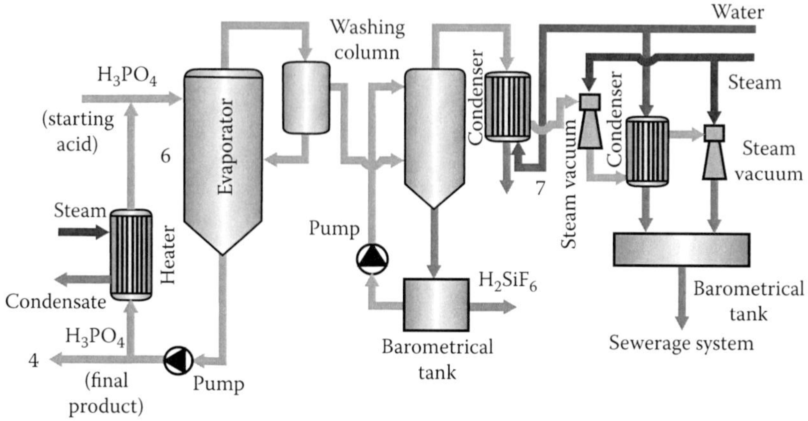

Figure 9.13 Flowsheet of the concentration unit.

acid industry are being proposed within the EC supported FP6 INCO ECOPHOS Project. It is designed to develop ecologically sustainable, environmentally friendly, resource- and energy-saving industrial process technologies for the production of a wide class of phosphorus-containing substances, as discussed by Seferlis et al. (2006). Considerable improvement in process energy saving and pollution reduction can be achieved with the use of pinch technology and process integration approach. With this technique considerable reduction in energy consumption of sodium hypophosphite production was shown by Tovazshnyansky et al. (2010). At the same time the importance of receiving economically optimal solutions with correct accounting for cost and energy prices can substantially limit the possibilities of energy saving, counting that highly expensive materials are required for heat transfer surface production. This can explain the fact that in a lot of phosphoric acid production plants, which are now in operation, few process integration and heat recuperation opportunities are used. As it is discussed by Kapustenko et al. (2009) on the example of PHEs, this obstacle can be overcome with the less material consuming, compact heat exchangers, which can be used instead of conventional tubular ones.

Plate heat exchanger is much more compact and requires much less material for heat transfer surface production; thus, there is a much smaller footprint than with conventional shell-and-tube units. It makes it feasible and economically efficient to produce plates from expensive, sophisticated materials. PHEs have a number of advantages over shell-and-tube heat exchangers, such as compactness, low total cost, less fouling, flexibility in changing the heat transfer surface area, accessibility and, what is very important for energy saving, close temperature approach – down to 1 °C. The main limitations for their use are pressure and temperature restrictions imposed by gaskets. The comparison of capital cost data for both PHE and shell-and-tube heat exchangers for a wide range of thermal duties is presented by Edwards and Stinchcombe (1977). They found that, for many cases, stainless steel plate units are even cheaper than shell-and-tube units made from lower-cost carbon steel. They also noted that, on average, the all-stainless-steel tubular unit would be about four times as expensive as PHE for the same duty. This advantage is magnified when specialised and expensive metals are used for PHE. Such material is needed for the heat transfer plates used in the phosphoric acid production process. To determine the kind of material needed and the possibility of using compact heat exchangers, the different possible combinations of streams in heat exchangers should be analysed.

9.2.2 Placement of Heat Exchangers in the Wet Process of Phosphoric Acid Production

The main placements of heat exchangers in the manufacturing of phosphoric acid by the wet method include the following, as shown in Figures 9.12 and 9.13:

Cooling of sulphuric acid solution 78%–98% H_2SO_4 (Position 1 in Figure 9.12)
Cooling of weakly concentrated phosphoric acid (3%) after washing of the filter sediments (Position 2 in Figure 9.12)
Heating of 30% phosphoric acid before sulphate sedimentation (Position 3 in Figure 9.12)
Cooling of phosphoric acid (final product) with 40%–42% or 50%–54% P_2O_5 concentration after evaporation (Position 4 in Figure 9.13)
Cooling of scrubber acid (8%–11% H_2SiF_6) (Position 5 in Figure 9.12)
Evaporation of phosphoric acid (Position 6 on Figure 9.13)
Cooling of the barometric condenser waters in the evaporating station (Position 7 in Figure 9.13)

The content and temperature conditions of streams, the types of heat exchangers that can be used and heat transfer surface materials are presented in Table 9.6. The materials for the gaskets for PHEs are also presented in Table 9.6. The pressures of all streams are under 10 bar, which is much lower than the upper limit for this type of heat exchanger (usually 25 bar). This means that the pressure is not limiting PHE application in this process.

9.2.3 Calculation of PHEs for Use in Phosphoric Acid Production

The following streams of the phosphoric acid production process require cooling or heating:

Heating of streams that contain 30% P_2O_5 from 20 °C to 40 °C
Cooling of streams with 54% P_2O_5 from 85 °C to 55 °C
Cooling of streams with 54% P_2O_5 from 55 °C to 25 °C

Table 9.6 Requirements for Heating and Cooling in Phosphoric Acid Production Process

Position, Figure 9.12, 9.13	Destination	Type of HE	Process Stream		Material	Gaskets
			T_{in}, °C	T_{out}, °C		
1	Taking the heat off dilution	PHE	Cooling 98–78% H_2SO_4		Diabon F100, NS1 or NS2	EPDM
			80	30		
2	Weak phosphoric acid cooling	PHE	Cooling 3% P_2O_5		AISI 316L, 254 SMO	EPDM
			50	25		
3	Heating of 30% phosphoric acid prior to sulphate precipitating	PHE	30% P_2O_5		Hastelloy G30, Grafite	EPDM
			20	40		
4	Cooling of 40–42% or 50–54% (final product) phosphoric acid after evaporation	PHE	54% P_2O_5		AISI 316L, 904L, 254 SMO, Sanicro 28, C-276, G30	MS2000 (Klingersil), PTFE
			85	55		
4	Cooling of 40–42% or 50–54% (final product) phosphoric acid after evaporation	PHE	54% P_2O_5		Hastelloy G30, Grafite	EPDM
			55	25		
5	Cooling of scrubber acid	PHE	H_2SiF_6, 8–11%		254 SMO and C-276	EPDM
			50	37		
6	Evaporation of phosphoric acid	Shell and tube	30–54% P_2O_5		Grafite	–
			85	90		
7	Cooling water of barometric condenser	PHE	Water		Hastelloy G30	EPDM
			38	25		

The important feature of stream heating or cooling is the high intensity of fouling on the heat transfer surfaces due to gypsum and fluoride precipitation from the solution. The intensity of fouling depends on the structure of the impurities of the product solution, the quantity and sizes of the particles, the flow velocities and the temperatures. In spite of these challenges and severe operating conditions, PHEs of different types can be safely used in these process phases. A detailed study of calcium sulphate deposition from phosphoric acid solution in tubular heat exchangers was reported by Jamialahmadi and Muller-Steinhagen (2007). They identified the key variables in deposition rates of fouling to be (1) the concentration of the fouling substances, (2) the surface temperatures and (3) the flow velocities. In the first-effect evaporators, the range of temperatures was from 83 °C to 97 °C and the predominant fouling mechanism is calcium sulphate dihydrate formation. A significant increase of fouling thermal resistance with temperature and fouling substance concentration, as well as inverse influence of flow velocity, was noticed. This is the reason why in this chapter PHEs are not considered for the first-effect evaporators. At lower temperatures (see Table 9.6) and cleaner acid solutions, PHEs can be used to benefit from their lower susceptibility to fouling.

Table 9.7 Composition of Alloys for Heat Transfer Surface Manufacturing

	Metals,%				
Alloy	**Cr**	**Nickel**	**Mo**	**Cu**	**Others**
AISI 316	17.0	12.0	2.0	–	–
Avesta 254 SMO	20.0	18.0	6.1	1.7	N 0, 2
Alloy C276	15.5	58.0	16.0	–	W
Hastelloy C22	21.0	44.0	17.5	–	W 3, Fe 2–6
Hastelloy G30	29.5	40.0	5.0	1.7	W 2, 5, Fe 18–21
Hastelloy D205	20.0	64.5	2.5	2.0	Si 5, Fe 6

Table 9.8 Composition of Aqueous Phosphoric Acid Solutions

Main Elements, %	30% P_2O_5	54% P_2O_5
P_2O_5	29.8	53.0
H_2SO_4	1.9	2.0
SO_3	1.6	2.5
Fe_2O_3	0.15	0.24
F	0.9	0.4
SiO_2	0.3	0.03
Cl, ppm	755	165

The presence of different impurities substantially influences the choice of heat transfer material. It should be selected depending on acid concentration, temperature range and presence of chlorine, iron and other impurities. The use of Hastelloy G30 or graphite can be recommended for 50% P_2O_5 at temperatures lower than 85 °C and in the presence of 1% of HF, Fe_2O_3, Al_2O_3, 4% of H_2SO_4 and 600 ppm HCl. The synthetic EPDM rubber (ethylene propylene diene monomer) can be used as a gasket material. The alloys of AISI 316L, 904L, 254 SMO, Sanicro 28, C-276 and G30 are also used as heat transfer surface materials for PHEs, depending on the specific conditions (see Table 9.7).

The cooling of the end product of 54% P_2O_5 concentration by a phosphoric acid of 30% P_2O_5 was considered for the calculation of PHEs with different plate geometries. The cooling is usually done in two stages. At the first stage the concentrated acid is cooled from 85 °C to 55 °C with the lower concentrated (30%) acid heated from 20 °C to 40 °C. The composition of phosphoric acid aqueous solution in both streams is presented in Table 9.8. Thermophysical properties for phosphoric acid of different concentrations at various temperatures are derived from literature data. Intermediate values were calculated by means of linear interpolation of adjacent values. Data for the calculation of the first stage are presented in Table 9.9.

Further, there is cooling of the concentrated phosphoric acid from 55 °C to 25 °C while 30% of the acid is heated from 20 °C to 40 °C. The physical properties of the streams of the 54% and the 30% acid for their respective temperature ranges are presented in Table 9.10. Data for calculation of the second-stage heat exchanger are presented in Table 9.11.

Table 9.9 Data for Calculation of the First-Stage Heat Exchanger

Heat Load	$Q = 866.8$ **kW**	
Stream	**Hot**	**Cold**
Medium	**Phosphoric Acid (54% P_2O_5)**	**Phosphoric Acid (30% P_2O_5)**
Working pressure, MPa	0.5	0.5
Flow rate, kg/h	40,000	48,000
Inlet temperature, °C	85	20
Outlet temperature, °C	55	39.93
Pressure drop, MPa	≤0.1	≤0.1

Table 9.10 Physical Properties of 30% and 54% Phosphoric Acid

Medium 1 (hot): Physical properties for three temperatures—phosphoric acid 54% P_2O_5			
Temperature, °C	85	70	55
Density, kg/m^3	1,335	1,346	1,357
Specific heat, kJ/(kg·°C)	2.569	2.605	2.612
Heat conductivity, W/(m·°C)	0.546	0.534	0.520
Dynamic viscosity, centipoises	1.122	1.729	2.753
Medium 2 (cold): Physical properties for three temperatures—phosphoric acid 30% P_2O_5			
Temperature, °C	20	30	40
Density, kg/m^3	1,181	1,176	1,171
Specific heat, kJ/(kg°C)	3.237	3.262	3.284
Heat conductivity, W/(m°C)	0.533	0.547	0.560
Dynamic viscosity, centipoises	2.996	2.310	1.738

Table 9.11 Data for Calculation of the Second-Stage Heat Exchanger

Heat Load	$Q = 861.2$ **kW**	
Stream	**Hot**	**Cold**
Medium	**Phosphoric Acid (54% P_2O_5)**	**Phosphoric Acid (30% P_2O_5)**
Working pressure, MPa	0.5	0.5
Flow rate, kg/h	40,000	48,000
Inlet temperature, °C	55	20
Outlet temperature, °C	25	39.81
Pressure drop, MPa	≤0.05	≤0.05

For corrosive media such as phosphoric acid, the Hastelloy G30 alloy is used as the heat exchanger plate material (see Table 9.7). The minimal thickness of the plates from the Hastelloy G30 is 0.6 mm. It is necessary to use synthetic rubber EPDM as the material for the gaskets. The direction of flow of heat carriers is countercurrent. The calculations were made, accounting a 10% margin for overall heat transfer coefficient.

It is possible to install M10B units with the diameter of port connections 100 mm on both stages, as the biggest velocity in connections equal to 1.438 m/s does not

Table 9.12 Results of Calculations for Heat Exchangers with Different Plate Type Combinations for the First Stage

Type of Plates	Arrangement of Channels	Number of Plates	Area, m^2	Pressure Drop (Hot Side), kPa	Pressure Drop (Cold Side), kPa
H–L	1 × (9MH + 4L)/1 × (9ML + 4L)	27	6.48	65.93	98.45
H–L	1 × 13MH/1 × 14ML	28	6.72	79.5	98.91
H–H	1 × 24H/1 × 25H	50	12.00	61.96	93.22
L–L	1 × 20L/1 × 21L	42	10.08	21.6	31.42

exceed permissible limit, 5 m/s. Results of calculations for the first stage with various combinations of plates in the unit are presented in Table 9.12. From the comparison of the table rows, the use of plates with different geometries of corrugation in one heat exchanger allows to reduce the heat transfer surface area in comparison with the use of only one plate type in the heat exchanger. The plates have corrugations with different angles to the vertical axis and to the main flow direction. Plates of H type have corrugations with bigger angles (about 60°) that form the H channels with higher efficiency of heat transfer and higher hydraulic resistance (see Figure 5.4c). Plates of L type have a lower angle (about 30°) and form the L channels with lower heat transfer and hydraulic resistance (see Figure 5.4a). Combined, these plates form channels MH or ML with intermediate characteristics (see Figure 5.4b).

From results presented in Table 9.12, it can be recommended to install one-pass unit (6.48 m^2, 27 plates, grouping 1 * (9MH + 4L)/1 * (9ML + 4L)) for cooling down the concentrated phosphoric acid at the first stage.

Results of calculations for the second stage with various combinations of plates in one unit are presented in Table 9.13. Advantages of using various plate types in one unit on the second stage are seen even more clearly, in spite of the fact that two-pass heat exchangers are needed due to operating conditions. From the results of calculations, it is possible to recommend installation of the two-pass units with an area of 50.16 m^2 (209 plates) and channel arrangement of 2 * (49H + 3MH)/2 * (49H + 3ML).

The analysis of data shows that the use of mixed grouping of plates in one heat exchanger (type H/MH–H/ML and MH/L–ML/L) allows satisfying the heat exchange requirements fully. Thus, the number of plates (the heat transfer area) is minimal and the condition of pressure drop on one of the streams is completely satisfied.

Table 9.13 Results of Calculation of Heat Exchangers with Various Plate Combinations for the Second Stage

Plate Combination	Arrangement of Channels	Number of Plates	Area, m^2	Pressure Drop Hot Side, kPa	Pressure Drop Cold Side, kPa
H–L	2 × (49H + 3MH)/2 × (49H + 3ML)	209	50.16	34.5	48.33
H–L	3 × 41MH/(2 × 41ML + 1 × 42ML)	248	59.52	37.45	46.84
H–H	2 × 53H/(1 × 53H + 1 × 54H)	214	51.36	35.25	49.27
L–L	(1 × 38L + 3 × 39L)/4 × 39L	312	74.88	40.44	49.29

To obtain an optimal solution in this work, the concept of PHE design with fixed allowable pressure is used. With this approach, the objective function is the heat transfer surface area, which for PHE corresponds to the minimal capital cost; which is mostly due to the fact that the PHE plates are in contact with aggressive media and should be manufactured from expensive alloys. The principle of combining two plates with different corrugation geometries provides a new dimension to PHE design. In effect, by selecting the proper proportion of H and L plates, it allows changing the thermal and hydraulic characteristics of the plate pack with the level of discreteness equal to one plate in a pack. This makes the mathematical problem for designing the PHEs essentially a continuous process that can be solved using the continuous algebraic equations derived from the mathematical model of heat exchangers (see Chapter 5). The advantages of this technique are obvious for HEN optimisation, where it allows to obtain a continuous mathematical model of the system and to optimise pressure drop at the target stage.

Application of the algorithm for calculation of mixed grouping PHEs makes it possible to achieve the following advantages:

- To intensify the process of heat transfer and to reduce the heat transfer area
- To select one-pass units
- To completely utilise the available pressure drop
- To eliminate unproductive pressure losses in comparison with multipass heat exchangers

9.2.4 Closed Circuit Circulation for Barometric Condenser

In operating evaporation plants the secondary steam should be condensed to create a vacuum in the evaporation units. With that steam the fluorine is released, which can be partly recovered as a commercial by-product (fluorosilicic acid 20%–25%), but its remainder passes to the condenser. It is producing liquid effluent that contains mainly fluoride and a small amount of phosphoric acid (see EFMA, 2000). These substances can cause severe damage when discharged to the environment. To avoid this, it is possible, (1) as recommended by EFMA (2000), to arrange a closed-loop water system with barometric mixing condenser or (2) to use surface condensers. Let us examine both options for the case of evaporation unit with production capacity of 120 t of 54% P_2O_5/d.

The basic flowsheet of barometric condenser with recycled water circulation is shown in Figure 9.14. In summer, the water inlet temperature in the mixing condenser is about 25 °C and the outlet temperature is 38 °C. Plate heat exchangers of Alfa Laval are calculated for the location of recycled water cooling. Plates are made from Hastelloy G30 with a thickness of 0.6 mm and gaskets made from EPDM synthetic rubber. The calculation was done accounting for a 20% margin for possible heat transfer surface fouling during process operation. The value of shear stress on the plate's wall was controlled at not less than 50 kPa. This ensures stable operation of the heat exchanger between the scheduled maintenances. Data for the calculations are listed in Table 9.14.

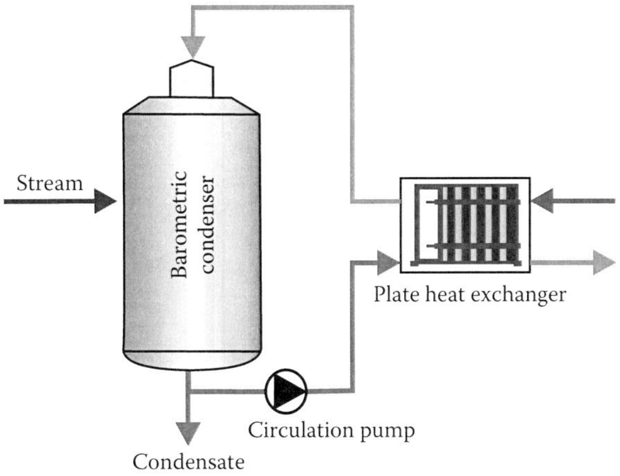

Figure 9.14 The barometric condenser unit with water cooled in external heat exchanger.

Table 9.14 Initial Data for Design of Heat Exchangers for Cooling of Recycled Water of Barometric Mixing Condensers

	Duty Parameters				
Flow	T_{inlet}, °C	T_{outlet}, °C	Flow Rate, m³/h	Pressure Drop, kPa	Margin %
Barometric condenser water	38	25	368	150	20
Cooling water	20	30	475	150	20

Three unit types were evaluated:

1. Traditional plate-and-frame PHEs
2. Spiral heat exchangers
3. Wide-gap PHEs (with the wider channel on the side of the cooled, recycled water of the barometric condenser)

Table 9.15 shows the results of calculations for the main parameters. Analysis of these data gives the following conclusions. The installation of one M30M unit or of two T20M heat exchangers in parallel is most suitable. Although the installation of two units is more expensive, it provides reserve capacity. It allows for increasing the cold heat carrier flow rate, to clean one unit without stopping the other one or to work with one unit at decreased capacity under emergency operation conditions. It is possible to install M15M units, but they are two-pass heat exchangers, which have a strong tendency for plate fouling and cleaning them is a more difficult process because both frame plate and pressure plate have connections. It requires dismounting of piping on moving the frame plate when disassembling PHE for mechanical cleaning. The same can be said about three-pass wide-gap heat exchanger WG200, which also has a much bigger heat transfer surface area and is about twice more expensive than M30M. Although use of SHE 1H-L-1T spiral heat exchanger provides stable operation with

Table 9.15 Data for Heat Exchangers Designed for Barometric Mixing Condenser Recycled Water Cooling (Figure 9.14), when Water is the Cooling Medium

Type of PHE	Number of Units	Channel Arrangement for One PHE Type or for Spiral Unit	Total Surface Area, m^2	Pressure Drop, kPa		Margin %
				Hot Side	Cold Side	
M15M	2	2 × (37MH + 7L)/ 2 × (37ML + 7L)	215.4	92	145	21
T20M	1	1 × 135H/1 × 135H	230.4	55.5	92.4	20
T20M	2	1 × 68H/1 × 68H	234.6	50.5	81.8	20
M30M	1	1 × (32H + 24ML)/ 1 × (32H + 24MH)	204.2	86.9	146	20
WG200	2	3 × 14SH + 4 × 15SH/ 1 × 33NH + 2 × 34NH	322.4	149	79	21
SHE	2	1H-L-1T	572.6	137	143.6	20

minimal heat transfer surface fouling, this type of unit has a considerable heat transfer surface area and, consequently, very high price (about four times more expensive than wide-gap heat exchanger and 8 times more expensive than traditional units).

The increase of cooling water temperature can occur during operation in the summer time; therefore, it is essential to take into account such situations during the design stage. It is possible to provide heat exchanger capacities at the design stage in two ways: first, to provide for cooling water flow rate increasing across the heat exchanger and, second, to design the unit with possible closest temperature approach. In this case the cooling water inlet temperature should be defined as near as possible to the outlet temperature of the barometric condenser, which equals 25 °C according to design conditions (Table 9.14). The decrease of the temperature approach leads to the increase of the heat transfer surface area of the unit. Eventually, the heat transfer surface increases so that such heat exchanger design becomes unreasonable in terms of cost and technical aspects. Figure 9.15 shows the dependence of the total heat transfer surface area F_{all} of M30M heat exchanger from the cooling water inlet temperature T_{inlet} , when other calculation parameters are constant (see Table 9.14) and the margin is equal to zero. When the cooling water inlet temperature is equal to 21 °C, then the total heat exchanger surface area must exceed 200 m^2. A further approach of T_{inlet} to 25 °C leads to a sharp increase of the heat transfer surface area. Figure 9.16 shows the change of the cooling water outlet temperature T_{out}, barometric condenser water outlet temperature T_{outlet} and heat transfer coefficient K dependence from the value of the external circuit cooling water inlet temperature.

Increasing the cooling water flow rate through the heat exchanger causes significant difficulties. First, phosphoric acid production enterprises are mostly located in regions with scarce water supply, especially in summer. Second, increasing the flow rate requires installing heat exchangers with headers and connections for high input. Taking this into account the selection of M30M heat exchanger or two T20M units in parallel is the right choice. But one M30M unit costs less. Third, whichever heat exchanger is installed, if the cooling water temperature exceeds 24 °C, then the requirement of cooling the barometric condenser water to the temperature of 25 °C cannot be satisfied.

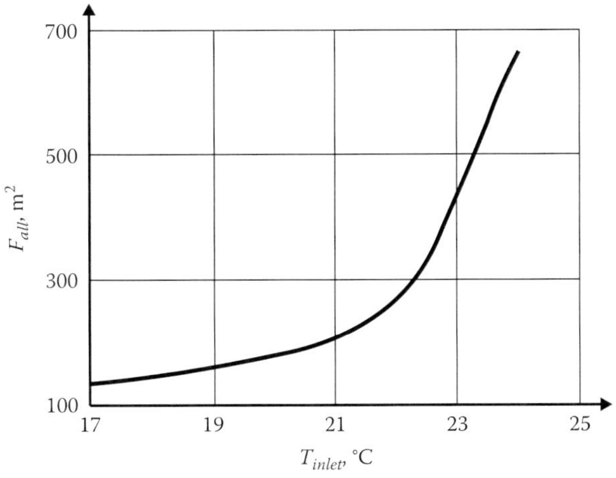

Figure 9.15 Dependence of total heat transfer surface area from the cooling water inlet temperature.

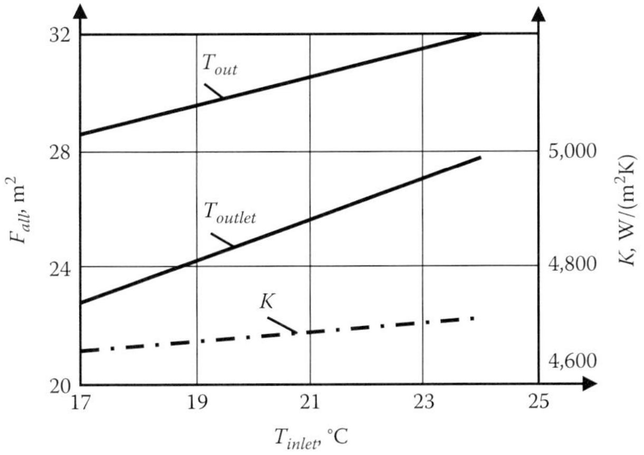

Figure 9.16 Dependence of cooling water temperature T_{out}, barometric condenser water outlet temperature T_{outlet} and heat transfer coefficient K from the value of the inlet temperature of the external circuit cooling water.

The current trend for chemical production plants is to replace mixing condensers by surface condensers, which is more economic with PHEs. It relates first of all to the following positive aspects of such replacements:

- Installation of plate condensers provides a faster response to changes in the barometric condenser system (for the comparison of PHE and shell-and-tube heat exchanger's dynamic behaviour in fouling conditions, see Georgiadis and Macchietto, 2000).
- It saves freshwater.
- Compactness of the unit, ease of maintenance and operation.

The disadvantages are that the plates will accumulate surface fouling during operation, removal of non-condensing gases and the requirement to provide low pressure drop in units operating at a vacuum.

The calculations of surface condenser are made for the same evaporation unit capacity and cooling water parameters as for the mixing barometric condenser (see Table 9.14). The steam pressure is 9.7 kPa, temperature 45 °C and flow rate 6,314 kg/h. The special condenser AlfaCond 600 with a steam connection diameter of 600 mm, for high steam throughput, has a total surface area of 72.8 m^2 and one-pass channel arrangement of 1 * 51SW/1 * 52NG. The pressure drop of steam is equal to 2.8 kPa and that of cooling water is 37.2 kPa. The plates from Hastelloy G30 are welded in pairs on the steam side for more reliable operation under a vacuum.

The comparison with the case (1) of closed-loop barometric mixing condenser shows significant reduction (up to 3 times) of the heat transfer surface area. These advantages of surface condenser application make this option preferable in both grass root and retrofit projects for evaporation units in phosphoric acid production.

9.2.5 Conclusion

The compact heat exchangers can be efficiently utilised in the phosphoric acid production process. The highly corrosive activity of the heat exchanging process streams requires the use of highly corrosion-resistant materials for heat transfer surface manufacturing. The cost of such materials makes it complicated to implement economically energy-saving methodologies including advanced Pinch Analysis and Process Integration. The use of such type of compact heat exchangers as PHEs, requiring much less material for their production than conventional shell-and-tube units, makes such solutions economically sound. It is possible to increase heat recuperation and to save up to 25%–30% of energy consumed in the process of phosphoric acid production.

To minimise the surface area of PHEs, reliable methods and algorithms for their design were developed. It is shown that one of the possibilities to minimise the heat transfer area is to use the principle of combining the plates with different geometries of corrugations on the heat transfer surfaces in one heat exchanger.

The current trend is to replace mixing condensers with surface condensers and especially with plate units. The simulation for the unit of recycled barometric mixing condenser water cooling with plate heat exchangers clearly shows the possibility of the application of surface condensers instead of mixing condensers. Proposed engineering solutions are suitable for phosphoric acid concentration stages for different processes of production. The use of PHEs in phosphoric acid production can increase energy and material savings and to make the production environmentally cleaner.

9.3 HEAT INTEGRATION OF AMMONIA REFRIGERATION CYCLE INTO BUILDINGS' HEATING SYSTEM

There are over 200 million buildings in the EU 28 alone, which account for about 40% of final energy demand and 36% of greenhouse gas emissions according to EC DIRECTIVE 2010/31/EU. Similar proportions in energy usage are observed in the United States with share of commercial and public buildings 45% according to US Department of Energy (USDOE, 2011) and in all OECD countries (IEA (2011)). The heating, cooling and hot water preparation are responsible for about 65% of that amount. The increase of energy efficiency in buildings is a very important task. One of the powerful tools to fulfil it is the use of process integration and modern energy-saving equipment for low-grade heat utilisation as discussed by Kapustenko and Arsenyeva (2013).

The refrigeration appliances of supermarkets use ammonia refrigeration machines in systems for freezing, cooling and storing food. Supermarkets are usually located near densely populated residential blocks. During the operation of similar ammonia units, a great deal of low potential heat is not used and is wasted into the atmosphere through the cooling system. On the other hand, there is a need to deliver heat in the heating, conditioning and water supply systems of buildings. However, the temperature regimes of consumers are often such that the utilisation of waste heat is frequently not feasible. It is difficult to find an optimal solution without a detailed analysis of the process stream system, heat exchange at different temperature levels and a heat recovery system. One of the most efficient methods of attaining optimal efficiency is Pinch analysis, which is based on the thermodynamic analysis of a process stream system and the creation of flowsheets with consideration for the cost of utilities and equipment as discussed by Boldyryev et al. (2013) who studied the possibility of utilising the condensation heat of a supermarket refrigeration machine for existing heat consumers. The Process Integration methodology, which allows to reduce the energy consumption and construct a heat exchange network based on compact heat exchangers, is described in Section 9.3. The objective function in the optimisation problem is to minimise total reduced cost.

9.3.1 System Description and Modelling

Supermarket refrigeration units are conventional ammonia refrigeration machines. The obtained process data allow to construct a computer model of the studied refrigeration unit using the UniSim Design software (Honeywell, 2015). This makes it possible to refine the temperature, pressure and flow rate of the cooling agent and obtain the unavailable thermophysical properties. The pressure of ammonia at the inlet and outlet of a compressor is equal to 196 and 1,200 kPa, and the unit's refrigerating capacity is 3,956 kJ/h. The technological streams of the studied process and their thermophysical properties are given in Table 9.16.

The computer model of the ammonia refrigeration cycle is illustrated in Figure 9.17. At this stage, the overheating and condensation heat of ammonia is withdrawn with

Table 9.16 Stream Data of the Existing Refrigeration Process and Heat Users

No.	Name of Stream	Type	TS, °C	TT, °C	*G*, t/h	*C*, kJ/(kg·°C)	*r*, kJ/kg	CP, kW/°C	ΔH, kW
1	Ammonia cooling	Hot	155	30	3.194	3.250		2.883	360.43
	Ammonia condensation	Hot	30	30	3.194		1,146		1,016.76
	Cooling of liquid ammonia	Hot	30	20	3.194	4.750		4.214	42.14
2	Water heating	Cold	15	60	15.000	4.190		17.458	785.63
3	Air preheating	Cold	10	30	50.000	1.005		13.958	279.17
4	Air to fans	Cold	10	55	25.000	1.005		6.979	314.06

Compressor		
Power	0.3210	kW
Feed pressure	195.9	kPa
Product pressure	1,200	kPa
Product temperature	155.2	°C
Duty	1,156	kJ/h

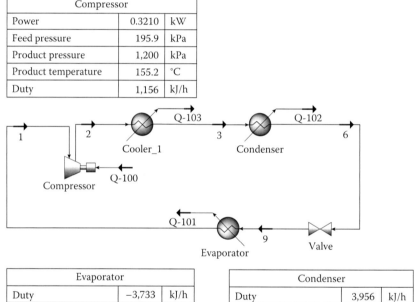

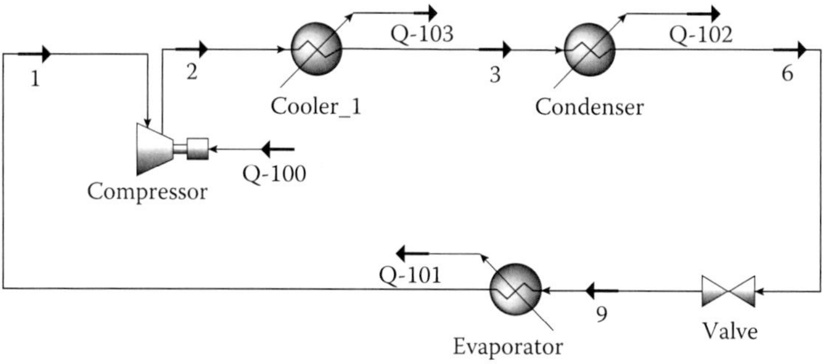

Evaporator		
Duty	−3,733	kJ/h
Feed temperature	5.318	°C
Product temperature	−19.00	°C

Condenser		
Duty	3,956	kJ/h
Feed temperature	31.00	°C
Product temperature	20.00	°C

Figure 9.17 UniSim model of the existing ammonia cooling cycle.

cooling water and wasted into the environment. This can clearly be seen from the Composite Curves of the existing process (Figure 9.18). The heating of all the cold streams requires 1,379 kW, the cooling of the hot streams represented by the hot composite curve needs 1,419 kW, and the heat recovery region is completely absent.

9.3.2 Heat Integration of Existing Refrigeration Cycle

The horizontal shift of Composite Curves towards each other up to the point of attaining the feasible minimum temperature difference (ΔT_{min}) between the process

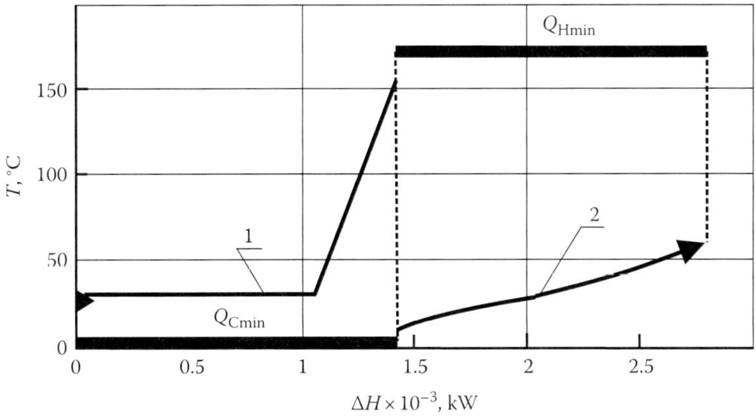

Figure 9.18 Composite Curves of the existing system of heat streams: (1) hot and (2) cold.

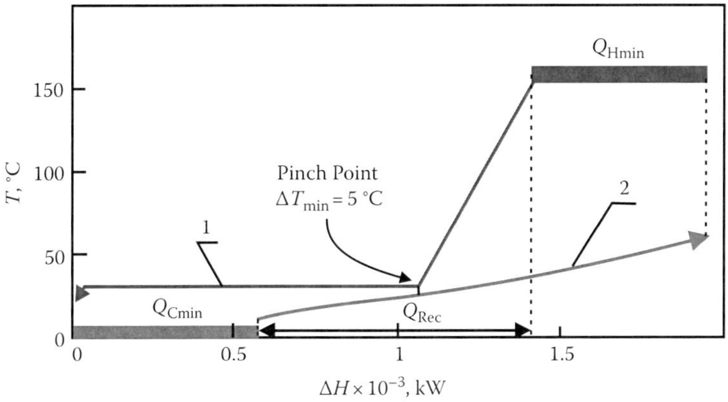

Figure 9.19 The Composite Curves for existing process for $\Delta T_{min} = 5$ °C. 1, Hot Composite Curve; 2, Cold Composite Curve; $Q_{Hmin} = 530$ kW, hot utilities; $Q_{Cmin} = 570$ kW, cold utility; $Q_{Rec} = 849$ kW, heat recovery.

streams makes it possible to obtain the maximum recovery of heat energy in the system. The feasible minimum difference between the temperatures of heat transfer agents for the existing heat exchangers is taken as 5 °C. The Composite Curves of the process with a $\Delta T_{min} = 5$ °C are shown in Figure 9.19.

The Composite Curves show that the heat recovery for the existing system of process streams is 849 kW for $\Delta T_{min} = 5$ °C. The required power of hot and cold utilities is 530 and 570 kW. The analysis of the graph shows that the overheating heat of ammonia and its condensation heat are utilised only partially. Because of the structure of the Composite Curves and the location of the Pinch Point, all condensation heat cannot be totally used. To achieve the target powers of hot and cold utilities and the best recovery of heat, it is necessary to use the network of heat exchangers. Based on the analysis of data, gathered from the constructed Composite Curves, the HEN is

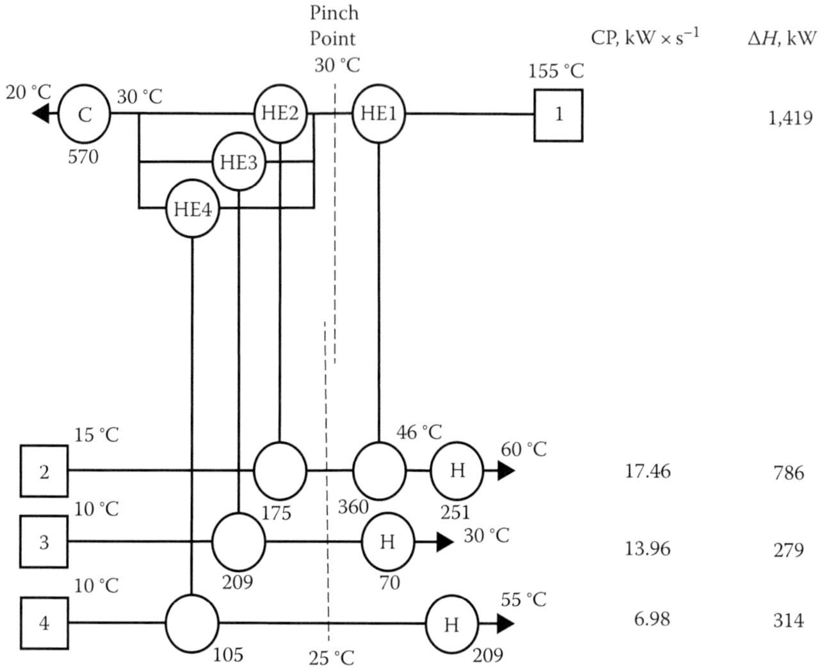

Figure 9.20 The Grid Diagram of the investigated stream system for $\Delta T_{min} = 5$ °C. HEi-4, recuperative heat exchangers; C, cooler; H, heater; CP, stream heat capacity; ΔH, stream enthalpy. Dashed lines indicate Hot (above) and Cold (below) Pinch Point.

built. For its development, the grid diagram representation is used. The Grid diagram of HEN for the investigated process streams is shown in Figure 9.20.

The Grid Diagram shows that the recovery of 849 kW of heat energy requires four heat exchangers, one of which must be located in the subsystem above the Pinch Point, while the others must be included into the subsystem below the Pinch Point. The power of hot and cold utilities amounts to 530 and 570 kW, which is in complete agreement with the Composite Curves. The temperature difference $\Delta T_{min} = 5$ °C is feasible to achieve in compact PHEs. In this case, heat exchangers has to be welded on the ammonia side due to its high toxicity. The estimated total area of the additional heat transfer surface will amount to 225 m^2. For the economy calculations it is assumed that the cost of installing a single heat exchanger and the cost of 1 m^2 of heat transfer area are equal to $5,000, and $500. The costs of hot and cold utilities were taken for Ukraine to be approximately $350 and $35 per 1 kW/y. In result the estimated payback period is 5 months.

9.3.3 System with Additional Compression of Ammonia

More detailed analysis of the system of streams shows that when increasing the condensation temperature changes the location of the Pinch Point and the shape of

Printed and bound by CPI Group (UK) Ltd, Croydon, CR0 4YY
22/10/2024
01777647-0004

Compressor_1		
Power	0.3210	kW
Feed pressure	195.9	kPa
Product pressure	1,200	kPa
Product temperature	155.2	°C
Duty	1156	kJ/h

Compressor_2		
Power	0.1347	kW
Feed pressure	1,190	kPa
Product pressure	2,604	kPa
Product temperature	111.4	°C
Duty	485.0	kJ/h

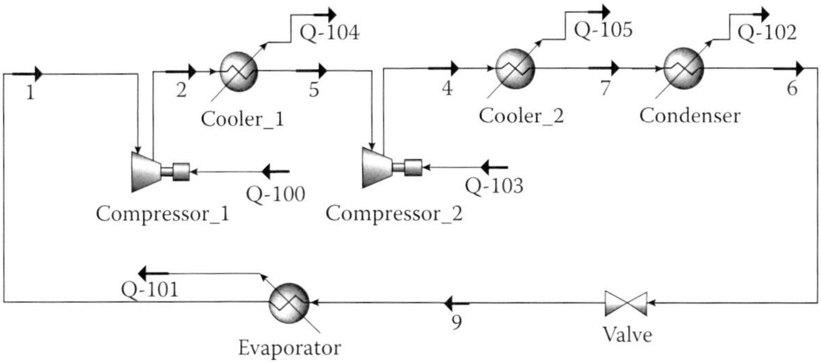

Evaporator		
Duty	−3,731	kJ/h
Feed temperature	20.06	°C
Product temperature	−19.00	°C

Condenser		
Duty	585.5	kJ/h
Feed temperature	55.00	°C
Product temperature	20.00	°C

Figure 9.21 UniSim model of ammonia refrigeration machine with additional compression.

the Composite Curves, the heat recovery potential can be raised. Using the UniSim Design software, a model has been constructed of a refrigeration machine with additional ammonia compression (Figure 9.21). In this new process system, the ammonia after the first compression stage is cooled down from 125 °C to 30 °C. Then the stream enters the second compression stage and leaves it with a pressure of 2,604 kPa and a temperature of 111 °C. The ammonia is delivered then from a compressor to cooling, then to condensation and afterwards to throttling and evaporation.

The calculations with a computer model allow determining of all characteristics of the ammonia stream in the refrigeration cycle with additional compression. This enables to form the table of the parameters of process streams to analyse the system of streams. The parameters of process streams that can be involved in Heat Integration are given in Table 9.17.

For determination of the minimal temperature difference for the HEN, the cost relations of the heat transfer area, utilities and total reduced cost depending on ΔT_{min} are plotted (Figure 9.22). The curve of total reduced cost has a minimum at ΔT_{min} = 23 °C. But on the curve's range from ΔT_{min} = 17 °C to ΔT_{min} = 27 °C, the value changes are insignificant. At the same time the cost for heat transfer area is less at ΔT_{min} = 17 °C. The minimum on the reduced utilities cost curve (curve 1 in Figure 9.22) is explained by the threshold problem of Composite Curves

Table 9.17 Parameters of Process Stream System with Additional Compression

No.	Name of Stream	Type	TS, °C	TT, °C	*G*, t/h	*C*, kJ/(kg·°C)	*r*, kJ/kg	CP, kW/°C	Δ*H*, kW
1	Ammonia cooling 1st stage	Hot	125	30	3.194	3.250		2.883	273.93
2	Ammonia cooling 2nd stage	Hot	111	60	3.194	4.275		3.793	193.44
	Ammonia condensation	Hot	60	60	3.194		986.2		874.98
	Cooling of liquid ammonia	Hot	60	20	3.194	4.935		4.378	175.14
3	Water heating	Cold	15	60	15.000	4.190		17.458	785.63
4	Air heating	Cold	10	30	50.000	1.005		13.958	279.17
5	Air for air curtains	Cold	10	55	25.000	1.005		6.979	314.06

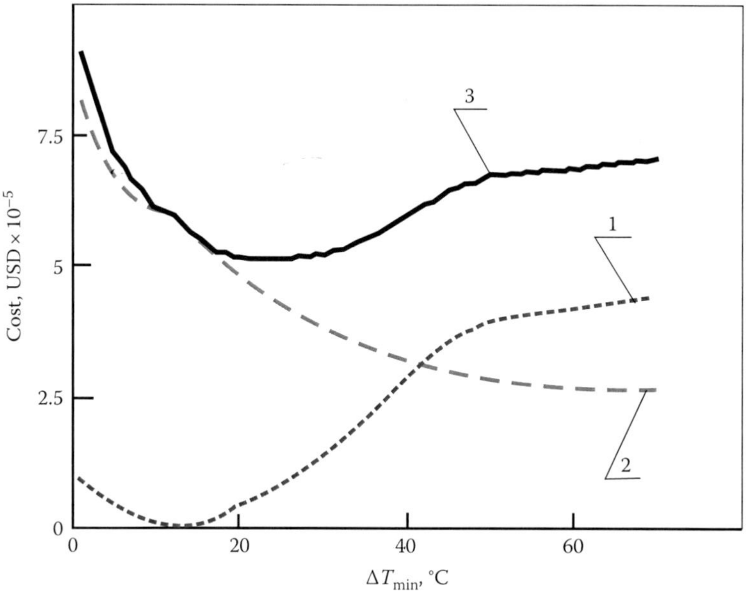

Figure 9.22 The costs curves: 1, reduced utilities cost; 2, the reduced capital cost; 3, total reduced cost.

(Klemeš et al, 2010). As a rule, for the Threshold problems, ΔT_{min} is selected equal to ΔT_{min} of the threshold. In our case this value is equal to 17 °C.

Accordingly, if ΔT_{min} = 17 °C is selected for the heat network design, then the total cost of the project is minimal and the value of ΔT_{min} conforms to the threshold value. The selection of ΔT_{min} = 17 °C for designing a HEN satisfies two requirements, namely, the total reduced cost of a design is close to its minimum possible value,

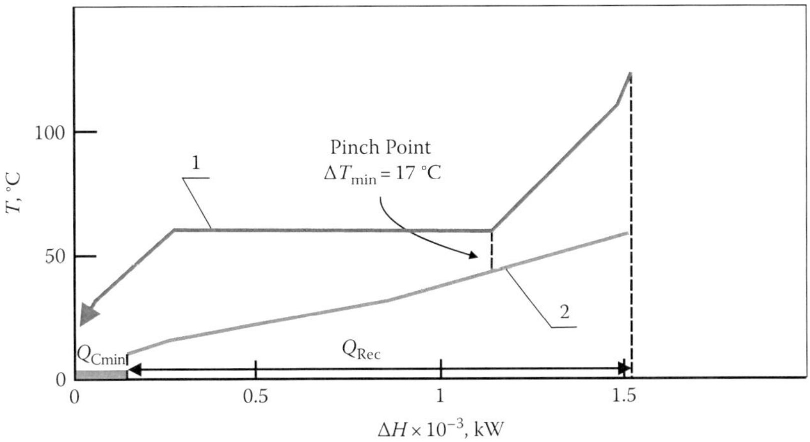

Figure 9.23 Composite Curves of the process with the additional ammonia compression for ΔT_{min} = 17 °C: 1, Hot Composite Curve; 2, Cold Composite Curve; Q_{Cmin} = 139 kW, cold utility; Q_{Rec} = 1,379 kW, heat recovery.

and ΔT_{min} corresponds to its threshold value. Moreover, if the prices on utilities will grow, the selection of a smaller temperature difference will lead to lower expenditures in the future.

The Composite Curves of process streams with additional compression for ΔT_{min} = 17 °C are plotted in Figure 9.23. The heat recovery amounts to 1,379 kW, the power of hot and cold utilities is reduced to 139 kW, and the need for hot utilities is no longer relevant, but the expenditures on the additional compression of ammonia are required. The application of the additional compression of ammonia allows us to satisfy the need for the heating of all the cold streams in the system with hot streams without utilities, but using the additional power of a compressor. The Composite Curves also show the location of the Pinch Point and its temperature that are required for constructing a network of heat exchangers with the use of a Grid diagram.

The developed Grid Diagram of a HEN for the integrated process with additional compression of ammonia is shown in Figure 9.24. It requires six recuperative heat exchangers with a total heat load of 1,379 kW and two coolers for the cooling and condensation of ammonia.

The total heat transfer area of the additional equipment is approximately 377 m^2, and it should be noted that heat exchangers of any design may be used in the system with additional compression due to the rather high minimum admissible temperature difference of 17 °C. To save space, compact heat exchangers are preferable choice. The operation of the system with additional compression also requires the installation of a compressor, which consumes 135 kW of electrical energy. The cost of a compressor was approximately \$50,000. The cost of installing heat exchange equipment, the specified cost of the heat transfer area and the cost of utilities were taken to be the same as for the system without additional compression. The estimated

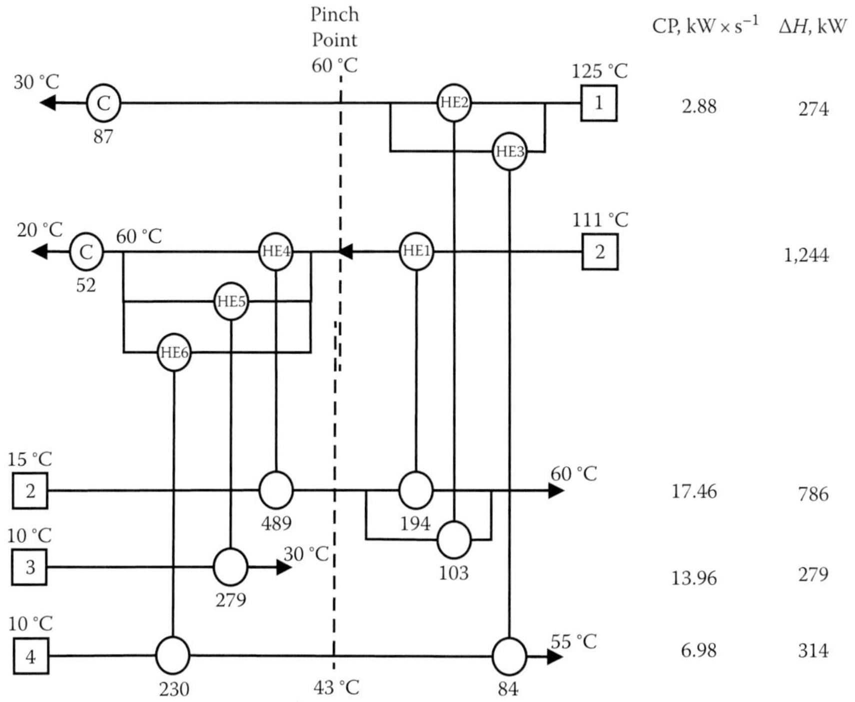

Figure 9.24 Grid Diagram of stream system with the additional ammonia compression for ΔT_{min} = 17 °C: HEi-6, recuperative heat exchangers; C, cooler; H, heater; CP, stream heat capacity; ΔH, stream enthalpy.

payback period of the system retrofit project with the additional compression of the ammonia stream amount for about 7 months.

9.3.4 Conclusion

An analysis of the operation of a supermarket ammonia refrigeration cycle shows the possibility of using the overheating and condensation heat of the ammonia stream. This heat may be used to heat water and air, thus considerably reducing the expenditures on utilities. The additional compression of the ammonia stream improves heat integration, but the implementation of this project requires bigger capital investments. However, the economic estimation of this retrofit project shows its reasonability. This methodology can be used to determine the capital cost of HENs within large industrial complexes that use intermediate utilities. The results can also be used to reconstruct existing supermarket ammonia refrigeration cycles and design new refrigeration cycles. When implementing these projects, it is necessary to perform an additional analysis of energy consumption systems, since each of them has its own set of streams, equipment features and process constraints.

REFERENCES

Abu-Eishah S.I., Abu-Jabal N.M. 2007. Parametric study on the production of phosphoric acid by dehydrate process. *Chemical Engineering Journal* 81(1–3): 231–250.

Boldyryev S.A., Garev A.O., Klemeš J.J., Tovazhnyansky L.L., Kapustenko P.O., Perevertaylenko O.Y., Arsenyeva O.P. 2013. Heat integration of ammonia refrigeration cycle into buildings heating systems in buildings. *Theoretical Foundations of Chemical Engineering* 47(1): 39–46.

Directive 2010/31/EU of the European Parliament and of the Council of 19 May 2010 on the energy performance of buildings. *Official Journal of the European Union*, 18 June 2010.

Edwards M.F., Stinchcombe R.A. 1977. Cost comparison of gasketed PHEs and conventional shell and tube units. *The Chemical Engineer* 5: 338–341.

EFMA. 2000. *Production of Phosphoric Acid*, Booklet No 4 of 8. European Fertilizer Manufacturers Association. Brussels, Belgium. www.productstewardship.eu/fileadmin/user_upload/user_upload_prodstew/documents/Booklet_nr_4_Production_of_Phosphoric_Acid.pdf (accessed 15 April 2015).

Georgiadis M.C., Macchietto S. 2000. Dynamic modelling and simulation of PHEs under milk fouling. *Chemical Engineering Science* 55: 1605–1619.

Honeywell. 2015. UniSim – Software for Process Design and Simulation. www.honeywellprocess.com/en-US/explore/products/advanced-applications/unisim (accessed 15 April 2015).

International Energy Agency (IEA). 2011. *Energy Balances of OECD Countries*, 2011 edn. IEA, Paris, France, 2011.

Jamialahmadi M., Muller-Steinhagen H. 2007. Heat exchanger fouling and cleaning in the dihydrate process for the production of phosphoric acid. *Transactions on IChemE, Part A, Chemical Engineering Research and Design* 85(A2): 245–255.

Kapustenko P., Boldyryev S., Arsenyeva O., Khavin G. 2009. The use of plate heat exchangers to improve energy efficiency in phosphoric acid production. *Journal of Cleaner Production* 17(10): 951–958.

Kapustenko P.O., Arsenyeva O.P. 2013. Process integration for energy saving in buildings and building complexes. In Klemeš J. (Ed.), *Handbook of Process Integration (PI): Minimisation of Energy and Water Use, Waste and Emissions*. Woodhead/Elsevier, Cambridge, UK, 938–965.

Kapustenko P.O., Ulyev L.M., Boldyryev S.A., Garev A.O. 2008. Integration of a heat pump into the heat supply system of a cheese production plant. *Energy* 33(6): 882–889.

Klemeš J., Friedler F., Bulatov I, Varbanov P. 2010. *Sustainability in the Process Industry – Integration and Optimization*. McGraw-Hill, New York, USA.

Klemeš J., Perry S.J. 2007. Process optimisation to minimise energy use in food processing. In Waldron K. (Ed.) *Handbook of Waste Management and Co-Product Recovery in Food Processing*, Vol. 1. Woodhead Publishing Limited, Cambridge, UK, 59–89.

Korfitsen E., Kristensen A.P.R. 1998. Ammonia high pressure heat pumps in food refrigeration applications. *International Journal of Refrigeration* 21(3): 212–218.

Reay D.A., Macmichael D.B.A. 1988. *Heat Pumps: Design and Applications*. Pergamon Press, Oxford, UK.

Seferlis P., Klemeš J., Bulatov I., Koltsova E., Kapustenko P., Soboleva I. 2006. Development of sustainable processes for waste utilisation in phosphoric acid industry. H8.03. In: *Proceedings of the Ninth Conference on Process Integration, Modelling and Optimisation for Energy Saving and Pollution Reduction – PRES 2006/CHISA 2006*, Prague, Czech Republic, Vol. 4, 1154.

Tovazhnyansky L., Kapustenko P., Ulyev L., Boldyryev S., Arsenyeva O. 2010. Process integration of sodium hypophosphite production. *Applied Thermal Engineering* 30(16): 2306–2314.

USDOE. 2011. 2010 Buildings Energy Data Book. D&R International Ltd, Silver Spring, MD, USA. buildingsdatabook.eren.doe.gov/ (accessed 10 November 2014).

Index

A

B

C

I

K

L

M

N

O

P

R

S